People, Land, and Community

P9-DNN-629

People, Land, and Community

Collected E. F. Schumacher Society Lectures

Edited by Hildegarde Hannum

With an Introduction and Comments by Nancy Jack Todd

Yale University Press / New Haven and London

Published with assistance from the Charles A. Coffin Fund.

Among the many contributions to the debate on economic policy, the works of E. F. Schumacher present a clear, integrated vision of a decentralized and sustainable economy that nourishes the earth and its inhabitants. The E. F. Schumacher Society, located in the southern Berkshire region of western Massachusetts, was founded in 1980 as a membership organization to implement Schumacher's ideas in practical programs for economic self-reliance. The Annual E. F. Schumacher Lectures, sponsored by the Society, and the library of the E. F. Schumacher Center are acquainting a growing number of people with the promise and viability of Schumacher's concepts.

Please address inquiries to 140 Jug End Road, Great Barrington, MA 01230. Contributions made to the E. F. Schumacher Society are tax-deductible.

The chapter by Wendell Berry, "People, Land, and Community," appeared in *Standing by Words: Essays* by Wendell Berry. Copyright © 1983 by Wendell Berry. Reprinted by permission of North Point Press, a division of Farrar, Straus & Giroux, Inc.

Copyright © 1997 E. F. Schumacher Society. All rights reserved. This book may not be reproduced, in whole or in part, including illustrations, in any form (beyond that copying permitted by Sections 107 and 108 of the U.S. Copyright Law and except by reviewers for the public press), without written permission from the publishers.

A catalogue record for this book is available from the British Library.

Library of Congress Cataloging-in-Publication Data
People, land, and community : E. F. Schumacher Society Lectures /
    edited by Hildegarde Hannum with an introduction and comments by
    Nancy Jack Todd.
    p.   cm.
    Includes bibliographical references and index.
    ISBN 0-300-06966-9 (cloth: alk. paper) 0-300-07173-6 (pbk.: alk. paper)
    1. Environmental economics.   2. Sustainable development.
    I. Hannum, Hildegarde.   II. E. F. Schumacher Society.
    HC79.E5P433   1997   338.9'27—dc20   96–46526
                                                     CIP

Printed in the United States of America on acid-free, recycled paper. The post-consumer waste content of the recycled paper is 20 percent.

The paper in this book meets the guidelines for permanence and durability of the Committee on Production Guidelines for Book Longevity of the Council on Library Resources.

10 9 8 7 6 5 4 3 2 1

Frontispiece: From left, E. F. Schumacher, Robert Swann (President, E. F. Schumacher Society), and George Wald (recipient of the 1967 Nobel Prize in physiology) at a gathering at the New Alchemy Institute, Cape Cod, Massachusetts, in 1974.

Perhaps we cannot raise the winds. But each of us can put up the sail so that when the wind comes we can catch it.

—E. F. Schumacher

Contents

Preface

Of the forty lectures sponsored by the E. F. Schumacher Society from 1981 to 1996, twenty are included in this collection (for a complete listing see page 309). Most of them were submitted in writing by the speakers—Thomas Berry's and David Orr's in substantially altered form—but several were transcribed from tape and therefore have a more conversational tone. I have made every effort in my editing to leave the original voice intact in the transition from the spoken to the written word.

With one exception the lectures selected for this volume have been published individually as pamphlets by the E. F. Schumacher Society (and have been edited for inclusion here): "People, Land, and Community," delivered by Wendell Berry as the first E. F. Schumacher Lecture, was revised for inclusion as a chapter in *Standing by Words* and is reprinted here as it appears in his book.

Robert Swann's 1988 lecture, "The Need for Local Currencies," has been expanded as an essay co-authored by Susan Witt. They are the president and executive director, respectively, of the Schumacher Society.

Bringing this collection to twenty-one is Hunter Hannum's "Wagner and the Fate of the Earth," an essay inspired by Thomas Berry's lecture, "The Ecozoic Era." It was published by the Society as a pamphlet and also appeared in the quarterly journal *Annals of Earth*.

The Introduction and the comments preceding each lecture are by Nancy Jack Todd, cofounder with John Todd of New Alchemy Institute and Ocean Arks International. She is the editor of *Annals of Earth* and serves as a board member of the Schumacher Society.

For three years, 1993–1995, the Annual E. F. Schumacher Lectures were held at Yale University, hosted by the Yale Student Environmental Coalition (YSEC). The Afterword is by Benjamin Strauss, who, as an undergraduate, was the YSEC coordinator for the 1993 Annual Lectures.

It has been a rewarding task to be the editor of this volume and to play a role in bringing these lectures, which have inspired many listeners, to a wider audience.

My appreciation goes to Schumacher Society staff Susan Witt, Sara Wilson Doyle, and Erika Levasseur for their help in preparing the book. Too numerous to name individually are all those staff, board members, and interns of the Schumacher Society whose enthusiastic efforts have helped to make the lecture series a success over the years.

—Hildegarde Hannum

Nancy Jack Todd
Introduction

The lectures in this volume reflect the enduring relevance of one of the most quietly influential philosophies of the second half of the twentieth century, the culmination of the lifework of economist E. F. Schumacher. The series of lectures that bears his name is given by thinkers and leaders from many disciplines and is presented annually by the E. F. Schumacher Society of Great Barrington, Massachusetts. They address such topics as the importance of community, locally based economies of scale, education, the dignity of good work, intelligence applied to the exigencies of everyday living, the necessity of balance between human needs and the health of the natural world, and the dangers of corporate globalization. Underlying all these is the foundation of hope for a sustainable future that was characteristic of the message Schumacher worked tirelessly to bring to the world.

According to *The London Times Literary Supplement* of October 6, 1995, Schumacher's *Small Is Beautiful: Economics as if People Mattered* ranks among the hundred most influential books published since World War II. The selection was made by a group of writers and scholars hoping to create "a common market of the mind" to bridge the cultural divisions of postwar Europe. Others to be so acknowledged were Simone de Beauvoir, André Malraux, Albert Camus, George Orwell, Jean-Paul Sartre, Hannah Arendt, Carl Gustav Jung, and Erik Erikson. Although less academic than many of the other works, *Small Is Beautiful* foreshadowed with extraordinary accuracy many of the major issues we would be struggling with at the end of the century. Since its publication in 1973, it has been translated into approximately twenty languages

Schumacher was a tall man who cast a long shadow—particularly, as the lectures attest, in the realm of ideas. At the time of his death in 1977 he was eulogized as "a prophet standing against the tide" and "a man who asks the right questions of his society and of all societies at a crucial time in their history." This assessment remains valid. From the perspective of those who have since witnessed rampant hostile corporate takeovers and the use of the term "downsizing"—which originally referred to reducing the size of cars—for firing employees, the compassion inherent in an "economics as if people mattered" is all the more compelling. Adamantly opposed to excessive material consumption, meaningless growth, corporate domination, and world-scale economic systems, Schumacher would have been gratified to see how his ideas, steadily gaining momentum over the years, have created a significant undertow to counter the 1990s global dynamic dominated by GATT, the General Agreement on Tariffs and Trade.

The particular genius of the thinking of E. F. Schumacher lay in his union of the theoretical and the practical, embodying a rare combination of sound epistemology and pragmatic common sense. He was also a deeply spiritual man with a strong love for and understanding of the natural world. Although he had been well-known in Europe since the end of World War II, it was not until the publication of Small Is Beautiful that his influence became widespread in North America. In the few remaining years of his life, he became recognized throughout the United States and Canada and had met with a number of national figures, including former President Jimmy Carter.

Schumacher's life mirrored directly the events of the first three-quarters of the twentieth century. He was, paradoxically, a man both of and very much ahead of his time. Like Lewis Mumford, he was critical of unthinking acceptance of technological innovation masquerading as progress, warning against individual and local loss of autonomy and quality of life. Schumacher, known as Fritz to his friends, was born to a traditional academic family in Bonn, Germany, in 1911. According to his daughter Barbara Wood in her biography, E. F. Schumacher: His Life and Thought, he proved a quick and talented student, and in 1930, having studied at the universities of Bonn and Berlin, he was chosen to represent Germany as a Rhodes scholar at New College, Oxford. Two years later he made his first trip to America, where he discovered an intellectual freedom he had never known before. In 1934, however, increasingly anxious about the rise of National Socialism in Germany, he left a promising career in New York and returned home. The situation there confirmed his worst fears. Many of the people he respected, understandably intimidated, were closing their eyes to the evils around them. At the core of his own opposition to the Nazis was his rejection of their habitual manipulation of information and their flagrant violation of the truth. With deep foreboding, in 1936 he left Germany with his new wife to settle in England, the country that was to become his home.

From then on, either directly or indirectly, he was part of the unfolding events of his time. With other German expatriates in England, he agonized over the fate of his country and of Europe. Once the war had begun, with anti-German feeling running high, Schumacher, the patrician intellectual, was relegated to the country to work as a farm laborer. At one point he was taken from his wife and infant son and interned for three months in a detention camp, where he initiated a number of practices to improve sanitation as well as the quality of the food. He later came to consider his time at the camp to have been his real university. It was there that the thinker became a doer.

After his release he returned to the farm, preoccupied with the search for economic prerequisites for a lasting peace in Europe. His writings on the subject brought him to the attention of a number of prominent people, and he was soon drawn into their discussions on the postwar economy. After becoming a British citizen in 1946, he was sent to Germany as a member of the British Control Commission. In thinking about the reconstruction of German industry, he began to formulate his ideas on what was appropriate—a word that came to be strongly associated with him—in terms of scale and ownership. As he studied the restructuring of Germany's economy, the strategic role of energy, soon to become a linchpin of his thinking, became apparent to him. He became equally convinced of the necessity for currency reform as a means of preventing the concentration of wealth among the few at the expense of the many, another of his legacies that remains alive in the movement of the last two decades of the twentieth century to establish local currencies.

In late 1949 Schumacher was asked to become an economic advisor to the National Coal Board of Britain. He accepted and remained in his post as chief economic advisor for the next twenty years. To accommodate his growing family, he bought a house with a large garden in Surrey. This proved to be another turning point for him; he became fascinated with his garden, joined the Soil Association, and became an enthusiastic exponent of organic gardening. Observing natural processes at work in his garden, he developed an insightful understanding of the interrelated complexity of living systems. In anticipation of advanced ecology and the Gaia theory—which views the Earth as a single, self-regulating, living entity, the sum of the interactions of the atmosphere, hydrosphere, surface sediments, and biota—he wrote then: "It makes sense that nature is an unbelievably complicated, self-balancing system in which the unconsidered use of partial knowledge can do more harm than good. As far as I can see, chemical agriculture has over-reached itself. It is working against nature instead of with her." Later, in *Small Is Beautiful*, he noted that "the wider human habitat, far from being humanized and ennobled by man's agricultural activities, becomes standardized to dreariness or even degraded to ugliness."

One of the major influences on Schumacher's thinking was the work of a little-known Austrian economist named Leopold Kohr. In *The Breakdown of Nations* Kohr treats the subject of scale, attributing the ills of the modern world to bigness. He writes: "If a society grows beyond its optimum size, its problems must eventually outrun the growth of those human faculties which are necessary for dealing with them." Schumacher referred to Kohr as "a teacher from whom I have learned more than from anyone else."

An inveterate reader, Schumacher studied the Buddhist and Taoist sages and was deeply impressed by the nonviolent message of Mahatma Gandhi. In 1955 he was offered a United Nations assignment in Burma and seized the opportunity to see Asia. The time he spent there had a lasting influence on his thinking. He found the land, the culture, and the people of Burma compellingly beautiful. He was moved by their freedom from materialism and their apparent happiness under conditions that would be deemed impoverished in developed countries. This reinforced his inclination to look beyond abstraction and theory to embrace the constants of health, beauty, and permanence. He wrote: "Economics means a certain ordering of life according to the philosophy inherent and implicit in economics. The science of economics does not stand on its own feet: it is derived from a view of the meaning and purpose of life."

Prophetically, he further noted: "A civilization built on renewable resources, such as the products of forestry and agriculture, is by this fact alone superior to one built on non-renewable resources, such as oil, coal, metal, etc. This is because the former can last, while the latter cannot last. The former cooperates with nature, while the latter robs nature. The former bears the sign of life, while the latter bears the sign of death." Later, in his most famous essay, "Buddhist Economics," he advocated a form of economics based on "Right Livelihood" as part of the Buddha's Noble Eightfold Path. Fundamental to such an economics are simplicity, nonviolence, the importance of community, and the necessity and dignity of work. Schumacher returned from Burma convinced that a sustainable form of economics must be found that would be appropriate as a path for the developing world, "a middle way between materialist heedlessness and traditionalist immobility." He spent the rest of his life seeking and advocating that path.

Schumacher was equally foresighted in his analysis of the industrial world. In 1958, before the founding of OPEC and to the disbelief of his colleagues, he warned that Western Europe would attain "a position of maximum dependence on the oil of the Middle East . . . The political implications of such a situation are too obvious to require discussion." Even greater than his concern about the conflicts that would ensue was his fear of the possibility of a nuclear exchange. He became adamantly opposed to the use of nuclear energy. The accumulation of large amounts of toxic substances, he claimed, "is a transgression against life itself, a transgression infinitely more serious than any crime ever perpetrated." Echoing the Gandhian philosophy of nonviolence, he wrote: "A way of life that ever more rapidly depletes the power of the Earth to sustain it and piles up ever more insoluble problems for each succeeding generation can only be called violent. . . . Non-violence must permeate the whole of man's activities, if mankind is to be secure against a war of annihilation."

In India in 1961 Schumacher had his first exposure to the grinding poverty of much of that subcontinent. It was in brooding over the lot of millions of people there that his pivotal idea and most lasting legacy was born. He came to believe that the cause of their profound moral and physical impoverishment lay in the demoralizing impact of the industrialized West on traditional, formerly self-sufficient cultures. What was needed, he urged, was a level of technology more productive and effective than that employed traditionally in rural areas but simpler and less capital intensive than Western technologies. His groundbreaking concept of intermediate technology had taken form.

This idea was wholeheartedly embraced by a younger colleague, a Scot named George McRobie, who became key to its development. The first of Schumacher's articles on intermediate technology appeared in *The Observer* in 1965 and was given an enthusiastic reception. He and McRobie responded by forming a small, self-financed group soon to become well-known as ITDG, the Intermediate Technology Development Group. They set up an office and began researching the kinds of equipment that could be made available to small-scale farmers and crafts people. This quickly developed into a buyer's guide; it too was well received, leading them to increase their efforts to fill the gaps in existing technologies. It was Schumacher's intent to combine traditional and advanced knowledge for creating new technologies to address questions of impact and scale. ITDG grew rapidly as a result of the deep chord it had sounded around the world. Intermediate technology was seized upon as an innovative tool for tackling the problems of poverty. It fell to Schumacher, as he became recognized as an international figure, to become intermediate technology's global ambassador and interpreter.

Schumacher believed that it was vital for poor people to be able to help themselves and that intermediate technology could enable them to do so. He traveled widely, advocating small-scale technologies as well as enterprises, workshops, and factories that would serve communities in such a way that no one need be exploited for another's gain. The technologies and community structures he envisioned would produce material sufficiency rather than surfeit and would be a source of work, which he considered everyone needed in order to become fully human. Again prophetic in his insights, he was convinced that the affluence of the West could not be maintained indefinitely and tried to teach that the hope of the powerless and the poor lay in becoming as independent as possible of the corporate dynamic.

With the publication of *Small Is Beautiful* in 1973, Schumacher's status was transformed to one approaching that of guru. He made several trips to North America, where, especially among young people, his words rang true. The non-violent message of intermediate technology and economies of scale was seen as just as applicable to the Western world as to nonindustrialized areas. Schumacher wrote: "The key words of violent economics are urbanization, industrialization, centralization, efficiency, quantity, speed. . . . The problem of evolving a non-violent way of economic life [in the West] and that of developing the under-developed countries may well turn out to be largely identical."

This observation is proving to be the case, as the accounts in many of the lectures that follow substantiate. A central theme advocates that people rise to the challenge of taking action themselves to create much-needed and appropriate changes in their lives. This kind of activism is occurring in many places in the developing world, in some cases in the face of appalling poverty and lack of resources. It is happening in industrial countries too as a response to widespread disillusionment with the remoteness of government and with corporate irre-sponsibility. Resourcefulness, applied intelligence, community, and appropriate action synergistically forge the road to change and concomitantly create a sense of hope. Schumacher was at heart a man of the people and a man of hope.

Of the many groups eager to carry on Schumacher's work, the leadership has fallen to two organizations, one in the United States and one in Great Britain.

The E. F. Schumacher Society in England—established in 1979 by Satish Kumar, editor of *Resurgence* magazine since 1973—sponsors a popular lecture series held annually in the city of Bristol. Kumar was also instrumental in the founding of Schumacher College in Devon. Its interdisciplinary curriculum examines the foundations of a sustainable, balanced, and harmonious worldview. The College practices as well as teaches Schumacherian principles of shared work, meditation, and intellectual inquiry. By bringing together an eclectic mix of instructors from around the world and international participants ranging in age from twenty to eighty, the College is creating a global network of people who share Schu-macher's vision and are implementing it in ways that justify his optimism.

The E. F. Schumacher Society in Great Barrington, Massachusetts, was founded in 1980 by Robert Swann and a group of Schumacher's American friends and colleagues. A lifelong pacifist and advocate of decentralism, Swann was drawn to

Schumacher's ideas through reading his articles in *Resurgence* magazine. In 1967 he went to England to meet Schumacher and suggested to him that these articles be published in book form. This led directly to the collection of essays that became *Small Is Beautiful*. Swann subsequently organized Schumacher's 1974 North America tour to promote the book, a trip that was also to have a catalytic effect on the sustainable energy movement in the United States. At the end of the tour Schumacher suggested that Swann establish a United States-based group to work at the interface of economics, land use, and applied technology. Six years later, at the urging of Ian Baldwin, David Ehrenfeld, Hazel Henderson, Satish Kumar, and John McClaughry, Robert Swann took on the challenge, and the E. F. Schumacher Society came into being. With Susan Witt as executive director, the Society has evolved programs that have grown increasingly effective in fulfilling the mission envisioned by Schumacher.

The Annual E. F. Schumacher Lectures provide a public forum for scholars and activists working in the Schumacher tradition, and they are becoming recognized as a resource presenting knowledge too valuable to be ignored.

The E. F. Schumacher Center, located in the Berkshire hills of western Massachusetts, has grown to a facility housing a five-thousand-volume, computer-indexed library of books, pamphlets, tapes, and specialized bibliographies. The subject matter focuses on decentralism, human-scale societies, regionally based economic systems, local currency experiments, and community land trusts. In 1995 Vreni Schumacher, Schumacher's widow, bequeathed his entire library to the Center. This asset infuses the Center, according to board member Kirkpatrick Sale, with "the essence of the man himself, in all his dimensions. I was especially pleased to see that we have all his book reviews and articles (published, and often with typescript, sometimes with manuscript) and loads of his speeches (manuscript and mimeographed), none of which has ever been gone through systematically and some of which I think no one even knew about or remembered. Plus those notebooks—who knows what rich veins may be there."

In addition to the resources of the Center, the annual lectures, and other educational programs, the Schumacher Society develops model projects that effectively put power back into the hands of the people. "Among material resources, the greatest, unquestionably," wrote Schumacher, "is the land. Study how a society uses its land, and you can come to pretty reliable conclusions as to what its future will be." One of the Society's goals has been to create new institutional forms that provide access to land based on social and ecological objectives rather than market forces. The community land trust model developed by Robert Swann provides just such a vehicle for decommoditizing land and places stewardship in the hands of a democratically structured, regional organization. The Schumacher Society, actively involved with its Berkshire community land trust, has published a handbook of legal documents that help others to organize community land trusts.

In light of the growing influence of the two Schumacher organizations and the ideas reflected in the lectures that follow, it is evident that the work of E. F. Schumacher lives on. The world has changed dramatically since his death but not in ways he would have found surprising. Many of the developments he cautioned against have accelerated. As David Orr maintains in his lecture, "Issues of human survival . . . will dominate the world of the twenty-first century." Corroborating Schumacher's worst fears, linguist and social critic Noam Chomsky has summarized the crisis as follows: "In the real world human rights, democracy, and free markets are all under serious attack in many countries, including the industrial societies. Power is increasingly concentrated in unaccountable institutions. . . . People's lives are being destroyed on an enormous scale through unemployment alone. . . . The economic system is a catastrophic failure."

The first section of the book, entitled "Beyond a Legacy of Domination," addresses the origins of this breakdown. Kirkpatrick Sale and Winona LaDuke contrast the worldview and the unsustainable ways of life of Western civilization with more enduring traditional and indigenous cultures. David Ehrenfeld and Frances Moore Lappé expand the analysis of how we have reached such an impasse, and Dana Jackson describes the impact that feminism and environmentalism have had by questioning the foundations of this now changing legacy.

In the second section, "Toward Decentralism and Community Revitalization," the speakers describe autonomous local alternatives to the thrall of corporate domination. From Hazel Henderson, Jane Jacobs, Robert Swann and Susan Witt, John McClaughry, Wendell Berry, Wes Jackson, John McKnight, and Cathrine Sneed we learn of ingenious programs in fields ranging from local economics to community activism to agriculture to inner-city garden projects—programs intended to undermine and free the present economic gridlock and reach out to the people who have become its victims.

The third and final section, "Toward a New Era in Human-Earth Relations," is a soul-searching exploration of the philosophies and practices that will awaken a new sense of human destiny. The section begins with Thomas Berry's sweeping vision of the evolving human place in the larger realms of planet and cosmos. Hunter Hannum traces the healing wisdom embedded in mythology and high culture. Kirkpatrick Sale, David Orr, Wes Jackson, John Todd, and Stephanie Mills variously construct the knowledge base and the codes that can lead to behavior more honoring of the uniqueness and preciousness of life in its myriad planetary forms. Finally, from the vantage point of a lifetime of conservation and stewardship, David Brower recounts what can—and must—be done if we are to preserve ecological viability and harmony for ourselves and the generations who will follow us.

Such thinking is hardly the subject of either mainstream debate or regular media coverage. Yet as cultural historian William Irwin Thompson has argued, the potential impact of a small group on the larger culture can be likened to that of an enzyme on a physical system. In *Small Is Beautiful* Schumacher wrote: "Our most important task is to get off our present collision course. And who is there to tackle such a task? I think every one of us, whether old or young, powerful or powerless, rich or poor, influential or uninfluential. To talk about the future is useful only if it leads to action *now*."

The kind of enzymatic activity he was urging is no longer confined, as it largely was in his day, to those considered outside the mainstream or to the so-called alternative movement. Toward the end of the twentieth century a shift akin to a grassroots sea change became perceptible. Across the North American continent and in Europe and Australia, communities and groups of citizens who see a need for greater economic autonomy and a healthy environment are tackling their problems with creativity, zeal, and common sense. Such coalitions may typically involve businesspeople, educators, local and state government officials, union members, farmers and growers, nature lovers, and others who care about where they live or who want to maintain open options for their children. Simultaneously breathing new life into democratic process, they are launching programs and projects to address their problems. It is in part Schumacher's legacy, no longer small but perennially beautiful, that many people now search beyond the plundered despoiled landscapes and the polluted air and waters of today envisioning a human/natural continuum, with human beings living sustainably in landscapes regenerating from the assaults of recent centuries.

Schumacher urged people to be constantly observing and questioning what goes on around them, and not merely with regard to technology or economics. He would have us examine honestly the foundation and scale and civility of our lives, our vitality, our integrity, and our spiritual wholeness. He was involved in a lifelong quest, rarely settling for a single answer beyond that of absolute honesty but instead seeking long and diligently for the most intelligent, adaptive, appropriate, and hopeful means for us to conduct our lives as if people—and nature—mattered.

In their various ways the Schumacher lecturers are continuing that quest.

Part I Beyond a Legacy of Domination

Kirkpatrick Sale
The Columbian Legacy and the Ecosterian Response

If our present global economic structure is as unsustainable as it is widely thought to be, it is important to examine the causes of this crisis. In this first section of the book, the speakers trace, from shifting perspectives, the paths by which we have reached an impasse characterized by a despoiled environment, growing poverty, and weakening democratic structures. In spite of this deadlock, each of them shares, in his or her respective way, E. F. Schumacher's vision of potential responses to the spectrum of problems before us. Building on the foundation of his work, they point to how we may yet piece together a mosaic of solutions.

Author Kirkpatrick Sale is among the most well-known of the Schumacher lecturers because of his radio broadcasts, lectures, and columns for *The Nation*. His books include *Human Scale, Dwellers in the Land: The Bioregional Vision, The Conquest of Paradise: Christopher Columbus and the Columbian Legacy*, and *Rebels Against the Future*. He is a board member of the E. F. Schumacher Society.

Sale dedicated *Dwellers in the Land* to Schumacher; as is apparent from his text, the thinking of the two men constitutes a paradigmatic continuum. Sale was the first to introduce the concept of bioregionalism, the subject of his second lecture in this volume, to the broader public. An exuberant iconoclast, he is part of the Neo-Luddite movement, whose adherents are critical of society's unquestioning acceptance of technological innovation that has little regard for its potentially negative impact on individuals, culture, and the environment.

In the following lecture, which became part of *The Conquest of Paradise*, Sale draws a parallel between the crisis of the late twentieth century and the corruption, human misery, and spiritual bankruptcy prevalent in Europe as the great voyages of exploration to the New World began—in an "age of decadence before the end." Christopher Columbus, a product of that culture, set sail with the original intent of converting the heathens to Christianity. Upon his arrival in the Western Hemisphere, so blinded was he by his cultural biases, he failed to recognize that he was, for the first time, witnessing cultures living, in the speaker's words, in "exquisite relationship to the natural world." Sale paints a fleeting but haunting portrait of natives of the New World and a paradise of timeless balance, irrevocably lost with the intrusion of Columbus and those that followed him.

It is vital, if humbling, for members of our culture, one more powerful and more technologically advanced than the world has ever known, to understand what it has destroyed in order to achieve this pinnacle. It is equally important to be reminded how transitory is such power, how shaky are its foundations, and how disempowered are all but an elite few. Many scholars offer similar critiques, but Kirkpatrick Sale does not rest with pointing out what is wrong. In his call for the creation of "ecosteries," he presents a strategy for the inevitable transition between the perils of the present and the building of a decentralized, democratic, and sustainable society.

In the fifteenth century the subcontinent of Europe, containing perhaps sixty million souls, was a society in crisis, of spirit as much as of substance, sickly, miserable, melancholic, anguished, without a faith to believe in, institutions to trust or values to rely on; it was the victim of such a series of calamities, and had been for well over a century, that violence had become the tenor of everyday life ("The citizens of Mons," writes the historian Huizinga, "bought a brigand, at far too high a price, for the pleasure of seeing him quartered, at which the people rejoiced more than if a new holy body had risen from the dead"). Disease had become a daily agent of death ("How can it even be called a life," cried Thomas à Kempis at the time, "which begets so many deaths and plagues?"), and starvation had become the regular alternative to scarcity ("Famine constantly visited the continent," historian Fernand Braudel notes, "laying it waste and destroying lives"). A French poet of the time, Eustache Deschamps, was moved to write:

> Time of melancholy, and of temptation,
> Age of tears, where envy and torment blend,
> Time of lassitude and of condemnation,
> Age of decadence before the end.

Before the end: yes, none doubted, as the Habsburg court historian then put it, that "the end of the world is near and the waters of affliction will flow over the whole of Christendom," such is "the miserable corruption and the wretchedness of all classes."

He was perfectly a man of his age, so it is no surprise that just at that time a middle-aged merchant seaman named Cristóbal Colón, a native of Genoa but then living in Cordoba, apparently without any other visible means of support than the ministrations of a generous mistress, was obsessed with the idea of Armageddon, the approaching end of the world, and spent countless hours with his Bible and religious tomes figuring out exactly how many years remained until the Final Judgment, eventually deciding that the world would end after its seven thousandth year—which, he reasoned, was about 155 years away, or less, depending on which authority to trust. And thus he was moved, so he later said, to think of sailing to new islands and mainlands in the unknown parts of the Ocean Sea, there to fulfill the two necessary conditions that must be met before the Final Judgment—the conversion of all heathen idolaters on earth to Christianity, including such heathen idolaters as might be found in those parts, and the military assault to free the holy city of Jerusalem from the infidel, payment for which was to be supplied by such treasure as might be found there. These, apparently, the basic elements of the proposal he was even then trying to put to the King and Queen, the plan that would bring Europe the salvation it so badly needed and so desperately sought in those dark days.

As we know, Cristóbal Colón (who has come down to those of us in the English-speaking world as Christopher Columbus) did indeed find heathen idolaters and considerable treasure—but he did not find salvation. Or, more accurately, he found salvation, and it was there among those idolaters and their exquisite relationship to the natural world, but he did not know it, did not even have the capacity to know it. And he began the long process by which those idolaters— some 100 or 120 million of them, I believe—were effectively destroyed and much of their culture annihilated and by which the treasures of their two continents—gold, silver, pearls, timber, fish, tobacco, potatoes, corn, and medicines, among much else—were discovered, exploited, and exported, with most of the land laid waste besides; the long process by which Europe, on the strength of that treasure and little else, was able to finance and forge the institutions that gave it the power to spill out from its borders and conquer not only those two vast continents but most of the rest of the world as well. To this day it is European culture and artifacts and technologies that exist everywhere, European languages that are spoken right around the world, European institutions and economies that dominate all countries of whatever land or longitude—the most successful domination by any civilization in the history of humanity and leading to, even more, the most successful domination by any single species in the history of life.

That, in all its glory and all its terror, is the Columbian Legacy. Today, after the five-hundred-year trajectory of their worldwide conquest, we can see it in fullest clarity and ponder what it has brought us to. I have isolated four of its essential characteristics, those that may be said to be the cornerstones of European civilization nascent in the fifteenth century and embedded somewhere in the soul of the Great Discoverer, who spread them across the Ocean Sea, and that, thanks to him, came to support the edifice we call the Modern Age, indeed modern civilization:

1. Humanism—the declaration and celebration of the human species as the most important species of all (and of men as the most important component of it), with a God-given right to conquer and destroy and manipulate and control in its service, to have "dominion over" the species, the elements, even the processes of the earth.

2. Rationalism—that bipolar, straight-line, reductive way of looking at the world, according to which all is knowable, and knowable by us, finding its apex in that branch of rationalism we call science, which is our method of asserting control over nature and (in Schiller's phrase) "de-godding" its constituent parts.

3. Materialism—the narrow perception and appreciation of the world in terms of the corporal and tangible, and the valuation of it in terms of accumulation and possession, a belief-system that becomes most overt in the economic arrangements known as capitalism, whose genius is to permit virtually no other consideration than the immediate goal of profit to interfere with the exchange of goods.

4. Nationalism—that bold invention by which various self-styled "royal" families forged political institutions that took on the shape of nation-states, becoming over the centuries the central institution in daily life, deposing church, guild, manor, city-state, community, and individual, and creating that by which they were sustained: the standing army and the philosophy of militarism.

Those four, then—humanism and its domination, rationalism and its science, materialism and its capitalism, nationalism and its militarism—were the characteristics that made Europe successful, that made Europe powerful, that made Europe Europe. Fueled by the treasure extracted from the New World and working synergistically in a unique and marvelous way, they allowed one small set of people to expand and spread out and ultimately dominate not only the other peoples but the other species of the world as well and to do so unremittingly for five centuries—a dominance of white male over dark, technics over sodality, mechanical over organic, and, above all, of human over nature.

Those characteristics may all seem natural and inevitable, yet we might remind ourselves that they are not eternal givens but rather constructs, inventions, of a particular time and place and people, and they have had a life of barely more than half a millennium. They also may seem desirable and invaluable—humanism, science, modernity, civilization; how could they be anything but good? But it is well to realize that this is so only because those who believe in them and profit by them declare them to be so, to realize, too, that there is a growing body of people beginning to question their merit and wondering if in fact they are not perhaps the cause of our modern multiple crises.

For there is no longer room to doubt that now, five hundred years later, the subcontinent of Europe—and all the continents it has peopled and all the cultures it has touched—represents a society in crisis, a crisis, like the previous one, of spirit as much as of substance. The industrial world, the European-culture world, of which this nation is a preeminent example, is sickly, miserable, melancholic, anguished, without a faith to believe in, institutions to trust, or values to rely on, victim of the disease I have called "affluenza," the frenzied amassment of packages and products to the point that they choke our lives and clutter our landscapes while at the same time we amass slums, crimes, drugs, prisoners, suicides, debts, diseases, and pollution on a scale without parallel in history—and now stand at the point where not only is the survival of the human animal in real question but the survival of all oxygen-dependent species and indeed the living earth itself. We have as a culture subscribed to the theory of progress—it is time to cancel that subscription.

The Columbian Legacy stands before us today as never before—that legacy which we know by the name of European civilization, brought from the Old World to the New by the man who, as Columbia, is the very personification of the United States, the hero and champion of progress—stands before us, I might say, in the dock, affording us a chance, before it is too late, to examine its record and assess its crimes and pass judgment and weigh its future. That above all should be the project of this nation in the next two years as we approach the much-ballyhooed Columbian Quincentennial—a project that I trust you have already begun upon and will, with me, intensify in the months to come. For we really have no choice. Our planet, we now know, is on the endangered species list.

It is a somber and sobering prospect, and it poses an especially complex problem for all of us who realize the peril we are in but realize also the immense power and pervasiveness of that Western civilization which is, in effect, the sea we swim in. What can we possibly do? What must we do? And how?

Those are the crucial questions of our time, and I am asked them often. I am afraid I must tell you, even after more than two decades of pondering this, that I have no sure, no easy answers. But I would like briefly to look at our options, our choices as a society.

We can, of course, ignore the signs of the apocalypse and bury our heads in the sands of materialism and mindlessness, drugging ourselves insensate on the palliatives offered by press and pulpit and politics. This, the easiest, is the way of our leaders so far, who see no special reason to change a system that has given them so much wealth and power, no reason despite the fact that it assures the virtual doom of their grandchildren.

Or we can acknowledge the crisis and give it over to the "experts," the lawyers and environmental lobbyists and professionals, to solve. This is a familiar solution as we approach the twenty-first century. We have even created, as John McKnight pointed out in this forum some years ago, professional "bereavement counselors" to replace family and friends in time of grief over the death of a loved one—but the record of the environmental experts over the past few decades as the world has gone to hell does not give much cause for comfort.

Or we can decide the crisis is real and turn it over to the scientists to quantify and calculate, trusting them to come up with the magical technofixes that will let us go right on with what we are doing and yet be miraculously clean and green. As if we haven't learned by now that every technofix creates a whole array of new problems, all unforeseen by these same scientific geniuses, and as if it wasn't that very technofix mentality and the blind worship of science that got us into the eco-mess in the first place.

Or we can give the crisis over to the government and the politicians and the responsible agencies of the bureaucracy—the ones who have been so successful, for example, in cutting the deficit, solving the savings-and-loan crisis, creating a responsible budget, and eliminating campaign improprieties, defense-contract corruption, and junk-bond scandals. The ones whose dedication to swift and meaningful environmental action is attested to, in a symbolic way, by the agreement of the solons of Congress definitely to end by the year 2030 the production of hydrochlorofluorocarbons, among the most lethal of the chemicals destroying the ozone layer.

Or we can try to do the job ourselves, to create a home-grown version of the process by which much of Eastern Europe was able, in the space provided by Gorbachev, to overthrow its multiple tyrannies and embark on those exciting experiments that, we may hope, do not partake of so much of the free-market cure that they will kill the patient. This process, little noticed over here, was one in which first a few intellectuals and then increasing numbers of ordinary people came to the understanding that meaningful changes occur only through organizing structures *outside* those of the state and political apparatus, in multifarious task-oriented social groupings designed not to *reform* the system but to *redefine* and *reconstitute* civil society itself. Hence what the Poles called "social self-defense" and the Czechs called "the parallel polis," the building of small-scale structures that avoid the traps of reformism and co-optation because they look elsewhere, build alternative sources of power, and devote themselves to real needs as expressed by real people, unmediated by the shell game of politics.

This last does seem like the most promising strategy for us today, and it is perfectly in keeping with the very values of decentralism, democracy, and small-scale empowerment that E. F. Schumacher expressed so well—although we have to acknowledge that the task here in America is far different and far harder, on a far larger scale against far more potent forces, and without the black hole of Russian powerlessness to help us out. Still, it does seem our only choice, certainly the only hopeful one.

The future is not easy to contemplate, but it is, obviously, where we are going to spend the rest of our lives, and if those lives are to be anything more than the nasty, brutish, and short passages we are experiencing at the close of the twentieth century, it has to be an ecological future. Now, it seems to me that there are only two possible paths to achieving such a future: either by design or by catastrophe. By design if during the next decade the hopelessly large institutions of our industrial world prove themselves utterly inept and bankrupt and the citizens begin looking around for alternative, responsive, eco-centered institutions to put in their place; or by catastrophe, some devastating global eco-catastrophe, which alters or eliminates all existing systems and structures and vastly reduces many species, including the human, assuming they survive at all.

In either case, I would argue that the challenge for us is the same: to start now to establish small, local, bioregionally guided alternative institutions that can provide the *information* by which human communities can live in harmony with nature, the *strategies* by which such communities would go about doing this, and the *model* of how it is actually to be carried out. Specifically, I mean institutions guided by three essential tasks: (1) to gather the scholarship and lore that teaches us the characteristics of the species and habitats of our specific local area, from eco-niche to bioregion; (2) to inaugurate projects of rehabilitation, chiefly by ecological restoration that returns specific areas of the land and its species to their natural, largely wild state, within which humans fit their social and cultural constructs; and (3) to develop human communities, small-scale and eco-centered, that will carry out these tasks and guide us toward living within our restored eco-niches on the species level.

I have in mind something that might be compared, within European history, to the time after the fall of Rome when there emerged a small-scale, community-based, agriculturally rooted society and along with it the invention of the monastery, the institution more than any other that kept alive the wisdom of the past, that provided models of a new way of living, that became the source of creation and invention, not to mention inspiration and dedication, for the next thousand years. I am suggesting that the most important institution we can begin to create right now is something we might think of as an *"ecostery"*—a small community of men and women living and working together to learn about and restore important, sacred, and fructive portions of the earth to their fullest complexity and productivity, living within and keeping holy and learning from these ecosystems, systems that are wild and free and know us as one more large mammalian species marked especially by a capacity to carry on knowledge through myth and ritual and by the ability, unique in humans, to blush. Ecosteries that, however odd they may look now, come to be understood as the only ecologically based way of human existence in the future, where there is kept alive for at least the next thousand years the minority sensibility that has existed for centuries, even as the Modern Age was forming and marginalizing it, from St. Francis to Aldo Leopold, from the Celtic witches to Rachel Carson, the sensibility that has always reminded us of the right method of living on the earth; where there is developed and spread the system of values that reminds us of the inherent tragedy of the modern industrial way and teaches us that though we may have—I suppose we will have—the knowledge of how to cross the oceans, to make war, to build skyscrapers, to construct atomic bombs, to splice genes, none will choose to do those things, because they transgress the will of Gaea, they bespeak an alien, violent, disregardful, and nature-hating culture.

I am suggesting, in sum, that we understand our tasks right now to be the pursuit of scholarship, restoration, and community—not as separate tasks, you see, but as interrelated—and that we understand our goal as the building of these model ecosteries, in urban settings as well as rural, working to reinhabit the land with the wisdom that the original peoples had who lived there first. I know, of course, even as I say all this that it seems daunting and slightly mad. I know these are not easy tasks, especially in our current world—I myself have been struggling for a year and more just to give birth to a restoration project in New York City—and the forces that resist them are great. But I also know that there are even now some suggestive models, glimpses of how this all might work. Ecological restoration is being done across the country; people in community are understanding ecosystems in a bioregional way; quasi think tanks like Sister Miriam McGillis's farm and the permaculture centers and the Ecostery Foundation and the E. F. Schumacher Library are assembling the wisdom and the lore. And I know these tasks are the ones that must be done, one way or another, starting now, starting wherever the vision can locate.

I have come to think of the ecostery as something like the extra horse. You may know that fable of the father who died owning seventeen horses, and his will decreed that half should go to his first son, a third to the second, and a ninth to the third. Well, it was a plainly insolvable problem, and try as they might the children could not put those horses into groups that would satisfy their father's wishes—there was no way to take a half or a third or a ninth of seventeen.

Eventually they took their problem to the local wise man, who said, "I understand your problem and your dilemma. Let me help you. I will give you one of my horses."

The sons were perplexed—what good would that do?

"Well," said the wise man, "then you will have eighteen horses, and the first son may have half, that's nine; the second a third, that's six, and the third son a ninth, that's two—so you will be able to do as your father asked."

The sons of course were delighted and sat beaming at the old man, shaking his hand in gratitude. "But then, of course," the old man added, "you will have seventeen horses—nine plus six is fifteen plus two is seventeen—and so you may give the extra horse back to me. As soon as you have finished with it, of course."

The problem of the late twentieth century appears to be insolvable. But it is just possible that in the ecostery, or something very much like it as an ecological model, we might have our extra horse—the small, appropriate, organic, living solution that will finally allow us to understand and become a cooperative part of nature in her fullest, which means to let ourselves be, as we were no doubt meant to be, connected inextricably to the infinite web of life.

—1990

Winona LaDuke

Voices from White Earth: Gaa-waabaabiganikaag

The following lecture, originally titled "Learning from Native Peoples," is, in a fundamental way, among the most important of this lecture series. In addition to several of the authors in the series—Kirkpatrick Sale, John McKnight, Wes Jackson—E. F. Schumacher himself wrote of a sensibility, a paradigm, a worldview, in which human beings might exist in long-lived intimacy and harmony with the natural world. But for them, as for most of us, that possibility remains a longing, an instinctual hope for a condition we have never known. For Winona LaDuke it is a living heritage, the beleaguered but surviving belief system and chosen way of life of her people, the Mississippi band of Anishinabeg of the White Earth Reservation.

Harvard graduate Winona LaDuke is a natural leader and a compelling spokesperson and interpreter of Native American views. In her lecture she contrasts what she calls the settler mentality with that of her people and their sense of place, of home. She examines first the expression of the worldview of settlers, then that of indigenous peoples. Breathing life and experience into Schumacher's invoking of "the truths revealed in nature's living processes," LaDuke states unequivocally: "Indigenous peoples believe fundamentally in natural law and a state of balance . . . Natural law . . . is the highest law . . . superior to the laws made by nations, states, and municipalities. It is the law to which we are all accountable."

In her lecture Winona LaDuke offers a sense of what it is like to live according to such law, within the cycles of an animate, mutually interactive world of matter and spirit. She speaks personally of the suffering of indigenous peoples and of the struggle she, her family, and the tribe of the White Earth Reservation have undertaken to reclaim their ancestral lands. She describes the sustainable economies they derive from the land, claiming rightly that "traditional ecological knowledge is absolutely essential for the future." Like Schumacher, she is ultimately convinced of the essential value of community, stating, "Community is the only thing in my experience that is sustainable." She concludes: "Keewaydahn. It's our way home."

Thank you for inviting me to come here and talk about some of the things that are important to the Anishinabeg and to the wider community of native peoples. Today I would like to talk about *keewaydahn*, which means "going home" in the Anishinabeg language. It's something like what Wes Jackson said in his lecture earlier in today's program about the process of going home and finding home. I think that is essentially what we need to be talking about. It is a challenge that people of this society face in belonging to a settler culture. They have been raised in this land, but they do not know its ceremony, its song, or its naming. Early settlers re-used names from other places, calling their settlements "New England," "New Haven," and "New York." But at the same time there are many indigenous names that co-exist with them. I think naming, as well as knowing *why* names are, is very important in restoring your relationship with the earth and finding your place. Restoring this relationship is our challenge.

To introduce myself, I'll tell you a little bit about my work and about where I come from. I'm basically a community organizer, like a lot of you. I returned to the White Earth Reservation about ten years ago after being raised off-reservation, which is a common circumstance for our people. I then began to work on the land issue, trying to win back or buy back our reservation lands. In our community I am identified as Muckwuck or Bear clan, Mississippi band, Anishinabeg. That's my place in the universe. The headwaters of the Mississippi are on our reservation; where the river starts is where we are in the world.

Anishinabeg is our name for ourselves in our own language; it means "people." We are called Ojibways in Canada and Chippewas in the United States. Our aboriginal territory, and where we live today, is in the northern part of five American states and the southern part of four Canadian provinces. It's in the center of the continent and is called the Wild Rice Bowl or the Great Lakes region. Today we are probably the single largest native population in North America: there are at least two hundred and fifty thousand of us. We're on both sides of the border, and most people don't know who we are or know much about us. That ignorance stems in part from the way Americans are taught about native people.

There are about seven hundred different native communities in North America. Roughly one hundred are Ojibway or Anishinabeg communities, but we're
different bands. In Alaska there are two hundred native communities; in California there are eighty. In Washington state there are fourteen different kinds of
Indian people living on the Yakima Reservation alone. All different kinds of
indigenous people live in North America—all culturally and historically diverse.
The same situation is found on a larger scale when you look at the entire continent, the Western Hemisphere, and the world. I want you to rethink the geography of North America in terms of cultural geography, in terms of land occupancy.

Now, if you look at the United States, about 4 percent of the land is held by
Indian people. That is the extent of today's Indian reservations. The Southwest has
the largest native population, and there's a significant population on the Great
Plains. In northern Minnesota there are seven big reservations, all Ojibway or
Anishinabeg. But if you go to Canada, about 85 percent of the population north
of the fiftieth parallel is native. So if you look at it in terms of land occupancy and
geography, in about two-thirds of Canada the majority of the population is native.
I'm not even including Nunevat, which is an Inuit-controlled area the size of
India in what used to be called the Northwest Territories.

If you look at the whole of North America, you find that the majority of the
population is native in about a third of the continent. Within this larger area,
indigenous people maintain their own ways of living and their cultural practices.
This is our view of the continent, and it is different from the view of most other
North Americans. When we look at the United States and Canada, we see our
reservations and reserves as islands in the continent. When Indian people talk
about their travels, they often mention reservations rather than cities: "I went to
Rosebud, and then I went over to North Cheyenne." This is the indigenous view
of North America.

Going beyond North America, I want to talk about the Western Hemisphere
and the world from an indigenous perspective. My intent is to present you with
an indigenous worldview and our perception of the world. There are a number of
countries in the Western Hemisphere in which native peoples are the majority of
the population: in Guatemala, Ecuador, Peru, Bolivia. In some South American
countries we control as much as 22 to 40 percent of the land. Overall, the
Western Hemisphere is not predominantly white. Indigenous people continue
their ways of living based on generations and generations of knowledge and
practice on the land.

On a worldwide scale there are about five thousand nations and a hundred and
seventy states. Nations are groups of indigenous peoples who share common
language, culture, history, territory, and government institutions. That is how

international law defines a nation. And that is who *we* are: nations of people who have existed for thousands of years. There are about a hundred and seventy—maybe more now, about a hundred and eighty-five—states that are recognized by the United Nations. For the most part, these states are the result of colonial empires or colonial demarcations. And whereas indigenous nations have existed for thousands of years, many of the states in existence at the end of the twentieth century have been around only since World War II. That is a big difference. Yet the dominant worldview of industrial society is determined by these young states, not by the five thousand ancient nations.

The estimated number of indigenous people in the world depends on how you define indigenous people. It is said that there are currently about five hundred million of us in the world today, including such peoples as the Tibetans, the Masai, the Wara Wara, and the Quechua. I define indigenous peoples as those who have continued their way of living for thousands of years according to their original instructions.

That is a quick background on indigenous people. It should help you understand that my perspective, the perspective of indigenous peoples, is entirely different from that of the dominant society in this country.

Indigenous peoples believe fundamentally in natural law and a state of balance. We believe that all societies and cultural practices must exist in accordance with natural law in order to be sustainable. We also believe that cultural diversity is as essential as biological diversity to maintaining sustainable societies. Indigenous peoples have lived on earth sustainably for thousands of years, and I suggest to you that indigenous ways of living are the only sustainable ways of living. Because of that, I believe there is something to be learned from indigenous thinking and indigenous ways. I don't think many of you would argue that industrial society is sustainable. I think that in two or three hundred years this society will be extinct because a society based on conquest cannot survive when there's nothing left to conquer.

Indigenous people have taken great care to fashion their societies in accordance with natural law, which is the highest law. It is superior to the laws made by nations, states, and municipalities. It is the law to which we are all accountable. There are no Twelve Commandments of natural law, but there are some things that I believe to be true about natural law. And this is my experience from listening to a lot of our older people. What I am telling you is not really my opinion; it's based on what has happened in our community, on what I've heard people say, and on their knowledge. We have noticed that much in nature is cyclical: the movements of moons, the tides, the seasons, our bodies. Time itself, in most indigenous worldviews, is cyclical. We also have experienced and believe that it is our essential nature and our need always to keep a balance in nature. Most indigenous ceremonies, if you look to their essence, are about the restoration of balance. That is our intent: to restore, and then to retain, balance. Nature itself continually tries to balance, to equalize.

According to our way of living and our way of looking at the world, most of the world is animate. This is reflected in our language, Anishinabemowin, in which most nouns are animate. Mandamin, the word for corn, is animate; mitig, the word for tree, is animate; so is the word for rice, manomin, and the word for rock or stone, asin. Looking at the world and seeing that most things are alive, we have come to believe, based on this perception, that they have spirit. They have standing on their own. Therefore, when I harvest wild rice on our reservation up north, I always offer asemah, tobacco, because when you take something, you must always give thanks to its spirit for giving itself to you, for it has a choice whether to give itself to you or not. In our cultural practice, for instance, it is not because of skill that a hunter can harvest a deer or a caribou; it is because he or she has been honorable and has given asemah. That is how you are able to harvest, not because you are a good hunter but because the animal gives itself to you. That is our perception.

And so we are always very careful when we harvest. Anthropologists call this reciprocity, which means something anthropological, I guess. But from our perspective it means that when you take, you always give. This is about balance and equalness. We also say that when you take, you must take only what you need and leave the rest. Because if you take more than you need, that means you are greedy. You have brought about imbalance, you have been selfish. To do this in our community is a very big disgrace. It is a violation of natural law, and it leaves you with no guarantee that you will be able to continue harvesting.

We have a word in our language that describes the practice of living in harmony with natural law: minobimaatisiiwin. This word describes how you live your life according to natural law, how you behave as an individual in relationship with other individuals and in relationship with the land and all the things that are animate on the land. Minobimaatisiiwin is our cultural practice; it is what you strive towards as an individual as well as collectively as a society.

We have tried to retain this way of living and of thinking in spite of all that has happened to us over the centuries. I believe we do retain most of these practices to a great extent in many of our societies. In our community they are overshadowed at times by industrialism, but they still exist.

I would like to contrast what I've told you about indigenous thinking with what I call "industrial thinking." I think the Lakota have the best term to describe it. It actually refers to white people, although they are not the only ones who think this way. Indigenous peoples have interesting terms for white people: they are usually not just words, they are descriptions encapsulated in a word. I will tell you about one: the Lakota word for a white person is *wasichu*. It derives from the first time the Lakota ever saw a white person. There was a white man out on the prairie in the Black Hills, and he was starving. He came into a Lakota camp in the middle of the night, and the Lakota of course were astonished to see him. They began to watch him to see what he was doing. He went over to the food, took something, and ran away. A little while later, the Lakota looked to see what he had taken: he had stolen a large amount of fat. So the Lakota word for a white person, *wasichu*, means "he who steals the fat." Now, that is a description that doesn't necessarily have to do with white people, but taking more than you need has to do with industrial society. He who steals the fat. That's what I'm talking about when I refer to the industrial worldview.

Industrial thinking is characterized by several ideas that run counter to indigenous ideas. First, instead of believing that natural law is preeminent, industrial society believes that humans are entitled to full dominion over nature. It believes that man—and it is usually man of course—has some God-given right to all that is around him, that he has been created superior to the rest.

Second, instead of modeling itself on the cyclical structure of nature, this society is patterned on linear thinking. I went all the way through its school system, and I remember how time, for example, is taught in this society. It's taught on a timeline, usually one that begins around 1492. It has some dates on it that were important to someone, although I could never figure out to whom. The timeline is a clear representation of this society's linear way of thinking. And certain values permeate this way of thinking, such as the concept of progress. Industrial society wants to keep making progress as it moves down the timeline, progress defined by things like technological advancement and economic growth. This value accompanies linear thinking.

Third, there is the attitude toward what is wild as opposed to what is cultivated or "tame." This society believes it must tame the wilderness. It also believes in the superiority of civilized over primitive peoples, a belief that also follows a linear model: that somehow, over time, people will become more civilized. Also related of course is the idea behind colonialism: that some people have the right to civilize other people. My experience is that people who are viewed as "primitive" are generally people of color, and people who are viewed as "civilized" are those of European descent. This prejudice still permeates industrial society and in fact even permeates "progressive" thinking. It holds that somehow people of European descent are smarter—they have some better knowledge of the world than the rest of us. I suggest that this is perhaps a racist worldview and that it has racist implications. That is, in fact, our experience.

Fourth, industrial society speaks a language of inanimate nouns. Even words for the land are becoming inanimate. Jerry Mander discusses this idea when he talks about the "commodification of the sacred." Industrial language has changed things from being animate, alive, and having spirit to being inanimate, mere objects and commodities of society. When things are inanimate, "man" can view them as his God-given right. He can take them, commodify them, and manipulate them in society. This behavior is also related to the linear way of thinking.

Fifth, the last aspect of industrial thinking I'm going to talk about (although it's always unpopular to question it in America), is the idea of capitalism itself. In this country we are taught that capitalism is a system that combines labor, capital, and resources for the purpose of accumulation. The capitalist goal is to use the least labor, capital, and resources to accumulate the most profit. The intent of capitalism is accumulation. So the capitalist's method is always to take more than is needed. Therefore, from an indigenous point of view capitalism is inherently out of harmony with natural law.

Based on this goal of accumulation, industrial society practices conspicuous consumption. Indigenous societies, on the other hand, practice what I would call "conspicuous distribution." We focus on the potlatch, the giveaway, an event that carries much more honor than accumulation does. In fact, the more you give away, the greater your honor. We make a great deal of these giveaways, and industrial society has something to learn from them.

Over the past five hundred years the indigenous experience has been one of conflict between the indigenous and the industrial worldviews. This conflict has manifested itself as holocaust. That is our experience. Indigenous people understand clearly that this society, which has caused the extinction of more species in the past hundred and fifty years than the total species extinction from the Ice Age to the mid-nineteenth century, is the same society that has caused the extinction of about two thousand different indigenous peoples in the Western Hemisphere alone. We understand intimately the relationship between extinction of species and extinction of peoples, because we experience both. And the extinction continues. Last year alone the Bureau of Indian Affairs, which has legal responsibility for people like myself—legally, I'm a ward of the federal government—declared nineteen different indigenous nations in North America extinct. The rate of extinction in the Amazon rainforest, for example, has been one indigenous people per year since 1900. And if you look at world maps showing cultural and biological distribution, you find that where there is the most cultural diversity, there is also the most biological diversity. A direct relationship exists between the two. That is why we argue that cultural diversity is as important to a sustainable global society as biological diversity.

Our greatest problem with all of this in America is that there has been no recognition of the cultural extinction, no owning up to it, no atonement for what happened, and no education about it. When I ask people how many different kinds of Indians they can identify, they can name scarcely any. America's mythology is based on the denial of the native—of native humanity, even of native existence. Nobody admits that the holocaust took place. This is because the white settlers believed they had a God-given right to the continent, and anyone with this right wouldn't recognize what happened as holocaust. Yet it was a holocaust of unparalleled proportions: Bartholomew de las Casas and other contemporaries of Columbus estimated that fifty million indigenous people in the Western Hemisphere perished in a sixty-year period. In terms of millions of people, this was probably the largest holocaust in world history.

Now, it is not appropriate for me to say that my holocaust was worse than someone else's. But it is absolutely correct for me to demand that my holocaust be recognized. And that has not happened in America. Instead, nobody knows anything about us, not even educated people. Why? Because this system is based on a denial of our existence. We are erased from the public consciousness because if you have no victim, you have no crime. As I said, most Americans can hardly name a single Indian nation. Those who can are only able to name those that have been featured in television Westerns: Comanche, Cheyenne, Navajo, Sioux, Crow. The only image of a native that is widely recognized in this society is the one shown in Westerns, which is a caricature. It is a portrayal created in Hollywood or in cartoons or more recently to a minimal degree in "New Age" paraphernalia. In this society we do not exist as full human beings with human rights, with the same rights to self-determination, to dignity, and to land—to territorial integrity—that other people have.

The challenge that people of conscience in this country face is to undo and debunk the mythology, to come clean, become honest, understand the validity of our demands, and recognize our demands. People must see the interlocking interests between their own ability to survive and indigenous peoples' continuing cultural sustainability. Indigenous peoples have lived sustainably in this land for thousands of years. I am absolutely sure that our societies could live without yours, but I'm not so sure that your society can continue to live without ours. This is why indigenous people need to be recognized now and included in the discussion of the issues affecting this country's future.

I'd like to tell you now about indigenous peoples' efforts to protect our land and restore our communities. All across this continent there are native peoples— in small communities with populations of one hundred, five hundred, even five thousand—who are trying to regain control of their community and their territory. I could tell you many stories of these different struggles, but I'll use my own community as an example. Here is our story.

The White Earth Reservation, located at the headwaters of the Mississippi, is thirty-six by thirty-six miles square, which is about 837,000 acres. It is very good land. A treaty reserved it for our people in 1867 in return for relinquishing a much larger area of northern Minnesota. Of all our territory, we chose this land for its richness and diversity. There are forty-seven lakes on the reservation. There's maple sugar; there are hardwoods and all the different medicine plants my people use—our reservation is called "the medicine chest of the Ojibways." We have wild rice; we have deer; we have beaver; we have fish—every food we need. On the eastern part of the reservation there are stands of white pine. On the part farthest west there used to be buffalo, but this area is now farmland, situated in the Red River Valley. That is our area, the land reserved to us under treaty.

Our traditional forms of land use and ownership are similar to those of a community land trust. The land is owned collectively, and we have individual or, more often, family-based usufruct rights: each family has traditional areas where it fishes and hunts. In our language the words Anishinabeg akiing describe the concept of land ownership. They translate as "the land of the people," which doesn't infer that we own our land but that we belong on it. Our definition doesn't stand up well in court, unfortunately, because this country's legal system upholds the concept of private property.

Our community enforces its traditional practices by adhering to minobimaatisii- win. Historically, this involved punishing people who transgressed these rules. For instance, in our community the worst punishment historically—we didn't have jails—was banishment. That still exists in our community to a certain extent. Just imagine if the worst punishment in industrial society were banishment! With us, each person wants to be part of the community.

We have also maintained our practices by means of careful management and observation. For example, we have "hunting bosses" and "rice chiefs," who make sure that resources are used sustainably in each region. Hunting bosses oversee trap-line rotation, a system by which people trap in an area for two years and then move to a different area to let the land rest. Rice chiefs coordinate wild rice harvesting. The rice on each lake is unique: each has its own taste and ripens at its own time. We also have a "tally man," who makes sure there are enough animals for each family in a given area. If a family can't sustain itself, the tally man moves them to a new place where animals are more plentiful. These practices are sustainable.

My children's grandfather, who is a trapper, lives on wild animals in the wintertime. When he intends to trap beavers, he reaches his hand into a beaver house and counts how many beavers are in there. (Beavers are not carnivorous; they won't bite.) By counting, he knows how many beavers he can take. Of course, he has to count only if he hasn't already been observing that beaver house for a long time. This is a very sustainable way to trap, one based on a kind of thorough observation that can come only with residency. Further, I suggest that this man knows more about his ecosystem than any Ph.D. scholar who studies it from the university.

As I have indicated, the White Earth Reservation is a rich place. And it is our experience that industrial society is not content to leave other peoples' riches alone. Wealth attracts colonialism: the more a native people has, the more colonizers are apt to covet that wealth and take it away—whether it is gold or, as in our case, pine stands and Red River Valley farmland. A Latin American scholar named Eduardo Galeano has written about colonialism in communities like mine. He says: "In the colonial to neo-colonial alchemy, gold changes to scrap metal and food to poison. We have become painfully aware of the mortality of wealth, which nature bestows and imperialism appropriates." For us, our wealth was the source of our poverty: industrial society could not leave us be.

Our reservation was created by treaty in 1867; in 1887 the General Allotment Act was passed on the national level, not only to teach Indians the concept of private property but also to facilitate the removal of more land from Indian Nations. The federal government divided our reservation into eighty-acre parcels of land and allotted each parcel to an individual Indian, hoping that through this change we would somehow become yeoman farmers, adopt the notion of progress, and become civilized. But the allotment system had no connection to our traditional land tenure patterns. In our society a person harvested rice in one place, trapped in another place, got medicines in a third place, and picked berries in a fourth. These locations depended on the ecosystem; they were not necessarily contiguous. But the government said to each Indian, "Here are your eighty acres; this is where you'll live." Then, after each Indian had received an allotment, the rest of the land was declared "surplus" and given to white people to homestead. On our reservation almost the entire land base was allotted except for some pinelands that were annexed by the state of Minnesota and sold to timber companies. What happened to my reservation happened to reservations all across the country.

The federal government was legally responsible for this; they turned our land into individual eighty-acre parcels, and then they looked the other way and let the state of Minnesota take some of our land and tax what was left. When the Indians couldn't pay the taxes, the state confiscated the land. How could these people pay taxes? In 1900 or 1910 they could not read or write English.

I'll tell you a story about how my great-grandma was cheated by a loan shark. She lived on Many Point Lake, where her allotment was. She had a bill at the local store, the Fairbanks grocery store, and she had run it up because she was waiting until fall when she could get some money from trapping or from a treaty annuity. So she went to a land speculator named Lucky Waller, and she said, "I need to pay this bill." She asked to borrow fifty bucks from him until treaty payment time, and he said: "Okay, you can do that. Just sign here and I'll loan you that fifty bucks." So she signed with her thumbprint and went back to her house on Many Point Lake. About three months later she came in to repay him the fifty bucks, and the loan shark said: "No, you keep that money; I bought land from you instead." He had purchased her eighty acres on Many Point Lake for fifty bucks. Today that location is a Boy Scout camp.

This story could be retold again and again in our communities. It is a story of land speculation, greed, and unconscionable contracts, and it exemplifies the process by which native peoples were dispossessed of their land. The White Earth Reservation lost two hundred and fifty thousand acres to the state of Minnesota because of unpaid taxes. And this was done to native peoples across the country: on a national average reservations lost a full two-thirds of their land this way.

By 1920, 99 percent of original White Earth Reservation lands were in non-Indian hands. By 1930 many of our people had died from tuberculosis and other diseases, and half of our remaining population lived off-reservation. Three generations of our people were forced into poverty, chased off our land, and made refugees in this society. Now a lot of our people live in Minneapolis. Of twenty thousand tribal members only four or five thousand live on reservation. That's because we're refugees, not unlike other people in this society.

Our struggle is to get our land back. That's what we've been trying to do for a hundred years. By 1980, 93 percent of our reservation was still held by non-Indians. That's the circumstance we are in at the end of the twentieth century. We have exhausted all legal recourse for getting back our land. If you look at the legal system in this country, you will find that it is based on the idea that Christians have a God-given right to dispossess heathens of their land. This attitude goes back to a papal bull of the fifteenth or sixteenth century declaring that Christians have a superior right to land over heathens. The implication for native people is that we have no legal right to our land in the United States or in Canada. The only legal recourse we have in the United States is the Indian Claims Commission, which pays you for land; it doesn't return land to you. It compensates you at the 1910 market value for land that was seized. The Black Hills Settlement is one example. It's lauded as a big settlement—one that gives all this money to the Indians—but it's only a hundred and six million dollars for five states. That's the full legal recourse for Indian people.

In the case of our own reservation, we had the same problem. The Supreme Court ruled that to regain their land Indian people had to have filed a lawsuit within seven years of the original time of taking. Now, legally we are all people who are wards of the federal government. I have a federal enrollment number. Anything to do with the internal matters of Indian governments is subject to the approval of the Secretary of the Interior. So the federal government, which is legally responsible for our land, watched its mismanagement and did not file any lawsuits on our behalf. The Courts are now declaring that the statute of limitations has expired for the Indian people, who, when their land was taken, could not read or write English, had no money or access to attorneys to file suit, and were the legal wards of the state. We have therefore, the courts claim, exhausted our legal recourse and have no legal standing in the court system. That is what has happened in this country with regard to Indian land issues.

We have fought federal legislation for a decade without success. Yet we look at the situation on our reservation and realize that we must get our land back. We do not really have any other place to go. That's why we started the White Earth Land Recovery Project.

The federal, state, and county governments are the largest landholders on the reservation. It is good land still, rich in many things; however, when you do not control your land, you do not control your destiny. That's our experience. What has happened is that two-thirds of the deer taken on our reservation are taken by non-Indians, mostly by sports hunters from Minneapolis. In the Tamarac National Wildlife Refuge nine times as many deer are taken by non-Indians as by Indians, because that's where sports hunters from Minneapolis come to hunt. Ninety percent of the fish taken on our reservation is taken by white people, and most of them are taken by people from Minneapolis who come to their summer cabins and fish on our reservation. Each year in our region, about ten thousand acres are being clear cut for paper and pulp in one county alone, mostly by the Potlatch Timber Company. We are watching the destruction of our ecosystem and the theft of our resources; in not controlling our land we are unable to control what is happening to our ecosystem. So we are struggling to regain control through the White Earth Land Recovery Project.

Our project is like several others in Indian communities. We are not trying to displace people who have settled there. A third of our land is held by the federal, state, and county governments. That land should just be returned to us. It certainly would not displace anyone. And then we have to ask the question about absentee land ownership. It is an ethical question that should be asked in this country. A third of the *privately* held land on our reservation is held by absentee landholders: they do not see that land, do not know it, and do not even know where it is. We ask these people how they feel about owning land on a reservation, hoping we can persuade them to return it.

Approximately sixty years ago in India the Gramdan movement dealt with similar issues. Some million acres were placed in village trust as a result of the moral influence of Vinoba Bhave. The whole issue of absentee land ownership needs to be addressed—particularly in America, where the idea of private property is so sacred, where somehow it is ethical to hold land that you never see. As Vinoba said, "It is highly inconsistent that those who possess land should not till it themselves, and those who cultivate should possess no land to do so."

Our project also acquires land. It owns about nine hundred acres right now. We bought some land as a site for a roundhouse, a building that holds one of our ceremonial drums. We bought back our burial grounds, which were on private land, because we believe that we should hold the land our ancestors lived on. These are all small parcels of land. We also just bought a farm, a fifty-eight-acre organic raspberry farm. In a couple of years we hope to get past the "You Pick" stage into jam production. It is a very slow process, but our strategy is based on this recovery of the land and of our cultural and economic practices.

We are a poor community. People look at our reservation and comment on the 85 percent unemployment—they do not realize what we do with our time. They have no way of valuing our cultural practices. For instance, 85 percent of our people hunt, taking at least one or two deer annually, probably in violation of federal game laws; 75 percent of our people hunt for small game and geese; 50 percent of our people fish by net; 50 percent of our people sugarbush and garden on our reservation. About the same percentage harvest wild rice, not just for themselves; they harvest it to sell. About half of our people produce handcrafts. There is no way to quantify this in America. It is called the "invisible economy" or the "domestic economy." Society views us as unemployed Indians who need wage-earning jobs. That is not how we view ourselves. Our work is about strengthening and restoring our traditional economy. I have seen our people trained and re-trained for off-reservation jobs that do not exist. I don't know how many Indians have gone through three or four carpenter and plumber training programs. It doesn't do any good if, after the third or fourth time, you still don't have a job.

Our strategy is to strengthen our own traditional economy (thereby strengthening our traditional culture as well) so that we can produce 50 percent or more of our own food, which we then won't need to buy elsewhere, and can eventually produce enough surplus to sell. In our case most of our surplus is in wild rice. We are rich in terms of wild rice. The Creator, Gitchi Manitu, gave us wild rice—said we should eat it and should share it; we have traded it for thousands of years. A lot of our political struggle is, I am absolutely sure, due to the fact that Gitchi Manitu did not give wild rice to Uncle Ben to grow in California. Commercial wild rice is totally different from the rice we harvest, and it decreases the value of our rice when marketed as authentic wild rice.

We've been working for several years now to increase the price of the rice we gather from fifty cents per pound to a dollar per pound, green. We are trying to market our rice ourselves. We try to capture the "value added" in our community by selling it ourselves. We went from about five thousand pounds of production on our reservation to about fifty thousand pounds last year. This is our strategy for economic recovery.

Other parts of our strategy include language immersion programs to restore our language and revival of drum ceremonies to restore our cultural practices. These are part of an integrated restoration process that is focused on the full human being.

In the larger picture, in Wisconsin and Minnesota our community is working hard to exercise specific treaty rights. Under the 1847 treaty, we have reserved-use rights to a much larger area than just our reservations. These are called extra-territorial treaty rights. We didn't say we were going to live there, we just said we wanted to keep the right to use that land in our usual and accustomed ways. This has led us to a larger political strategy, for although our harvesting practices are sustainable, they require an almost pristine ecosystem in order to take as much fish and grow as much rice as we need. To achieve this condition the tribes are entering into a co-management agreement in northern Wisconsin and northern Minnesota to prevent further environmental degradation as a first step toward preserving an extra-territorial area in accordance with treaty rights.

There are many similar stories all across North America. A lot can be learned from these stories, and we can share a great deal in terms of your strategies and what you're trying to do in your own communities. I see this as a relationship among people who share common issues, common ground, and common agendas. It is absolutely crucial, however, that our struggle for territorial integrity and economic and political control of our lands not be regarded as a threat by this society. Deep-set in settler minds I know there's fear of the Indian having control. I've seen it on my own reservation: white people who live there are deathly afraid of our gaining control over half our land base, which is all we're trying to do. I'm sure they are afraid we will treat them as badly as they have treated us. I ask you to shake off your fear.

There's something valuable to be learned from our experiences, from the James Bay hydroelectric project in Quebec, for example, and from the Shoshone sisters in Nevada fighting the missile siting. Our stories are about people with a great deal of tenacity and courage, people who have been resisting for centuries. We are sure that if we do not resist, we will not survive. Our resistance will guarantee our children a future. In our society we think ahead to the seventh generation; however, we know that the ability of the seventh generation to sustain itself will be dependent on our ability to resist now.

Another important consideration is that traditional ecological knowledge is unheard knowledge in this country's institutions. Nor is it something an anthropologist can extract by mere research. Traditional ecological knowledge is passed from generation to generation; it is not an appropriate subject for a Ph.D. dissertation. We who live by this knowledge have the intellectual property rights to it, and we have the right to tell our stories ourselves. There is a lot to be learned from our knowledge, but you need us in order to learn it, whether it is the story of my children's grandfather reaching his hand into that beaver house or of the Haida up on the Northwest coast, who make totem poles and plank houses. The Haida say they can take a plank off a tree and still leave the tree standing. If Weyerhaeuser could do that, I might listen to them, but they cannot.

Traditional ecological knowledge is absolutely essential for the future. Crafting a relationship between us is absolutely essential. Native people are not quite at the table in the environmental movement—for example, in the management of the Great Plains. Environmental groups and state governors sat down and talked about how to manage the Great Plains, and nobody asked the Indians to come to the table. Nobody even noticed that there are about fifty million acres of Indian land out there in the middle of the Great Plains, land that according to history and law has never yet had a drink of water—that is, reservations have been denied water all these years because of water diversion projects. When water allocations are being discussed, someone needs to talk about how the tribes need a drink.

One proposal for the Great Plains is a Buffalo Commons, which would include one hundred and ten prairie counties that are now financially bankrupt and are continuing to lose people. The intent is to restore these lands ecologically, bringing back the buffalo, the perennial crops, and indigenous prairie grasses that Wes Jackson is experimenting with. I think we need to broaden the idea, though, because I don't think it should be just a Buffalo Commons; I think it should be an Indigenous Commons. If you look at the 1993 population in the area, you'll find that the majority are indigenous peoples who already hold at least fifty million acres of the land. We know this land of our ancestors, and we should rightly be part of a sustainable future for it.

Another thing I want to touch on is the necessity of shifting our perception. There is no such thing as sustainable development. Community is the only thing in my experience that is sustainable. We all need to be involved in building sustainable communities. We can each do that in our own way—whether it is European-American communities or Dené communities or Anishinabeg communities—returning to and restoring the way of life that is based on the land. To achieve this restoration we need to reintegrate with cultural traditions informed by the land. That is something I don't know how to tell you to do, but it is something you're going to need to do. Garrett Hardin and others are saying that the only way you can manage a commons is if you share enough cultural experiences and cultural values so that you can keep your practices in order and in check: minobimaatisiiwin. The reason we have remained sustainable for all these centuries is that we are cohesive communities. A common set of values is needed to live together sustainably on the land.

Finally, I believe the issues deep in this society that need to be addressed are structural. This is a society that continues to consume too much of the world's resources. You know, when you consume this much in resources, it means constant intervention in other peoples' land and countries, whether it is mine or whether it is the Crees' up in James Bay or someone else's. It is meaningless to talk about human rights unless you talk about consumption. And that's a structural change we all need to address. It is clear that in order for native communities to live, the dominant society must change, because if this society continues in the direction it is going, our reservations and our way of life will continue to bear the consequences. This society has to be changed! We have to be able to put aside its cultural baggage, which is industrial baggage. It's not sustainable. Do not be afraid of discarding it. That's the only way we're going to make peace between the settler and the native.

Miigwech. I want to thank you for your time. Keewaydahn. It's our way home.

—1993

Dana Lee Jackson

Women and the Challenge of the Ecological Era

While remarkably ahead of his time in many areas, E. F. Schumacher
was little acquainted with a movement, largely North American based,
that was rapidly gaining momentum at the same time his own work
with intermediate technology was finding international recognition.
This oversight is addressed here by Dana Jackson as she traces the integration of the environmental and feminist movements that began in
the early seventies and continue to gain in strength, as the world witnessed in the U.N. Women's Conference in Beijing in 1995.

Dana Jackson is an educator, writer, and administrator. She is also
the cofounder of The Land Institute in Salina, Kansas, a research and
education institute that has played a pioneering role in conducting research to develop a sustainable agriculture modeled on a prairie ecosystem. There she designed the intern education program and public
and community programs, such as the annual Prairie Festival and the
biennial organic gardening workshop. She also created a curriculum
called "Considerations for a Sustainable Society."

She is currently associate director of the Land Stewardship Project,
a Minnesota-based organization dedicated to the renewal of a stewardship ethic for farmland and the promotion of sustainable agriculture
and communities.

Jackson begins her Schumacher lecture with a brief review of what
she calls the ecological era before turning to her analysis of the parallel
and related feminist era. The implications of this most recent phase of

feminist studies have been extraordinary. As she points out, "Our goal must not be to change who dominates but to get rid of the model that justifies and promotes domination." At the crux of her argument is the thesis that our multiple environmental crises are the result of human domination over the natural world and that the hierarchical model of dominance is not appropriate in an ecological age.

To transform this destructive pattern, Jackson suggests that some of the feminine qualities formerly thought inappropriate in public life be introduced into that arena. The experience of women as nurturers, as compromisers in a relationship, as sustainers of the details of everyday life, and as negotiators in their approach to problem solving should be seen as assets. This is beginning to be evident in government at many levels. It is Jackson's theory that from an expanding mutual understanding and balance of power between men and women, as well as from reexamining our attitudes to the natural world, will come the partnerships that will guide us toward the next phase of the ecological era.

When E. F. Schumacher visited The Land Institute in March 1977, we were only seven months old. We were small, but our place was not beautiful. Three partially paid staff members and eight students worked and studied in a building under construction. Outside, in the mud around the building, there were piles of scavenged materials: lumber, 220 patio doors bought for a bargain price, scrap iron, and more. But Dr. Schumacher affirmed our efforts, and his ideas influenced the developing mission of this newest nonprofit devoted to sustainable alternatives.

Now we are fourteen years old with thirteen staff members and nine student interns. The Land Institute is still small by most standards but, we think, much more beautiful with our gardens, boardwalks, trees, research plots, and nearly one hundred acres of never-plowed native prairie. We have an office building in addition to the classroom we were constructing in 1977 plus barns and farm equipment. The two Carpathian walnut trees we planted in memory of Schumacher are thriving.

Many of the problems we discuss at The Land Institute would have interested Schumacher, but I doubt if my topic is one he would have chosen. While he obviously respected and quoted prominent women of his day, like the rest of the society he considered that the main role for women was predetermined by their sex. They were to care for their children at home. In the important essay on Buddhist economics in Small Is Beautiful, he described the three functions of work: to give a man a chance to utilize and develop his faculties, to enable him to overcome his ego-centeredness by joining with others in a common task, and to bring forth the goods and services needed for a becoming existence. When he said "a man," he meant a man. Women were not part of this description.

"Women on the whole," he said, "do not need an outside job." The legitimate need for women to have fulfilling work in addition to parenting responsibilities was not part of his range of considerations.

Attitudes about women's roles were already undergoing change in 1977 when Schumacher visited us, but I think he would be surprised at how actively women participate in mainstream society now. In general, society approves when they "utilize and develop their faculties" and engage in meaningful work, and it does not expect them to provide all the support services (cooking, cleaning, and laundry) that free their husbands to pursue careers. And just in time. Women's perspectives, values, and skills are needed as we respond to the complex problems making up the environmental crisis that is the theme of this year's lectures. We are in an age of ecology, what we might call the ecological era.

The challenge of the first part of the ecological era has largely been to recognize and understand our dependence upon nature, upon ecological systems, and to realize that humans cannot continue to ruthlessly exploit the nonhuman world to satisfy our needs and greeds.

The challenge of the second part of the ecological era—from 1990 on—is to transform our society so we can act on our ecological knowledge, change destructive patterns, and develop a sustainable society. Women must have a large part in this. I shall describe here the ecological era and, occurring in the same time frame, the feminist era. In my view, and the view of a growing number, the coming together of the ecological and feminist movements gives us a greater opportunity to change patterns that not only lead to the extinction of countless other species but also destroy what supports humans. We must change the underlying conceptual framework of Western society: a hierarchy that ranks white, heterosexual, male values, ideas, and work above that of women, people of color, and all other life forms. Certain attributes of women's culture must be employed to help us adapt to sustainable, ecological living patterns. What we might call a feminization of the culture will come about in response to the environmental crisis in the most decentralist social organizations of all, our families and partnerships. Let me begin by describing the ecological era.

The Ecological Era

In *Nature's Economy* Donald Worster writes that the Age of Ecology began "on the desert outside Almagordo, New Mexico, on July 16, 1945, with a dazzling fireball of light and a swelling mushroom cloud of radioactive gases." We had created a force capable of destroying the planet. Before Worster, in 1948 Fairfield Osborne, in his book *This Plundered Planet*, said he had come to understand towards the end of World War II that humans were involved in *another* war, one against nature. This was not the age-old literary theme of conflict with nature, in which nature was a worthy opponent. Osborne and Worster both recognized that nature was now the *victim* of our aggressive actions.

But I think we did not actually start the ecological era until this understanding became a part of the general public awareness, something that came about in the mid-1960s. By the time of the first Earth Day in 1970 the thinking, reading public in this country had become acquainted with several ecological concepts and had extended them to the human experience. We could see ecological damage that humans had created; we began to understand that what we do to nature, we do to ourselves.

Rachel Carson's *Silent Spring*, published in 1962, called attention to the flagrant misuse of persistent pesticides such as DDT, chlordane, heptachlor, and dieldrin and to their devastating effect on species other than insects. The chapter "Earth's Green Mantle" explained connections between plants and animals and described the concept of the web. This ecological concept was popularized around the time of the first Earth Day. And though, as David Ehrenfeld pointed out in *The Arrogance of Humanism*, "we greatly exaggerated the fragility of that web in developing our economic arguments for preserving natural resources," for those who had never thought much about our dependence upon nature, the concept was an eye-opener.

Next it was "carrying capacity" that became widely discussed when Paul Ehrlich's book *The Population Bomb* was published in 1968. The book stimulated people to ask questions like the following: How many people can the earth support? At what standard of living? By replacing what other life forms?

The idea of cycles in nature—nutrient cycles, life cycles, reproductive cycles—also became part of the public's awareness in the early stage of the ecological era.

The awakening of our minds to these and other ecological concepts led to an active grass-roots environmental movement in the early 1970s and the founding of alternative organizations such as the New Alchemy Institute, already established in 1969, and later The Land Institute in 1976. National environmental organizations had the support of their local letter-writing constituencies when promoting the significant national legislation of the period: the National Environmental Quality Act requiring environmental impact statements, establishment of the Environmental Protection Agency to enforce new laws regulating air and water quality, and legislation protecting wilderness. The Arab oil embargo of 1973 led to a growing awareness that natural resources were finite. We began to think about the fuel "cycles" of power plants, about net energy balances, and we experimented with conservation and renewable energy.

The ecological era continued into the 1980s in spite of Ronald Reagan and James Watt. Perhaps because of James Watt, the national environmental organizations became stronger and their leadership more professional but, sadly, not in proportion to the destructive forces that increased their membership. Three Mile Island, Bhopal, Chernobyl, Prince William Sound, Alaska—these plus other environmental catastrophes caused the public to acknowledge increasingly the inherent dangers of large-scale industrial technology. In the 1980s we wrestled with the clean-up and safe storage of toxic and radioactive wastes. We realized that both capitalism and socialism externalized the environmental costs of industrial growth and acted in ignorance of the second law of thermodynamics.

The relevance of ecology to farming became clearer in the 1980s as we studied the negative consequences of industrialized agriculture: soil loss, groundwater contamination, and the demise of family farm communities. In 1980 The Land Institute launched a research program based on nature as the teacher and the measure in agriculture. We set out to learn the wisdom of the prairie, a self-sustaining ecosystem that produced the soil that made Midwest corn and wheat fields productive. We continue our efforts to bring ecology and agriculture together as we attempt to develop prairie-like mixtures of perennial plants that produce seeds for people and livestock. We expect these domestic prairies to replace conventional crops on highly erodible soil. They will require no pesticides or herbicides and will use little or no chemical fertilizer. Tractor fuel consumption will be lower as we eliminate annual tillage for planting.

The conservation provisions in the 1985 farm bill were put there by the American people—city and country folk alike—to protect soil and water. Farmers were told that they must comply with conservation requirements on highly erodible land or be ineligible for subsidies. The discussion of ecology in agriculture increased as some researchers studied predator-prey relationships to control insect pests. We began hearing about studies of the role of legumes in the nitrogen fertilizer cycle and research on decomposition of organic material and release of soil nutrients.

In 1990 we are in the second stage of the ecological era. The second Earth Day celebration on April 22, 1990, focused on global environmental problems such as acid rain, the greenhouse effect, and ozone depletion. But everyone was reminded that solutions should be carried out *locally*, in particular places. One wonders if we will ever go beyond the Earth Day T-shirt level of consciousness. In this next stage of our ecological era we must do more than pick away at the symptoms of ecological disruption. It is essential to gain control over that which is responsible for our aggression against the earth, the hierarchal framework for society that is the basis of interaction between humans and the non-human world.

The Feminist Era

The ecological era appeared simultaneously with the feminist era. Changes in the status of women, in women's perception of themselves, in the opportunities and challenges they face have never before been so widespread nor so widely recognized as they have been in the period from the mid-sixties until now. I speak, of course, as a white, middle-class, married woman with three grown children, someone for whom the last twenty-five years have been a period of slow awakening. My own experience and the influence of women writers and feminist activists make me see the world in an entirely different way than I once did. My daughters and the young women interns at The Land Institute continue to enlighten me. Though I share much with women of color and lesbians, I am aware that their experiences have been in different contexts, and I know they have other opinions and proposals for change that I cannot adequately express.

Though I choose to describe the feminist era as beginning in the mid-sixties, parallel to the ecological era, I know that the groundwork of feminist philosophy was laid by women in the nineteenth century and feminist activism was born of the suffragettes in the early twentieth century. How much we owe to our fore-mothers—Sojourner Truth, Elizabeth Cady Stanton, Charlotte Perkins Gilman, Margaret Sanger, and others—for revealing the realities of sexist oppression. And how deprived our whole culture has been by the suppression of such books as *Woman, Church and State*, written by Matilda Joslyn Gage, first published in 1893 and reprinted in 1980.

I did not learn about these women when I was in school, nor in college. I was not taught that women abolitionists who attended the international anti-slavery convention in 1840 were not allowed to sit in the convention hall with the delegates or participate in the deliberations. Many women abolitionists were among the early advocates for women's rights: they spoke against the injustice of the common-law doctrine that considered wives to be chattel of husbands, that denied women the right to own property, that would not allow women to vote or hold office. When African-Americans were given the right to vote, white men still denied women—of all races—the same right. Why did the textbooks clearly teach the immorality of African-American slavery but not the immorality of women's oppression?

Women becoming adults in the 1950s, like me, did not question much. Textbook stereotypes, women's magazines, and movies all reinforced the belief that "a woman's place is in the home." This attitude kept women from competing for jobs with men who were World War II veterans, and it stimulated the consumer economy as women made a career of purchasing household goods for life in suburbia. Those who went to college and earned academic honors were not as much of a success by societal standards as those who dropped out after "catching a husband" and became adept at home decorating, dinner parties, and raising well-scrubbed children.

When I was twenty, I read Henrik Ibsen's play *A Doll's House*, first performed in 1879, and suddenly recognized the tyranny of patriarchy. I was horrified and depressed. But there was no one to talk to about my feelings, so I soon was again absorbed by the culture, became engaged, bought a copy of one of those glossy bride magazines and a cedar chest, and got married.

My mother-in-law approved of my temporary teaching job, saying that I was earning a PHT degree, Putting Hubby Through (a graduate program). She fully expected me to become a full-time homemaker again and stay home with our two preschool children when my husband finished his Ph.D. I expected this of myself. Friends in the Graduate Student Wives Club and I discussed our dissatisfactions, our buried intellectual interests, and the conflicts we felt because of our belief that it was the mother's duty to be home with the children.

The involvement of many women in the environmental movement has been an extension of the motherhood role. Women have always been involved in reform movements that they see as related to the welfare of their home and children. From promoting spittoons in the streets of frontier towns to prohibition of alcohol to working for air-pollution abatement and the safe disposal of hazardous or radioactive wastes, family well-being has been the impetus for action.

Friends and I arranged a "teach in" for homemakers at the Salina, Kansas, YWCA on Earth Day 1970, which then led to the organization of the Salina Consumers for a Better Environment. We lobbied grocery stores for less plastic packaging and more recyclable containers. We promoted tree planting. We also set up a speakers' bureau with self-educated women available to speak to clubs and service organizations about all the environmental issues of the day, everything from overpopulation to pesticides to declining fossil fuel supplies. My own interest in the issues—the personal environmental crusade I took on—kept me busy for most of the next decade as a professional citizen. The challenge of learning about many new subjects so we could give speeches and lobby lawmakers eased some of the suburban homemaker dissatisfaction Betty Friedan described in *The Feminine Mystique*, which had hit me hard when I read it in 1966. While still loyal homemakers for our husbands and children, we could also spend our days working in common cause with other women environmentalists. Undoubtedly we did some good, but from another perspective our involvement in working for the public good and our children's future sidetracked us from seeking personal fulfillment and independence.

Women became important in local chapters of mainstream environmental organizations in the 1970s. Too often they left the leadership to men and fell into the housekeeping chores: telephoning, licking stamps, baking cookies, and writing letters. This volunteer work force declined in the 1980s as more women took full-time jobs and professionals did more of the lobbying and office work. If we are to think globally and act locally, if we are to develop decentralist responses to the environmental crisis, we need to revitalize grass-roots organizations. But who will do the important volunteer work?

This brings us to the challenge for the next phase of the feminist era. The women's movement to this point has affected the way women work, how they relate to their families, and especially what they think about themselves and their aspirations. Contrasted to the 1950s, women expect to find careers in business, medicine, politics, and law. But greater equity in the workplace has not led to greater equity in the home. Studies show that women do a disproportionate share of the housework, take most of the responsibility for the children, and get less personal support from men than they give. Women develop their faculties on the job but still provide most of the physical and emotional support for the family. We have made great strides in social justice—except in our basic social unit, the home.

This inequity persists because the underlying conceptual framework for society is a hierarchy with white heterosexual men at the top. Men are considered to be more important than women and to pursue more important work. Men's patterns of thinking and making moral choices, of organizing ideas or work, of determining justice, of judging esthetics have been the *standards* for Western culture and the Judeo-Christian tradition.

These standards, carried out through industrialization, have molded the workplace in our country. Industrial values dominate: high production for profit is the bottom line, growth and more growth is the major goal, bigger or more is better. People must work at least forty hours a week—or fifty or sixty if they are aspiring corporate executives or academics—in order to hold jobs with decent pay. We cannot ignore the fact that many women are employed outside the home not because they have fulfilling work but because the family needs two salaries. Men—and now women—on that treadmill cannot reestablish a proper relationship with the earth, let alone an adequate relationship with their families and communities.

Women increasingly understand that working for their own release from male dominance cannot succeed unless they work to eliminate the dominance of one race over another, one age group over another. Sexism, heterosexism, ageism, racism, classism, and naturism are all the same problem. Women are not out to replace male dominance with female dominance but to correct the problem of dominance. Our goal must not be to change who dominates but to get rid of the model that justifies and promotes domination.

The environmental crises we face are the result of human domination over the natural environment. Humans have exploited the nonhuman world, treating other life forms as "the other" just as the dominant race has treated people of color as "the other" and just as men have treated women as the "other." The difference is that the consequence of our subjugating nature could be the destruction of ecosystems and the extinction of people of all races as well as many other species.

Now we must conclude that to live within the limits of natural ecosystems, to live sustainably, will require an entirely different way of relating to one another and to the earth. Rather than set up hierarchies of value, we must learn to deal with differences (gender, race, and species) by a process Riane Eisler, in *The Chalice and the Blade*, calls "linking" instead of "ranking." We must reestablish relationships with the natural world that will make us sensitive to the needs of other species.

A Synthesis of the Ecological Era and the Feminist Era

In this next phase of the ecological era and the feminist era, we must learn from nature and from women in order to transform our destructive patterns, but we cannot learn in a system that oppresses nature and women. The first step away from this system is to cultivate and elevate in importance some of the qualities and values most generally associated with women that can help us abandon our suicidal patterns. These are not to be considered innate characteristics, and they are not universally found in women, but they *are* identified more often with women than men, even though men express them also. Until recently, men have been criticized for exhibiting such qualities because that identified them with the inferior gender. But as we face a large number of environmental threats, not the least of which is still nuclear annihilation, we desperately need new standards of behavior. The feminization of our culture has already produced beneficial results in many workplaces, which leads us to believe that women's cultural patterns can benefit society on a broader scale also. These qualities are described in slightly different ways by different people. I've grouped them in four general categories, each of which includes a number of related traits.

First, women are considered to be nurturers. They take care of the physical and emotional needs of their families, but their strong nurturing impulse extends to all living things. They tend to place individual growth and fulfillment above abstractions. Women are attentive to the needs of nonhuman growing things such as pets, garden flowers and vegetables, and houseplants. Some claim that women are closer to nature, perhaps because their monthly cycle and their capacity to give birth and produce milk make them more tuned in to the world around them.

Second, women see themselves in relationship to others. Psychologists say that men are more likely to think of themselves as individuals who must accomplish things independently, while women tend to exhibit cooperative individualism. They see themselves as wives, mothers, friends, and members of groups and communities. Women empathize with others and are more adaptable and cooperative in group situations. They tend to integrate rather than separate, preferring networks to hierarchies.

Third, women have an attachment to the day-to-day process of sustaining life. They are used to taking care of many details at the same time. They will do the nitty-gritty work necessary to keep the household in good condition, complete projects, and organize events.

Fourth, women have a preference for negotiation as a means of problem solving that springs from an antipathy to violence. They tend to make moral choices based on causing the minimum of hurt, while men will tend to make moral choices based on rights and justice. Carol Gilligan points out in her book In a Different Voice that women do not like to make moral decisions based on dichotomies: either this is right or that is right. They prefer to look at the context of a problem and find a way out that causes the least hurt for those involved.

Why are these qualities associated with women? I think they are the consequences of the history of women's position. We've had to learn these patterns as subordinates, in some cases using them as techniques for survival. They are needed to do parenting and housekeeping, tasks traditionally relegated to women and passed on from mother to daughter. Society at large has benefited from the display or expression of women's qualities that help groups to work harmoniously, and society would benefit more if these qualities could replace some qualities of the dominant gender, such as aggressiveness and competitiveness and the tendency to prefer large and sweeping solutions or generalizations.

Women's culture has generally been disdained, and many women have forsaken much of it to become "honorary men" and succeed in the corporate business world. If gender differences are wiped out by women becoming men, then the earth will get a double whammy. But in a recent article in Working Women Thomas J. Peters, co-author of In Search of Excellence, says that women are feminizing corporate offices and being praised by their employers for introducing different ways of organizing work and relating to other employees.

Now we seem to be in a new trap. Because women's culture could be a dose of good medicine for society, are women responsible for solving our problems? Does this mean that women must be earthkeepers and work out the truce in the human war against nature? This sounds as if women are expected to clean up after men again. But we don't want to do that anymore. We want to share the clean-up jobs as partners and equals.

In a conversation I had with the organizer Byron Kennard in 1980, he referred to women activists as the "conscience of the community." They feel personally responsible for righting the wrongs in a community and volunteer their time for the organizing of causes. Women cannot carry this role alone, however, if they seek fulfilling work outside the home.

For progress to be made in the new stages of the ecological era and the feminist era, men and women must cooperate. It is time for the old domination structure to crumble, time for men to share the housekeeping and earthkeeping tasks, unglamorous as they may be. Instead of women sacrificing their talents and goals to enable their husbands to succeed, it is time for husbands and wives to help each other. But partnership and sharing must be extended beyond the household. Just as more working men now do share the laundry and shopping on evenings and weekends with their working wives, more men must share the tasks of community building, of earthkeeping.

We have problems even with this arrangement, however, if children are in the household and the parents' full-time work and volunteer schedules deny children loving attention. Varying the industrial model in the workplace by means of shared jobs, part-time work, and flexible working hours would enable us to express our nurturing natures more adequately.

Many fathers now do take care of their children. They are not babysitting; they are parenting. As men consider childcare a shared responsibility and are able to be with their children, a pattern unfolds that benefits men and the whole society as well as women.

To make progress in the second stage of the ecological era, humans must remake the relationship between nature and culture. The notion that women were closer to nature, thus wild, led men in the past to believe that women were not to be trusted but must be controlled and tamed. Our challenge is to learn from nature, from the wild, to study nature as the standard for agriculture and nature's economy as the basis of our human economy. As Ynestra King said, "Freedom lies in becoming natural beings in the deepest sense, rather than beings against nature."

Our Next Steps

The scope of our environmental problems is enormous: we must address our excessive faith in and dependence on technology, the overconsumption and waste of resources, overreliance on nonrenewable energy, destruction of habitats, and above all the question of how many people the earth can support. We must redefine national security and subdue costly militarism. We have so far been unsuccessful in turning around our bent for destruction through state and national legislation. (I do not mean by this that we should abandon protective laws such as the Clean Air Act.) And we will not be successful until we stop believing that an increasing domination of nature is a measure of human progress.

In his book *Envisioning a Sustainable Society: Learning Our Way Out* Lester Milbrath discusses the concept of social learning. He says that social learning, which is impossible to define in a phrase, comes about in different ways, generally recognizable only after it has happened. One way to explain it is to say that social learning occurs when society comes to understand something sufficiently for one dominant institution or practice to be replaced by another. We are in the midst of social learning about the relationship between men and women and our human relationship with nature, which has created the environmental crisis, but we have not reached the point where significant social change is imminent.

Milbrath suggests that we might open up our collective mind for social learning to take place because of a "slowly accelerating cascade of unfortunate developments." More industrial accidents like those at Bhopal and Chernobyl, increased ozone depletion, cancer threats, contaminated drinking water, population growth, and famines will finally convince us. As stories accumulate that show us the world is not working, he says, we may finally come to our senses. Then social learning will soak in, and we will have the potential for a sudden shift from the dominant social paradigm to a new paradigm.

David Ehrenfeld concludes the chapter "The Conservation Dilemma" in *The Arrogance of Humanism* with a similar position: "Non-humanistic arguments [for conservation] will carry full and deserved weight only after prevailing cultural attitudes have changed." The change may come only by a miracle, that is (he quotes Lewis Mumford), " 'not something outside the order of nature but something occurring so infrequently and bringing about such a radical change that one cannot include it in any statistical prediction.' " He reassures the reader that those who have considered the nonhumanistic arguments for conservation of nature will be ready to take advantage of the favorable circumstances. And in a broader context, those of us who have considered the advantages of a world in which patriarchy is no longer the conceptual framework and in which people understand and value our linkage with the natural world, will be ready for the paradigm shift or the miracle, and we can help it to happen. In the meantime, we should continue our work to effect social learning.

Social learning is underway in many parts of our culture. I am encouraged by what I see as an effort to integrate ecology and feminism into agriculture, religion, the arts, and community development.

Most people begin to understand their connection to the natural world when they start learning about food production. I think social learning has begun in agriculture as the environmental consequences of industrial farming have become public information. The connections between heavy use of agricultural chemicals and drinking water contamination led to revolutionary groundwater protection legislation in Iowa. The treatment of animals raised in confinement now really troubles consumers. Farmers themselves are looking for a way out of the costly input treadmill, and a transition from conventional agribusiness to sustainable farming has begun for many. In the grain/livestock agriculture of the Midwest, farmers like Dick and Sharon Thompson and their fellow members of Practical Farmers of Iowa talk about practices that will prevent soil erosion and nurture soil organisms. The Thompsons emphasize care and attention in their livestock programs, which are unlike large-scale hog confinement operations in Iowa. Each year several hundred farmers attend the Thompson field days and learn about their crop rotations and, most importantly, about their philosophy of working with nature. Similar learning goes on in sustainable agriculture organizations and on-farm experiments in Wisconsin, Minnesota, and Nebraska. The U.S. Department of Agriculture has not provided leadership; it is the decentralist organizations that are guiding us to a more ecological agriculture and making social learning happen.

The Land Institute goes further than other organizations in aiming for long-term sustainability in agriculture. We want to develop crops that can feed us when cheap fossil fuel is no longer available. This means creating a partnership with nature in which elements of the ecosystem contribute to soil fertility and to insect and disease control. Our focus on bringing ecology and agriculture together naturally embraces some of the feminine qualities I mentioned. For example, we must think about how the species of our planet will relate to one another and the places where they grow. Researchers must pay close attention to the growth habits and particular needs of each species. Our ecological model is not a pyramid with humans at the top of the food chain but a network of organisms linked together. Though the mixtures of grain crops we develop will mimic the prairie of the plains and Midwest states, we think the ecological principles we learn can apply to other ecosystems around the world.

The historical connections among food, religion, and women may be revitalized as churches recognize a new role in the ecological era. The newly appointed spiritual leader of the Church of England and seventy million Anglicans worldwide say that "God is Green." Although churches and synagogues in the United States mostly ignored the first Earth Day in 1970, they actively participated on April 22, 1990. From a concern about hunger and rural justice, religious leaders saw connections to soil and water stewardship. Now they regularly preach environmental messages from the pulpit and urge a higher ecological consciousness among their members. Jewish and Christian theologians are trying to renew a philosophy of nature and ethical guidelines for the human-nature relationship. Ten years ago women clergy were scarce, but now women are spiritual leaders throughout several Protestant denominations, and gender-inclusive language is found in church ritual. Books on theology by women have opened up discussions on the similarity between domination over women and domination over nature, and social learning is taking place.

The arts are also contributing to social learning and will help us change our conceptual framework for human-earth relations. The emotional response that the arts elicit affects social learning. Of course, the arts should not be forced to serve a particular moral vision or political position; artists respond to the world as they experience it, and the ecological era and feminist era have been a part of their experience. I think it is significant that more women artists are writing and exhibiting their works and that ecological themes are used more often by both women and men.

Our arts associate at The Land Institute is Terry Evans, a photographer whose book *Prairie: Images of Ground and Sky* has taught an esthetic understanding of prairie to many people. She is collaborating with nine other landscape photographers, three women and six men, in a special "Water in the West" project, which will depict many kinds of water use in western states. All the artists are concerned about problems resulting from the use of land and water. The project is unique in the way these independent artists learn from one another and work together developing the project.

Simone de Beauvoir said: "Representation of the world, like the world itself, is the work of men. They describe it from their own point of view, which they mistake for the absolute truth." Representation of the world from the perspective of women and their ethical sense of relatedness to living things and their propensity to nurture life are needed now. More than ever, we should encourage the feminization of the arts. We need an alternative to what Ynestra King calls the "androcentric master narrative."

Last is *praxis*, defined by David Orr as "the science of effective action." Action can contribute to social learning, and social learning must prepare us for action— that is, political action. William Ophuls, in *Ecology and the Politics of Scarcity*, describes politics as "the art of creating new possibilities for human progress." New possibilities and effective action seem most likely to materialize within the context of communities.

Faith in the small-scale, local approach turns into action in projects such as those begun through the Schumacher Society. The activities of community-based programs—land trusts, organic farm marketing cooperatives, and economic renewal projects such as Self-Help Association for a Regional Economy—are models of local politics as defined by Ophuls that open our minds to social learning. As communities stop trying to solve their economic problems by bringing in large polluting industries or expecting federal government assistance, they can improve their support of what is already there and develop more locally owned and locally controlled enterprises.

The most local situation of all, the smallest political unit, is the home. Here is where effective action toward social transformation begins—with partners in households helping each other develop as individuals, as individuals in relationships with others and the natural world. By caring for each other and sharing in the tasks of living, we undermine the old hierarchical framework that keeps us on the path of environmental destruction.

I conclude with three stanzas of a poem, a prayer, by Barbara Deming:

> Spirit of love
> That flows against our flesh
> Sets it trembling
> Moves across it as across grass
> Erasing every boundary that we accept
> And swings the doors of our lives wide—
> This is a prayer I sing:
> Save our perishing earth!
>
> Spirit that cracks our single selves—
> Eyes fall down eyes,
> Hearts escape through the bars of our ribs
> To dart into other bodies—
> Save this earth!
> The earth is perishing.
> This is a prayer I sing.
>
> Spirit that hears each one of us,
> Hears all that is—
> Listens, listens, hears us out—
> Inspire us now!
> Our own pulse beats in every stranger's throat,
> And also there within the flowered
>   ground beneath our feet,
> And—teach us to listen!—
> We can hear it in water, in wood, and even in stone.
> We are earth of this earth,
>   and we are bone of its bone.
> This is a prayer I sing, for we have
>   forgotten this and so
> The earth is perishing.

—1990

David Ehrenfeld

The Management Explosion and the
Next Environmental Crisis

David Ehrenfeld, professor of biology at Rutgers University, is a founding board member of the E. F. Schumacher Society. As a writer he follows in the tradition of such authors as Lewis Thomas and Loren Eiseley
in his ability to translate scientific information into poetic, memorable
prose. An outstanding scientist, he is specifically an informed and eloquent naturalist-biologist, the intellectual heir of the best scientific
minds of the nineteenth century. The author of *The Arrogance of Humanism*
and *Beginning Again: People and Nature in the New Millennium*, he is also a col-
umnist for *Orion* magazine and was the founding editor of the international scientific journal *Conservation Biology*. Currently he is working on a
book with David Orr about technology, society, and the environment in
the twenty-first century.

In the lecture that follows, Ehrenfeld examines our present dilemma from yet another angle, tracing the exponential explosion in
management that, he maintains, has distanced decision makers from
the reality and the processes that gave rise to their position in the first
place. It is the belief in control, he maintains, that has enabled administrators to gain such widespread power. This concentration of power,
a classical case of positive feedback, has been exacerbated by industrial
technology. As theory substitutes for experience, it not only weakens
the overall productive process but it undermines accountability and
demoralizes workers, depriving them of the dignity of work—something that E. F. Schumacher considered fundamental to becoming fully
human.

The administration explosion affects not only industry and commerce but also science and the universities. Nuclear power is only one example of the pursuit of theory and wealth at the expense of good science and the public welfare. As a matter of course, valuable knowledge and skills are rechanneled through the manipulation of research funding and support to those favored by management. The most serious instance of this trend in the universities, Ehrenfeld observes, is the disappearance of "our long-accumulated knowledge of the natural world." Like Schumacher, Ehrenfeld urges that we become aware and critical of docile acceptance of power and regain greater control of our lives.

An earlier version of this lecture appeared in England in *The Modern Churchman* (32, 2, 1990). A revised version was published as two chapters of Ehrenfeld's 1993 book *Beginning Again: People and Nature in the New Millennium.*

My subject is a problem that, unlike the greenhouse effect, one doesn't normally associate with the environment, although it is the indirect cause of many environmental crises and is critically involved in the exacerbation of all of them. Perhaps the reason we don't associate my topic with the environment is because it plays a causative role in so many other societal ills of our time. This lecture is about the extraordinary proliferation of administration, of bureaucracy, and of management—that is, the increasing percentage of people in our society who control events but do not themselves produce anything real. Please note at the beginning that I used the term "proliferation of management," not just management. The dreadful abuse I will be describing is an excess of something that is perfectly all right, even necessary, in reasonable quantities.

Management and managers are needed in any modern society, yet management, instead of being a service, can become a raison d'être and take on a life of its own. Swelling beyond all reasonable size, it appropriates and stifles the life of society: at this point it becomes utterly counterproductive and destructive. Unfortunately, I know of no good word in English to convey this bloated condition of management. "Bureaucracy" is not quite right because it usually refers to agencies of government and because it makes people think primarily of clerks and other low-level personnel. "Overmanagement" is better but cumbersome. I will use both terms, along with the words "management" and "administration." In each instance I am referring to the abuse of the process, not the process itself.

I also draw a distinction between producers and nonproducers, the latter being the administrators. This probably conjures up the revolutionary Marxist dichotomy of workers versus exploiters, with all of its pejorative connotations. I don't mind the comparison. Managers are playing a critical role in the destruction of our world and our children's world—we ought to hate them for it, or at least hate the process that has given them power. But at the risk of being considered a weak revolutionary, I caution that we shouldn't throw out the baby with the bathwater; when we find management that has resisted the tendency to grow out of control, it is deserving of both praise and study.

There is one modern belief that has enabled managers to take over much of our society, to direct our lives, and even to manipulate our goals. It is the belief in control. The theme of my book *The Arrogance of Humanism* is the misplaced faith in control that characterizes much of our century. The widespread idea that we are or ought to be in total control of the world and of our own destinies within the world is familiar to most of us, as is the accompanying idea that anything we don't control now, we will tomorrow or the next day.

To be sure, this faith has been severely shaken by comparatively recent events: the explosion of the space shuttle Challenger and the catastrophic partial meltdown of the nuclear reactor at Chernobyl are the two most conspicuous examples of a genre of disaster that promises to become much more familiar as time goes on and as the scope of our efforts to control everything in our environment increases. The Aral Sea is drying up and turning into a salty desert, rain forests and atmospheric ozone are declining, and urban sprawl is growing, yet the peddlers of the myth of control are still busily at work, as the promotional literature for Artificial Intelligence or for the genetic engineering of crop plants plainly shows. Nevertheless, the myth has been disturbed by recent events and promises to be disturbed even more as the enormously elaborate, interlinked, and ponderous system that has arisen to promote and maintain control begins to shake itself to pieces. Centralized administration is at the core of this system.

Early in the history of ideas it was observed that there is a relationship between centralized administrative control on the one hand and urbanization and the development of technology on the other. Genesis (10:9,10) describes Nimrod, the King of Babel, and the tower that was built under his rule. In biblical history Nimrod is the first city-builder, and although Genesis gives Nimrod fewer than forty words, the medieval rabbinic commentators Rashi and Nachmanides made it quite clear that to them Nimrod was also the first person to control the lives of many others.

By the nineteenth century, as the relationship between the technologies of power and urban population growth acquired its modern character, it became widely realized that bureaucracy persisted and developed even in the absence of all-powerful kings and dictators. This was most vividly described by Charles Dickens in *Little Dorrit*. In this great novel Dickens invented a government agency called the Circumlocution Office, "the most important Department under Government":

"No public business of any kind could possibly be done at any time without the acquiescence of the Circumlocution Office. Its finger was in the largest public pie, and the smallest public tart. . . . Whatever was required to be done, the Circumlocution Office was beforehand with all the public departments in the art of perceiving HOW NOT TO DO IT."

How did the Circumlocution Office keep things from getting done? Simple. Here are some instructions from a friendly official at the Office to an ordinary citizen who is trying to obtain a bit of public information:

"We shall have to refer it right and left; and when we refer it anywhere, then you'll have to look it up. When it comes back to us at any time, then you had better look us up. When it sticks anywhere, you'll have to try to give it a jog. When you write to another Department about it, and then to this Department about it, and don't hear anything satisfactory about it, why then you had better—keep on writing. . . . Try the thing and see how you like it. It will be in your power to give it up at any time if you don't like it. You had better take a lot of forms away with you. Give him a lot of forms!"

The first edition of Little Dorrit appeared in 1857. At that time the phenomenon of bureaucracy and centralized administration was already well-developed, even without today's huge urban populations, increase in technological complexity, and the invention of the computer, the copier, and the fax machine, which are such an important part of our own explosion of management. Although Dickens was a genius at describing administration, I don't think he really understood why it came about or realized that it was an organic outgrowth of a power-worshipping society dedicated to a belief in control. In Little Dorrit he told of the unhappy fate of an inventor stymied by the Circumlocution Office. To Dickens, an inventor was a noble person who ought not to be thwarted in his work. What he didn't realize was that invention and technological cleverness, in the context of England's growing cities, were responsible for the Circumlocution Office in the first place. The sudden access of power and concentration of power brought on by industrial technology came hand in hand with management. New kinds and levels of production were inevitably accompanied by squadrons of people who produced nothing in the way of tangible goods but managed the factories, distribution networks, and government taxation and regulatory offices that constituted the new production system. Not until approximately three-quarters of a century later did this idea occur to Lewis Mumford and George Orwell.

Orwell saw it quite clearly, although he died before his own insights into technology and administration could mature. In a book review written in 1945 he wrote, "The processes involved in making, say, an aeroplane are so complex as to be only possible in a planned, centralised society, with all the repressive apparatus that that implies." In other words, a technologically complex society

requires a good deal of management. Orwell associated this "repressive apparatus" especially with the manufacture of expensive, sophisticated weapons systems, but I think we could say that a society that produces antibiotics and computers requires more administrators than one that does not.

Nevertheless, the process has gotten out of hand. Having initially expanded in
response to a real need to organize complex processes, management continues to
expand according to its own, self-generated imperatives. Like a cancer, it has
become uncoupled from the society that produced it. Moreover, it has spread
beyond production and government services to nearly all walks of life; from
hospitals to universities, the hand of the manager is increasingly felt.

I am now going to concentrate on the mechanism by which administration
spreads and gains control, on the long-term consequences of this spread, with
particular reference to science, and—what Orwell never really got around to
thinking about—on the instabilities in this process that may ultimately serve to
derail, and possibly even stop, the administrative juggernaut.

Management spreads because its methods and output automatically create an
environment conducive to its increase. Each manager doing his or her job brings
forth more and more managers, as C. Northcote Parkinson was one of the first to
show. There need be nothing conspiratorial or even purposeful about the process,
which is what makes it so difficult to stop. It is a classical case of positive
feedback, with several feedback loops involved.

There are two elements in the job of management that fuel this feedback. First
and most obvious is the increasing prevalence of the habit of documenting
everything. This is not a new idea; only the perception of who is doing the
documenting has changed. In the Jewish *Ethics of the Fathers*, written down not
quite two thousand years ago, it says, "Know what is above you: an eye that sees,
an ear that hears, and a book in which all your deeds are recorded." And in
Matthew (10:29,30) it says that not a sparrow falls without God's knowledge,
and even "the very hairs of your head are all numbered." Disquieting as the idea
of this kind of minute, celestial documentation is for all of us at one time or
another, God as Recorder has the two great advantages of, first, being accurate
and, second, not charging for the service. Nowadays we have no such luck,
because there is a new recorder in town. Anyone who has ever had the misfortune
of compiling a university promotion packet or filling out tax forms knows that it
isn't only God who numbers the hairs of our heads. I will only mention in passing
that in a humanistic society in which religious faith in human infallibility and
control has replaced an older faith in divine control, it is sadly appropriate that
people now take over the job of counting and numbering.

The true purposes of this modern tendency to document everything are plain.

It serves to provide a source of undemanding work for bureaucrats who might otherwise not be terribly busy—thus justifying the call for yet more bureaucrats and conferring a kind of spurious legitimacy on the perpetual growth that is the hallmark of administration. More important, compliance with administrative demands for ever more and more minute personal and other information reinforces the desired belief that the provider of the information is subordinate to the recipient of the information, just as mere mortals are subordinate to God, the Judge. If you don't believe this, just ask an administrator—regardless of whether he or she is from government, industry, or academia—to fill out a questionnaire for you on his or her work habits and personal history.

Most of us in nonmanagerial positions have become so accustomed to this routine violation of privacy that we think nothing of it, except perhaps as an annoyance. It is easy to forget that just as in some code systems, the data that are being transferred are often themselves irrelevant, a blind. The real information is symbolic, and it resides in the context of the act of transfer. This explains why the data are often filed away unread. The fact that one has filled out a form is important; what is on the form is not.

Finally, in a more practical sense, managers use questionnaires, forms, plans, schedules, and protocols not just to intimidate but to gain control over other people's time, which in this society is probably the most effective form of control over others.

Associated with administrative paperwork, especially with the official justifications of it, is the use of various words of judgment—such as "evaluation" and "accountability." This practice further establishes the idea that the administrative evaluators are fit to judge, even though this is often not the case. Here, the use of such words, frequently abstract nouns, is analogous to the use of names in advertising; it bears little relationship to the highest use of language, which is to convey meaning directly through words. Designer words—such as Exxon, the oil company, or Camry, Sentra, and Cresida, the car models—either have no meaning at all or have meanings that have nothing to do with what they are attached to. They simply have a nice sound or the right, vague connotations. Similarly, when managers use such words as "excellence" and "vision," what is being conveyed is a feeling, not a meaning. This manipulation of language is hard to fight because it operates at a primitive level of human function not easily accessible to higher thought processes.

The second way in which administration expands involves a far less subtle mechanism than the proliferation of paperwork. It is the direct appropriation of power through control of the money supply and hiring and firing. We see this at its highest degree of refinement in the modern university, which has come to be dominated by priorities determined by a cash flow controlled by administrators.

In the words of the biochemist Erwin Chargaff, universities "have been turned into huge corporations whose only business is to lose money." But universities are not the only places where management has diverted and augmented the money supply to provide for its own continued growth at the expense of other parts of the institution. There are also charities, foundations, hospitals, and the many branches of government, especially the military.

What makes bureaucracies spiral out of control in their demand for institutional funds are the positive feedback loops that so many bureaucracies participate in and encourage. Universities again provide an excellent example. In the decades immediately following World War II, university administrators in the United States discovered a vast new source of unregulated cash, the so-called overhead on research grants. Administrators could set the overhead at virtually any percentage they liked—figures in excess of 50 percent are now common—and the incoming funds could disappear into an administrative black box. Pressure on scientists to obtain more grants (and patents) has increased steadily, more overhead dollars flow in, more administrators are hired, more money is needed. A positive feedback loop is thus established. The purposes of grants become irrelevant; only the amounts of the grants and their likelihood of attracting other grants matter. The fact that all hiring and promotion of faculty members, who are the ones receiving the grants, must be approved by the administration, while the hiring and promotion of administrators needs the approval only of senior administrators and not of faculty, greatly enhances the positive feedback. In other words, there is no negative feedback, or brake, built into the system. This changes the character of the faculty and the entire university. In other kinds of organizations, governmental and industrial, there are similar kinds of positive feedback loops. What they have in common is administrative control of the money flow and of hiring and firing.

The consequences of managerial proliferation are numerous and pervasive. I will discuss a few of the more important ones. All the problems are interrelated, so there is a certain amount of overlap.

First, there is the problem of bad decisions. As administrative control increases, as administrators become more numerous, and as the power gulf between administrators and producers widens, more and more of the critical decisions for an institution are made by administrators only on the basis of second- or third-hand information and in accordance with purely administrative priorities. The people who know, the people on the front lines, are shut out of the decision-making process. The result is bad decisions.

Most bad decisions are never brought to light, but there are so many that a few of the most acutely horrendous ones are finding their way into the newspapers. One example was given in a front page story in The New York Times, December 26,

1988, entitled "Nuclear Arms Industry Eroded As Science Lost Leading Role."
Times reporter Fox Butterfield wrote:

Many veterans of the bomb industry trace the problems of ageing equipment, shoddy
management and pollution and safety concerns to a withering of scientific and technical
expertise in the Government agency that runs the system.

"We have a big problem with competitiveness, and I think three-quarters of that is that
the guys making the decisions don't understand the technical things any more," said Dr.
Harold Agnew, a physicist who worked with Dr. Fermi on the world's first chain reaction at
the University of Chicago in 1942. "You can't run the bomb factories with a bunch of
lawyers and administrators" (Dr. Agnew said).

Of the Energy Department's eight regional office managers, only two have graduate
degrees in science and engineering. Several, including Jo Ann Elferink, the manager of the
San Francisco regional office, which supervises the Lawrence Livermore National Laboratory, have no academic training in science. Lawrence Livermore does research on new
warheads and the plan to build an anti-missile system in space.

Of course, my first thought on reading this article was, "How nice, the bomb
factories are breaking down." But this is not a very realistic attitude. When a
bomb factory falls apart, the potential consequences are different than, say, for a
bicycle plant, as the Russians learned at Chelyabinsk in 1957.

That article was mostly about the Hanford facility in the state of Washington.
In another front page article in the The New York Times, dated January 16, 1989,
reporter Keith Schneider wrote the following about the Savannah River nuclear
weapons plant: "Internal memorandums prepared by DuPont, which built and
operated the vast weapons plant for nearly four decades, show that company
scientists amassed volumes of research on key weaknesses in equipment." Yet, so
effectively was this vital information buried by the corporate and governmental
bureaucracies that even Westinghouse, which was scheduled to take over the
plant from DuPont in April 1989 and which is certainly privy to all the top-secret
information about it, said that it knew nothing about the seriousness of the
problems at Savannah River.

Anyone looking in the newspaper for articles about the consequences of
overmanagement does not have far to look. Here are some other examples. In a
New York Times opinion piece dated December 9, 1988, and entitled "To Revive
Schools, Dump Bureaucrats," John E. Chubb, a senior fellow at the Brookings
Institution, said:

New York City, as I discovered after a 35 minute phone call with nine different
bureaucrats at the Board of Education, has 6,622 full time employees in its public school
headquarters. That's one external administrator for every 150 students. (This does not
include administrators located at the schools.)

By comparison, the Roman Catholic Archdiocese of New York has so few employees in its headquarters that the first one I called simply offered to count them for me—30 central administrators; no more than one for every 4000 students.

   . . . In a new study of more than 400 American high schools, a Stanford political scientist, Terry Moe, and I conclude that the more centralized a school system is, the worse the achievement of its students, public or private.

In my own state, New Jersey, the Board of Higher Education has recently ruled that from now on, principals of public schools need not have any teaching experience at all, only a degree in management.

Or take this extract from a column by John Russell, art critic for the *The New York Times*, which appeared on May 14, 1989:

Every so often, there surfaces in the international art world a truly terrible idea. An idea of precisely that order was approved unanimously not long ago by the trustees of the Victoria and Albert Museum in London. The essence of it was that henceforth the curatorial staff of the museum were to concern themselves entirely with scholarly pursuits, leaving the day-to-day work of the museum—the "housekeeping," as it is now called—to a chain of command newly set up for the purpose and attuned to modern managerial methods.

I won't pain you with a catalogue of the disasters reported by Mr. Russell in his article—the brutalization of the curators, the sleazy ads in the London underground by Saatchi & Saatchi—but I will quote one more sentence: "When [the new Director] Mrs. Esteve-Coll appeared on television in December she is reported to have said—as if it were a handicap, like toothache—'One of our problems is that we deal in historic artifacts.'"

Bad decisions are inevitable in a process that is so utterly divorced from reality. Here is an example from the world of business. It is taken from another opinion article in *The New York Times*, written by a corporate executive named Herbert L. Kahn, dated July 25, 1988, and titled "My Years in Meetings." Mr. Kahn wrote that he spent much of his time in management meetings dealing with such questions as:

"What is the five-year trend of orders per square foot of branch sales offices, and what is the variation from region to region?" Or, "what is the output of your machine shop, both in dollar volume and in weight of metal, per gallon of lubricating oil, and does it vary seasonally?" . . .

There seemed to be only one thing missing [from the meetings]: A connection between our efforts and the company's real business. My division was supposed to design, make and sell high technology products. My superiors were presumably charged with guiding that enterprise. None of us, however, ever designed, made or sold anything, and we rarely even met anybody who did.

In this same regard I quote from a lead letter to the *Times* by Ron Szary, published on November 7, 1988, under the title "Management Mentality Is

Killing U.S. Industry." As his first reason for why the United States trails the Japanese in the superconductor race, Szary said:

> Unlike the Japanese, American companies are top-heavy with layers of management . . . all of whom want final say on products and production, but have little if any comprehension. . . . In the United States, it would be embarrassing to have the top man in most companies talk to the people on the floor: he doesn't know the product; he doesn't know the process. . . . Our layers of management are calculated to buffer the top levels. On the other hand, it is not unusual to see top Japanese management in direct, daily contact with workers. Who will outperform whom?

The next consequence of overmanagement, one that I will mention briefly, is the widespread problem of demoralization of the producers. The techniques that keep bureaucrats in control and help them expand their power base—the barrage of paperwork that everyone knows is worthless, the deluge of conflicting, often arbitrary memoranda, the insistence on "accountability" without any standards of reference for performance, the requirement of doing more things than are possible, some of which are in conflict with one another, and other forms of control practiced by management—create numerous double binds in the daily lives of each producer. Some of the techniques resemble scaled-down versions of the practices used to destroy the spirit of prisoners in concentration camps. Not surprisingly, workers in the most seriously overadministered institutions suffer from low morale and despondency. Some of them become physically ill. In such institutions, another positive feedback loop is created: excessive administration leads to demoralization, which leads to poor performance, which leads to yet more stringent and pervasive administrative control. For a few producers the only way to resolve the problem is by dying. They are soon replaced.

I come now to the last consequence of the management explosion that I am going to discuss: the destruction of science, a process that we are just beginning to witness. This too has profound environmental consequences, and I will give some examples, but most of the connections will have to be left unexplored.

At the outset I want to make plain that although administration is the immediate cause of the decline of science, the ultimate fault lies with science itself. The late René Dubos, a great scientist who was also a great environmentalist, saw what was happening to science and argued eloquently but futilely against the change:

> It seems to me unwise and ambiguous for scientists to affirm on the one hand that they are primarily searchers for truth and to claim on the other hand that everything they do is ultimately of practical importance. . . . Important as they are, the technological and other practical applications of science have been oversold. . . . Of probably greater usefulness would be the development of knowledge and attitudes that would help man to examine objectively, rationally, and creatively the problems that are emerging as a result of social evolution.

Although he perceived the problem, Dubos was too much in love with science ever to admit how corrupted it would become by its discovery of power. He cherished a Wellsian dream that science would somehow learn to abjure power, recover its purity, and reestablish a harmonious relationship with nature. This was naive. Perhaps only someone outside science can see the whole picture with prophetic clarity—can see where science came from and where it is going. In his novel *The French Lieutenant's Woman* John Fowles wrote: "We can trace the Victorian gentleman's best qualities back to the parfit knights and preux chevaliers of the Middle Ages; and trace them forward into the modern gentleman, that breed we call scientists, since that is where the river undoubtedly has run. In other words, every culture, however undemocratic, or however egalitarian, needs a kind of self-questioning, ethical elite, and one that is bound by certain rules of conduct, some of which may be very unethical, and so account for the eventual death of the form."

Fowles saw that between the medieval knight, the Victorian gentleman, and the modern scientist, each an "ethical elite" of its time, "there is a link: they all rejected or reject the notion of possession as the purpose of life, whether it be of a woman's body, or of high profit at all costs, or of the right to dictate the speed of progress. The scientist is but one more form; and will be superseded."

It was "trade" that killed the Victorian gentleman, and it is the lust for power that is killing the modern scientist. But the mere lust for power does not kill: the actual instrument of death is the administrative process created to handle all that power. Before explaining the mechanism of the destruction, I will give four examples, each of them from biology, which is what I know best, and each of them British.

First was the recent sale by Margaret Thatcher's government of much of the Plant Breeding Laboratory in Cambridge to the Unilever Corporation. Whatever the merits of the sale, and I can't think of any, it diminished the laboratory as a creative force in science.

Second was the recent edict by the Universities Funding Council (formerly the University Grants Committee) to condense all biology departments with fewer than twenty faculty members into larger units of only two types: a "B" type, which would concentrate on cells, organisms, and ecology, and an "M" type, which would study only molecular topics. This action will fairly quickly ruin much of what is left of British biology, especially ecology.

Third was the British government's decision to close the Brogdale Experimental Horticultural Station in Kent by April 1990. Brogdale has probably the world's best collection of living varieties of apple, not to mention pear, plum, cherry, and bush fruits. The government promises to maintain the collections. In the United States we have seen what happens when bureaucrats promise to take care of our precious heritage of fruits and vegetables and grains. Parenthetically, I should add that, in addition to closing Brogdale, the Ministry of Agriculture, Fisheries and Food will also close the Rosewarne Station in Cornwall, which *New Scientist* describes as a "unique site for research into early winter vegetables," and is considering closing the vegetable gene bank in Wellesbourne in Warwickshire.

Fourth was the administrative termination of research on birds, arachnids, and coelenterates at Tring and the gutting of ornithology elsewhere in the British Museum, events similar in character and motivation to what is happening at the Victoria & Albert Museum.

Each of these four examples was brought to pass because management was invited in to help control and enhance the flow of power emanating from modern science. It should not surprise anyone when the goals of administration are substituted for the societal goals that science once appeared to accept.

But the administratively imposed death of science is being caused by more tangible factors than the alteration of its objectives. There are direct causes, including, most obviously, its cost. Science is pricing itself out of existence. The sky-rocketing cost of equipment, supplies, and salaries is not an inherent part of modern science. Projects with high operating costs are carefully selected by administrators because the administration receives a percentage of the operating costs. Research that is spare and economical is automatically suspect—because it costs little money, it brings in little money. Such research, by administrative definition, is not "world class." In justification of this policy we hear self-serving arguments that scientific research creates more societal wealth than it consumes. Apart from the fact that there is no proof of this assertion, there is also no reason why expensive research should prove more beneficial than inexpensive research; often the opposite is true.

With dwindling resources and mounting debt, only the most stubbornly unobservant can expect the scientific gravy train to keep on running much longer. What we can expect in the coming years is a stream of theories and inventions that are supposed to create wealth and power out of inexhaustible commodities such as seawater. Nuclear power was an early example, now curdled and gone sour. Superconductivity and cold fusion are recent illustrations. All will promise much for very little, all would maintain the power and growth of the scientific / technical / managerial system that created them. They are a form of magical thinking grafted on to real physical, chemical, and biological processes bred by a mixture of greed and desperation.

There are other ways in which administration is killing science directly—for example, by consuming the creative energies of scientists in paperwork—but I want to go on to an indirect cause of death, one that has major environmental implications. For want of a better term, I call this the problem of de-skilling.

De-skilling is an ugly new word that denotes an even uglier process, the progressive draining of practical knowledge from a culture, a loss of skills by virtue of the loss of skilled practitioners to use them. When it was decided to begin work again on New York's vast Cathedral of St. John the Divine after a lapse of decades, a few old men in England were the only stonemasons left in the world who knew how to work the giant blocks from which a cathedral is built. If they hadn't been able to train young apprentices, there might have been no choice but to abandon the project.

I think that our concept of progress prevents us from being aware that skills and knowledge can vanish from the world. Most of us picture knowledge as cumulative: each advance is built on prior discoveries, block piled on block in an ever-growing edifice. We don't think of the blocks underneath as crumbling away or, worse yet, simply vanishing. Our worldview doesn't prepare us for that, although it is hardly a new phenomenon in the world. Oliver Goldsmith was writing about the loss of peasant agricultural skills when he penned these lines in "The Deserted Village," published in 1770:

> Ill fares the land, to hastening ills a prey,
> Where wealth accumulates, and men decay;
> Princes or lords may flourish, or may fade;
> A breath can make them, as a breath has made;
> But a bold peasantry, their country's pride,
> When once destroyed, can never be supplied.

Loss of knowledge and skills is now a big problem in our universities, and no subject is in greater danger of disappearing than our long-accumulated knowledge of the natural world. The problem is so serious that I have called it "the next environmental crisis." We are on the verge of losing our ability to tell one plant or animal from another and of forgetting how the known species interact among themselves and with their environment. What is happening is that the teaching of the part of biology devoted to the study of biological diversity—taxonomy, natural history, ethology, comparative physiology, biogeography, and allied subjects— is disappearing from the curriculum. The modern administrative climate is hostile to this kind of biology and to those who study and teach it. Our functional understanding of biodiversity is still in its late infancy, and never has the subject been so important to our existence, yet these riches have no value to university administrators. Grants given in the fields of taxonomy, biogeography, and comparative biochemistry are few and insignificant in amount. These subjects simply do not support enough administrators to make them worth keeping.

Here is another ominous positive feedback loop. The fewer research programs and courses there are in these subjects, the fewer people there will be to teach the next generation of students. Where will we be when there isn't a soul left who can tell one kind of grass from another or anyone who knows the habits of grasshoppers?

It cannot be denied that the environmental consequences of the management glut are both deep and pervasive. Modern management, far removed from the actions it regulates, damages the environment through countless bad decisions. For similar reasons, it hinders the efforts to rectify its own mistakes. And as I have indicated, its actions are draining the reservoirs of skills and knowledge that we need now and will need even more in the future.

Can we resolve the crisis of overadministration? Can anything bring the bureaucracy to heel? I think the answer to this question is readily apparent, but it may not be satisfactory to everyone. Management, like anything undergoing perpetual growth, will eventually bring itself under control by running out of resources. It will self-destruct. This will be hastened by the tendency of management to cripple the producers, the people who provide the wealth in the first place. Because it incorporates so many positive feedback loops, modern management is inherently unstable.

But the breakdown of the bureaucracy may take longer than we care to wait, and its loss may be small comfort if the rest of society is a bankrupt ruin. Many of the problems the world faces—nuclear, chemical, and biological weapons, the hole in the ozone layer, the greenhouse effect, the international agricultural crisis, deforestation, loss of species, loss of human cultures—need to be addressed quickly and effectively. Bureaucracy is slow and ineffective. Can we do anything selective to put it back in its proper place before it is too late?

Before attempting to give a constructive answer, I must at least mention the issue of overpopulation. Overpopulation is partly responsible for all the environmental problems I listed, but is it also responsible for the management glut? I said in my introduction that management, historically, appears to have been an urban phenomenon; it depends on high densities of people and the concentrations of complex activities and power that they generate. Now that we have sprawling conurbations rather than cities, with computers to help overcome the problems of control brought about by distance, bureaucracy has been liberated from its urban context. Moreover, with so many people interacting in so many ways in a technologically complex world, we have to ask whether intensive management is an unavoidable accompaniment to overpopulation. Does it provide a short-term stability that is the only alternative to chaos? I don't know the answer, but I don't think we ought to assume the game is lost before we start to play. Simple, deterministic ideas of the future are usually wrong; let us hope that the notion that overpopulation makes overmanagement inevitable is no exception to the rule.

To curb managerial excess, the first thing that comes to mind is shutting off the money tap, or at least slowing the flow. Bureaucratic growth consumes expensive office space and is labor intensive; it therefore costs a great deal of money. But because the administrators control institutional budgets, it is hard to reduce their funding. I know of only a few cases where this has been done. For example, the Internal Revenue Service now effectively limits the percentage of income that a non-profit foundation can spend on the administration of its grants. In this instance, one organization is limiting the administrative spending of another. The limit is imposed from outside. One can imagine informed taxpayers putting pressure on governmental granting agencies to reduce sharply the allowable overhead given to institutions as part of research and service grants. If adopted by the major granting agencies, such a policy would do much to restore teaching of all subjects as well as research in unglamorous fields such as taxonomy and natural history, because managers would no longer be preoccupied with the pursuit of huge grants for their overhead. Nor would there be so many managers.

Besides reducing the money supply for management, other methods of controlling the controllers suggest themselves. Positive feedback loops can be eliminated if producers can participate, at some level, in the hiring and firing of administrators. Even more effective would be the breaking down of the work barriers between bureaucrats and producers through rotation of jobs; this sort of system was pioneered by the U.S. Geological Survey. If most administrative desk jobs were limited to a tenure of two to five years, to be followed by an equal stint of service in the field, administrative abuses would be sharply curtailed.

Widespread implementation of these kinds of reforms to shift power away from management will be very difficult without a revolutionary revision of the relationship between producers and nonproducers. The central assumption of bureaucracy is that the individual producer is subordinate to a larger system rather than to a fallible person or persons. Only by rejecting this central assumption can producers hope to regain their share of the world.

Personal anonymity of the managers and institutional anonymity concerning the internal structure and procedures of the management are essential features of what we could call hypermanagerial control. To the managee, a manager is always defined by his or her title, not by personal attributes. It is much more difficult to argue with a position than a person, especially if the position is itself in shadow. Nevertheless, this anonymity is potentially one of the most vulnerable parts of the bureaucratic hegemony. All that is needed to begin to fight the bureaucrats is some bright light: lists of major and minor administrators, with brief biographical sketches and accurate job descriptions; pamphlets showing the salaries of administrators and their staffs and, if possible, office expenses; organizational charts of the bureaucracy, with graphs showing its growth during the past ten or twenty years. In the Columbia University protests of 1968 the most effective single device in bringing down the university administration was a little pamphlet entitled "Who Rules Columbia?" Related strategies suggest themselves. Why not send questionnaires to the managers, demanding the kinds of professional and personal information that the producers are continually being asked to supply? If the questionnaires are not returned, no matter: the act of sending them redefines the relationship between the managers and the managed.

Of course, such spotlight strategies are not applicable to all bureaucracies, and there is always the problem of who will bell the cat? In some cases of managerial explosion, especially governmental, relief must come from outside. This relief has never been organized, although I have no doubt it will be. The force of angry producers is considerable. Both Presidents Jimmy Carter and Ronald Reagan won election by promising to curb governmental bureaucracy, although both ended up by making it worse. There is still no serious strategy for this task, no group of activists uniquely dedicated to its completion. One problem is that overmanagement presents a revolutionary challenge: the old political dichotomies of liberal versus conservative and labor versus capital do not help us understand the present realities of our condition. New battle lines are not yet drawn, but the fight is coming.

To survive with the many good features of our society intact and with our environment in a livable condition, we must solve the problem of bureaucracy before it solves itself. Surely, once the problem is widely identified, we will make some progress toward a solution. Nevertheless, honesty requires me to state that as long as we continue to be a power-worshipping society dominated by the myth of control, we doom ourselves to an excess of administration and all the misery this entails. Our task is to find a way to convert a "power economy to a life economy," in Lewis Mumford's words. In the language of Deuteronomy (30:19) the injunction is even more direct: "I have set before you life and death, the blessing and the curse; therefore choose life that you may live, you and your seed."

—1990

Frances Moore Lappé
Toward a Politics of Hope:
Lessons from a Hungry World

Frances Moore Lappé is in the second phase of a unique and extraordinarily effective career. In 1971, as a very young woman, she wrote *Diet for a Small Planet*. Like *Small Is Beautiful* it had an impact that was diametrically opposite to small. Beyond the three million copies sold and the translation into six languages, it changed the way countless thousands of people thought about the growing of food and altered the eating habits of countless more. Until *Diet for a Small Planet* was published, few realized, for example, the extent to which the widespread consumption of beef contributes to human suffering. Most people simply did not know that the land used for cattle grazing and the grain used in feedstocks are directly depriving millions around the world who live racked by hunger and threatened by famine.

In this lecture Lappé examines the economic system underlying the politics of hunger. Again we hear an echo of E. F. Schumacher's analysis: "At the surface layer," she claims, "we can identify the root of hunger in powerlessness imposed by the increasing concentration of decision-making power over all that it takes to grow and distribute food. On a deeper level the root of hunger, I believe, lies in our self-imposed powerlessness before economic rules that, if taken dogmatically, create that concentration to begin with."

In 1971 Lappé cofounded the Institute for Food and Development Policy. The Institute has won worldwide recognition as an education and research center and has reshaped the international debate on hunger. More recently she has devoted her efforts to what she refers to as the *root causes* of poverty and hunger. Her 1989 book, *Rediscovering America's Values*, created a resource for the democratic renewal she considers fundamental to challenging our apparent social and political ineffectiveness.

In 1993, with her husband, Paul Martin Du Bois, she cofounded the Center for Living Democracy in Brattleboro, Vermont, a national organization intended to act as a catalyst in transforming democracy from theory to practice. It does so by actively engaging citizens in public problem-solving and by enabling them to acquire the skills needed for public life and the arts of democracy. Their most recent book, *The Quickening of America: Rebuilding our Nation, Remaking our Lives*, also teaches those skills and tells the success stories of hundreds of people across the country who are doing just that. Lappé and Du Bois believe that "the old measures aren't really adequate anymore in capturing the richness of citizen engagement in public life." As Lappé indicates here, in questioning ingrained economic dogma lie the seeds of a politics of hope.

Thinking back on when I began my present work in the early 1970s, now close to fifteen years ago, I recall the bewildered looks of my friends. Why would anyone choose to spend all day, every day, thinking about the most depressing subject in the world—hunger? What they failed to grasp, and what I want to share with you, is that world hunger, rather than being simply a depressing subject to be avoided, can give us a powerful tool for making sense out of our increasingly complex world. World hunger, I learned, holds the key to discovering where our own legitimate interests lie: they lie in common with the hungry.

Hunger: What Is It?

Today's headlines cry out the news of famine, now threatening thirty million people in Africa. Already hundreds of thousands have died. This is hunger in its acute form.

There is another form, however. It is less visible. It is the chronic day-in-day-out hunger that afflicts from five hundred million to as many as eight hundred million people. While chronic hunger rarely makes the evening news, it is just as deadly. Each year it kills as many as eighteen million people, more than twice the number who died annually during World War II.

These statistics are staggering. They shock and alarm us. But several years ago I began to doubt the usefulness of such numbers. Numbers can numb. They can distance us from what is actually very close to us. So I asked myself, What is hunger, really? Is it the gnawing pain in the stomach when we try to stay on that new diet? Is it the physical depletion that comes with chronic undernutrition?

Yes, but it is more, and I became convinced that as long as we conceive of hunger only in physical measures, we will never truly understand it, certainly not its roots. What, I asked myself, would it mean to think of hunger in terms of universal human feelings, feelings that all of us have experienced at some time in our lives?

I'll mention only four such emotions, to give you an idea of what I mean.

To begin with, being hungry means making choices that no human being should ever have to make. In Guatemala today, many poor Indian families send a son to join the army. Yes, many know that this same army is responsible for killing tens of thousands of civilians, mostly the Indians themselves. But the $25 a month the army pays each soldier's family—half the total income of a typical poor family in Guatemala—may be the only means the family has to keep the rest of the children alive.

A friend of mine, Dr. Charles Clements, a former Air Force pilot and Vietnam veteran, who as a medical doctor spent a year treating peasants in El Salvador, writes in *Witness to War* of a family he treated there whose son and daughter had died from fever and diarrhea. "Both had been lost," he writes, "in the years when Camila and her husband had chosen to pay their mortgage, a sum equal to half the value of their crop, rather than keep the money to feed their children. Each year, the choice was always the same: if they paid, their children's lives were endangered. If they didn't, their land could be repossessed."

Being hungry thus means anguish, the anguish of making impossible choices. But it is more than that.

Two years ago in Nicaragua I met Amanda Espinoza, who until then had never had enough to feed her family. She told me that she had endured five stillbirths and watched six of her children die before the age of one. To her, being hungry means watching people you love die. Therefore, hunger means grief.

In this country and throughout the world, the poor are made to blame themselves for their poverty. Walking into a home in the rural Philippines, the first words I heard were an apology for the poverty of the dwelling. Being hungry also means living in humiliation.

Anguish, grief, and humiliation are part of what hunger means. But increasingly throughout the world, hunger has a fourth dimension.

In Guatemala, in 1978, I met two highland peasants. With the help of a U.S.-based voluntary aid group, they were teaching other poor peasants to make "contour ditches" to reduce the erosion on the steep slopes to which they had been pushed by wealthy landowners in the valley. Two years later the friend who had introduced us visited me at the Institute for Food and Development Policy in San Francisco. I learned that one of the peasants I had met had been forced underground; the other had been killed. Their crime was teaching their neighbors better farming techniques, for any change that might make the poor less dependent on low-paying plantation jobs threatens Guatemala's oligarchy. Increasingly, then, the fourth dimension of hunger is fear.

What if we were simply to refuse to count the hungry? What if we instead tried to understand hunger in terms of four universal emotions: anguish, grief, humiliation, and fear? We would discover, I believe, that how we understand hunger determines what we think its solutions are.

If we think of hunger in terms of numbers—numbers of people with too few calories—the solution also appears to us in terms of numbers—numbers of tons of food aid or numbers of dollars in economic assistance. But once we begin to understand hunger as real families coping with the most painful of human emotions, we can perceive its roots. We need only ask when we have experienced any of these emotions ourselves. Hasn't it been when we felt out of control of our lives, powerless to protect ourselves and those we love?

Hunger has thus become for me the ultimate symbol of powerlessness.

The Causes Of Powerlessness

We must go further. We must pull back the layers of misunderstanding that hide the roots of hunger. The first step is to ask, What are the causes of this powerlessness that lies at the very root of hunger?

Certainly it is not scarcity: not when the world is awash with grain (reserves are at record highs); not when the world produces five pounds of food every day for every woman, man, and child alive; not when a mere 2 percent of the world's grain output would eliminate the food deficit of all the world's 800 million hungry people.

No, we cannot blame nature—not even in Africa. Even there, experts tell us, the continent could well be food self-sufficient. Neither can we blame natural disasters, droughts, or floods. Between the 1960s and the 1970s, deaths from so-called natural disasters leapt sixfold, but climatologists tell us that no weather changes can account for this drastic increase. Instead, increased deaths from drought and flood reflect a social breakdown in the structures protecting people from nature's vagaries.

If it is not people's powerlessness before nature's scarcity and her unpredictability, what is the cause of growing hunger? On one level we can answer that the root cause lies not in a scarcity of food or land but in a scarcity of democracy. I mean by this the increasing concentration of decision-making power over all that it takes to grow and distribute food—from the village level to the national level to the level of international commerce and finance.

Let's look briefly at these three levels. First, at the village level, fewer and fewer people control more and more land. A United Nations study of eighty-three countries showed that less than 5 percent of the rural landholders control three-quarters of the land. For example, in El Salvador six families have come to control as much land as do 300,000 peasant producers. With fewer families controlling an ever greater share, more and more people have no land at all. Since 1960 the number of landless in Central America has multiplied fourfold. By the mid-1970s, in twenty Third World countries 50 percent or more of the rural people were effectively landless, deprived of the most basic resource needed to feed their families.

The village is but one level of concentrated power. There is a second level: the concentration of decision-making in the hands of national governments that are unaccountable to their people. Such governments answer only to a small elite, lavishing credit and other help on them and on a military force that protects their privileges.

On average, Third World governments devote less than 10 percent of their budgets to agriculture. Their expenditures on arms, however, leapt fourfold in the decade ending in 1980. In Ethiopia the government is spending 50 percent more on its military than on all other budget categories combined. Such governments actively—and with increasing brutality—resist genuine reforms that would make the distribution of control over food resources more equitable.

They love terms like "land reform," but we must not be deceived. In El Salvador the "land to the tiller" program failed to distribute land to two-thirds of the eligible recipients because poor peasants were too afraid of retaliation to apply for land or for this same reason relinquished the land they were given. This U.S.A.I.D.-designed reform left the most powerful—the owners of the big coffee estates—untouched and the most powerless—the 60 percent or more of rural people—with no land at all. By selling land, on time, to tenant farmers unable to afford it, all the "reform" accomplished was to lock poor families into perpetual debt for plots too infertile and too small to support them.

Similarly, in the Philippines, Ferdinand Marcos promised land reform as soon as he took over in 1972. But under this one-family rule, land ownership in the Philippines has become more concentrated, not less, and the Filipino people are among the hungriest in all of Asia, according to the World Health Organization.

Honduras provides yet another example. In that country, pressure from the peasants has shifted the land reform from paper to action, but at the present pace of government action, it will take over one hundred years to achieve the stated goals of the reform. As one Honduran explained so clearly, "Waiting for the government to give you land is like waiting for the Second Coming." Meanwhile, since 1980 aid to Honduras has increased 500 percent. Thus, the second level on which we can document the increasing concentration of decision-making at the root of hunger is that of national governments beholden to self-serving elites.

There is yet a third level on which democracy is scarce: the international arena of commerce and finance. A handful of corporations dominate world trade in most of the raw commodities that are the lifeblood of Third World economies. According to the United Nations, of the approximately $200 billion that consumers in the industrial countries pay for agricultural products from the Third World, only 15 percent returns to the Third World countries, and of course only a fraction of that returns to the producers themselves.

Dependent on international markets over which they have no control, Third World producing countries have seen the price of every single one of their commodities—with the exception of cocoa—fall in real terms over the past thirty years.

So far, however, we have pulled off only one layer in our effort to grasp the roots of hunger. We have identified the problem: it exists not in the scarcity of resources but in the scarcity of democracy, reflected in the tightening control over economic resources.

But we must dig deeper. We must ask why. Why have we allowed this to happen at the cost of millions of needless deaths each year? Why do we rationalize, condone, and indeed shore up with our tax dollars systems that generate such needless suffering? Even here in our own country, according to the recent Harvard physicians' study, one in ten of us is so poor that we are at risk of hunger.

I have asked myself these questions many times. I have thought long and hard, and I have come to an important realization.

Powerlessness Self-Imposed by Economic Dogma

Peeling off another layer, I have concluded that at the root of hunger lies our own self-imposed powerlessness before economic dogma.

Eighteenth-century intellectual advances forced us to relinquish our ever-so-comforting notion of an interventionist God who would put the house aright. We then faced a frightening void. We have desperately sought a substitute concept—something, anything, to relieve us of the responsibility of moral reasoning. With Newton's discovery of laws governing the physical world and with Darwin's discovery of laws governing nature, we seized upon the notion of parallel laws governing the social world, absolutes that we could place above human intervention.

These absolutes I call our false gods, precisely because, though they be human inventions, we have made them sacred. Placing ourselves at the mercy of dogma, we acquiesce in hunger. This is the tragedy.

I will mention only two tenets of the economic dogma now ruling the West, the market and property rights, and the consequences of making them absolutes rather than simply devices to serve our values.

We certainly hear a lot about the free market's virtues these days, and who can deny that it is a handy device for distributing goods? Any society that has tried to do away with it has faced some mighty serious headaches.

The problem arises when we convert a useful device into an absolute. We become blind to its pitfalls. Unfortunately, this has been the Reagan administration's response to hunger. What are the central pitfalls of the market that directly relate to the causes of hunger?

I recently had the dubious pleasure of debating Milton Friedman. Nobel Laureate Friedman insists that the greatest virtue of the free market is that it responds to individual preferences. "But wait," I said to Dr. Friedman, "I thought that the preference of most individuals is to eat when they are hungry. Yet more than half a billion people living in market economies are not eating." The lesson is unmistakable: the market does *not* respond to individual preferences; it responds to money.

Nowhere is this obvious truth clearer than in the flow of food in world trade. While we think of the Third World as dependent on imports, in fact we in the industrial countries are the largest importers of agricultural commodities, importing almost 70 percent of all farm commodities traded. The United States, known for its cowboys and sixteen-ounce steaks, is actually one of the world's largest importers of beef. This flow from the hungry to the overfed is simply the market at work.

As the Third World poor are increasingly pushed from the land and must compete for jobs as day laborers, they are less able to make their demand for food register in the market. With a stagnant or shrinking domestic market for basic foods, naturally those who remain in control of the land orient their production to the highest-paying consumers, and these consumers are abroad. Voilà! We have the Global Supermarket, in which even Fido the dog and Felix the cat in North America can outbid the hungry in the Third World.

I had my first glimpse of the Global Supermarket when I was driving in northwestern Mexico in the late 1970s. I wound through land made productive by means of expensive irrigation systems, paid for with billions of pesos from the Mexican government, purportedly installed to grow food for hungry Mexican peasants. But I didn't see corn or beans growing. Instead I saw mile after mile of cotton and then tomatoes, cucumbers, and peppers—all destined for North America!

Stopping at a government agricultural research station, I asked, "Why are these farmers growing tomatoes and specialty vegetables for North American tables when Mexicans are going hungry?" The agronomist sat down and scratched out a few numbers. "It's quite simple," he said. "An entrepreneur here can make twenty times more growing tomatoes for export than growing the basic foods of our people."

I recall landing in the rural Philippines several years later. From my airplane window I noticed banana trees growing as far as the eye could see. Only a few years earlier that same land had grown a variety of crops, many for local consumption. Then transnational firms, including Del Monte and Dole, offered contracts to the biggest local land owners to produce bananas for the lucrative Japanese market. It was not difficult to push the peasants from the land. After all, who in the Third World has a legal title to the land? And what poor peasant can afford a lawyer's defense? So within ten years, fifty thousand acres were taken over by banana trees.

Overall since 1970, while hunger deepens, export crop production throughout the world has grown two-and-one-half times faster than the production of basic foods.

To me, the most dramatically telling consequence of the market's distribution within a world of gross inequalities is the disposition of the world's grain supply. When I wrote *Diet for a Small Planet* in 1971, I was shocked to learn that about one third of the world's grain was going to feed livestock. When I did further research for the tenth anniversary of the book, I learned that fully one half of the world's grain was now going to feed livestock. Even in famine-stricken Africa the demand for feed (that is, the demand that the market can register) is growing twice as fast as the demand for food for human consumption.

The growth of the Global Supermarket is a reflection of the problem, not the problem itself. It reflects the increasing gap between rich and poor, between those few who can live by their wealth and the many who are unable to live by their work. The lesson is clear: left to its own devices, the market simply reflects and reinforces the deadly wealth gap in our world. Thus, it must be seen for what it is, a useful device and nothing more. We must no longer delude ourselves into thinking that it registers the needs and wishes of real people; it measures the power of wealth.

The second pitfall of the market is that it is blind and therefore misleads us. It is blind to the human and resource costs of the productive impetus it claims to foster.

Let us return to an example close to home. Throughout the 1970s our agricultural exports boomed, growing sixfold in value in only a decade. In one year our agricultural exports brought in over $40 billion in foreign exchange. What a bonanza, the market told us. All that grain we exported could pay for imported oil.

What did the market fail to tell us? The market could not tell us that producing so much grain required an energy expenditure equivalent to at least one third of what we earn by exporting it. Neither could the market tell us about the topsoil eroded from prime farmland at an accelerated rate, up by 39 percent in just the first three years of the export boom. Nor could it tell us that the push to export means that groundwater is being pumped from the earth much faster than nature can replenish it.

Neither did, neither could, the market inform us of the social cost of all-out production: the tens of thousands of good farmers pushed from the land and the hundreds of rural communities destroyed. In theory the market rewards hard work and production. But that is theory. In reality the market demands hard work and production but rewards only those who can expand. And who are those whom the market rewards? Just as in the Third World, they are those with considerable equity, which gives them access to credit. They can therefore expand to make up in volume what they are losing in profits per acre as the production push leads to price-depressing gluts. From the social cost of the devastation of livelihoods to the increased rural landlessness to the shocking phenomenon of farmers, even entire rural communities, on food stamps—to all of this the market is blind.

The market, left to its own devices, has yet a third fatal drawback, which undermines deeply held values. It leads to a concentration of economic power that directly contributes to hunger and makes genuine political democracy impossible. This situation is apparent when we look at the Third World. The connection between hunger in El Salvador and six families controlling as much land as 300,000 peasants, mentioned earlier, is obvious. Why can't we see this connection at home between the growing concentration of economic power and needless human suffering?

Left more fully to the mercy of international markets, in the 1970s American farmers experienced perhaps the most dramatic concentration of reward in our history: in just one decade the top 1 percent of farmers in terms of sales increased their share of net farm income from 16 to 60 percent.

This third pitfall of the market, its tendency to concentrate economic power beyond anything efficiency justifies, draws our attention to the fourth and final point I want to make about the market as a dogma. Clinging with blind faith to the ideology of the market as price setter and allocator of resources hides the truth that nowhere are markets free. Whereas ideologues view the market as the interplay of impersonal, automatic forces, in fact all markets—because they lead to concentration—reflect the disproportionate power of a relatively few actors. Nowhere is this more true than in world agricultural trade, as we have seen.

Facing unflinchingly the pitfalls of the market does not mean that we throw out the market in favor of another dogma such as top-down state planning. No, it means that we approach the market as a useful device and nothing more. We ask ourselves, "Under what circumstances can the market serve our values?" And then we work to ensure those conditions.

Let us pause to ask, Under what conditions could market distribution serve to reduce hunger? Under what conditions could the market respond to human preferences, as Milton Friedman would have it? I put forth this simple proposition for your consideration: the more widely dispersed is purchasing power, the more the market will respond to actual human preferences.

As we have already seen, the opposite is true. That is, where incomes are highly skewed, the preferences of the majority are ignored by the market, whether it be in the Philippines, where bananas are grown in the midst of hunger for the Japanese market, or in Central America, where beef is raised in the midst of hunger for our tables.

What can we say in a positive vein? Is there evidence of relative equality of income allowing the market to eliminate hunger? We can say that those few market economies in the world that have successfully eliminated hunger—the Scandinavian countries, for example—enjoy a more even distribution of income than we do here. In the Third World we see some indicative examples. Compare the Indian state of Kerala with other states in India. In Kerala the death rate of babies (a good measure of nutrition) is half the all-India average, in part because land reform and a strong union movement there resulted in a wider distribution of economic power.

If we truly believe in the value of the market in enhancing human freedom, then the challenge should be clear: to work for all politics that reduce rather than reinforce the concentration of income and wealth.

But within a market system where land, food, and human skills are bought and sold with no restriction, how can we work toward a more equal distribution of buying power? The answer is that we cannot. For the historical record shows that the market leads in the opposite direction—toward concentration. In other words, the market, left to its own devices, undermines the very condition so obviously necessary for it to serve human needs.

If we agree that tossing out the market would be foolish, yet we want to let go of rigid dogma and take our rightful responsibility as moral agents, what do we do? Unfortunately, to answer this question we must face the second major stumbling block posed by our economic dogma—the absolute notion of unlimited private control over productive property.

The dogma of property rights allows us to accept as fair and inevitable the accelerating consolidation of our farmland in fewer hands and in absentee ownership, just as we have long seen happening in the Third World. In Iowa, the very symbol of family-farm America, half or more of the land is now rented, not owner operated. Similarly, we accept the accelerating concentration of corporate power: one-tenth of one percent of U.S. corporations control two-thirds of corporate assets.

Believing that our very nation was built on the right to unlimited private control of productive property, many Americans view this right—certainly President Ronald Reagan does—as the most basic protector of our freedom. But Yale economic philosopher Charles Lindblom points out what we often overlook. He has written, "Income producing property is the bulwark of liberty only for those who have it!" And, I will add, most Americans do not have it. Eighty percent of us own no stock at all. Not only do most of us have no income-producing property, the majority of Americans have no net savings.

While President Reagan and many other Americans may believe that the right to unlimited private control over productive property is the essence of "the American Way," this was certainly not the vision of our nation's founders. It was not their understanding of property. In their eyes property rights were not absolutes but were linked to the concept of the common good. Dismayed by the misery caused by land concentration in Europe, Thomas Jefferson wrote to James Madison in 1785, "Legislators cannot invent too many devices for subdividing property." Indeed, Jefferson wanted to redistribute land every generation.

In the view of the founders, property could serve liberty only when widely dispersed. The right to property was legitimate only when it served a useful function in society—that is, when it did not interfere with all people's need to own property. Benjamin Franklin, for example, argued that society had the right to reclaim "superfluous property" and use it as deemed best for the common good.

Thus, central to my concept of "a politics of hope," one that breaks free from constraints of dogma, is a fundamental rethinking of the meaning of ownership, certainly ownership of resources on which all humanity depends. Indeed, we see a worldwide movement toward such rethinking already underway, with ownership of productive resources as a cluster of rights and responsibilities in the service of our values, not as an absolute to place above other values. It is neither the rigid capitalist concept of private ownership nor the rigid statist concept of public ownership.

Where do I see movement toward such rethinking? In 1982 I visited one of the most productive industrial complexes in Europe, Mondragon in the Basque region of Spain. There, ninety or so enterprises—integrated into their own banking system, technical training school, and social services—are entirely owned and governed by the workers themselves. This noncapitalist, nonstatist form of ownership results in very different priorities, in another set of values. For example, during the recession of the early 1980s when Spain suffered 15 percent unemployment, virtually no one in Mondragon was laid off. Worker-owners were retrained to meet the needs of the changing economy.

We can detect a values-first approach to ownership in the Third World too. Since 1979 our Institute has served as an unpaid advisor to the Nicaraguan agrarian reform. Nicaragua's flexible, nondogmatic approach to reform has impressed us. The keystone of the agrarian reform is not the elimination of private property; indeed, many more landowners have been generated by the reform. The keystone is attaching an obligation to the right of ownership.

If you know anything about Latin America, you are undoubtedly aware that historically the large land owners, in control of the best land, have left most of it unplanted, preferring to graze cattle or simply let it sit idle. A study of Central America in the 1970s showed that only 14 percent of the land held by the biggest land owners was actually planted. In this context the theme of the Nicaraguan reform is simple: "Idle land to working hands."

If you are directly working your land, not renting it out, and you are making it produce, there is no ceiling to the amount of land you can own (in contrast to so many reforms that have tried to enforce a rigid limit). If you are not making the land produce, you will have it taken away and given to those who have gone hungry for want of land. The land is given free of charge, again in contrast to so many other so-called reforms that leave peasants as indebted after the reform as they were before. The concept of ownership is thus protected but not above a higher value—life itself, the right of all human beings to eat.

Do these examples sound far away, irrelevant, even alien to our own experience? Closer to home, consider the recent decision of Nebraskans on this very question of farmland ownership. A few years ago they passed an amendment to their state's constitution that said, in effect, you have to be a farmer to own farmland. Corporations like Prudential Insurance, which had been buying up Nebraska farmland, could buy no more. In their overwhelming support for this amendment, Nebraskans put the value of dispersed ownership of family farms above the absolute notion of the right to buy whatever one's dollars can pay for.

You may recall that I introduced my comments on property rights in response to the question, What would be required to achieve such a dispersion of economic power that the market could actually reflect human needs rather than the demands of wealth? Part of the answer, I have suggested, lies in rethinking property rights by regarding them as a device to serve higher values, not as ends in themselves. There is an additional approach worthy of consideration.

Given that the movement toward more fair distribution of buying power requires time and that even under the best of circumstances the market by its very nature has its ups and downs, many civilized people have simply decided that what is necessary to life itself should not be left to the vagaries of the market.

In Sweden the people decided that family-farm agriculture is too precious to be left to the market. Therefore, wholesale food prices are set by negotiation, not by the market; representatives of the government, food companies, retail food co-ops, and farmers themselves sit down periodically at the bargaining table. Retail food prices, however, are determined by the market. (Contrast this to the experience of American farmers, who do not know from one day to the next what price their commodities will bring.)

We should not overlook the fact that all Western industrialized countries have concluded that health care is too important to life itself to be left to the market—all countries except ours, that is.

I do not present these examples as the final word; I present them as signs of growing courage to confront the rigid isms in which we have trapped ourselves, courage to put our deepest values first and judge economic policies according to how they serve those values—not the other way around.

Let me take a moment to recap before the final portion of my talk. I have been attempting to peel away the layers of cause and misunderstanding that surround world hunger. At the surface layer we can identify the root of hunger in power-lessness imposed by the increasing concentration of decision-making power over all that it takes to grow and distribute food. On a deeper level the root of hunger, I believe, lies in our self-imposed powerlessness before economic rules that, if taken dogmatically, create that concentration to begin with. It is thus this rule of dogma that must be challenged if we are to end hunger.

The Impact of Our Dogma on the Hungry Abroad

The rule of dogma—the war of the giant isms, capitalism versus statism—has its perhaps most devastating impact on the hungry through the way it directs our country's foreign policy. Viewing the world as divided between two competing isms, our government becomes blind to hunger. Worse, it willingly abets the concentration of power at the very roots of hunger.

Our foreign aid becomes, therefore, not a channel through which we can put ourselves on the side of hungry people but a weapon our government uses to make the world conform to its dogma. As a result, the direction of our foreign aid has nothing to do with need. High-income recipients of U.S. foreign aid receive almost $12 per capita, but the low-income countries are given 50 cents per person, a twenty-fourfold difference. Today Central American allies receive per capita six times more food aid than sub-Saharan Africa, despite the terrible famine there.

Military aid and general budgetary support to our allies have grown from about half to almost three-fourths of our foreign aid under the Reagan administration. Militarization is dramatic: from 1980 to 1985 the number of African countries receiving U.S. military assistance has leapt from sixteen to thirty-seven. While food aid to Africa has fallen, military assistance in the proposed 1986 allocations to that continent has shot up threefold compared to 1980.

The role of U.S. foreign aid—lavished on anti-democratic regimes in countries like El Salvador (right behind Israel as the second biggest per-capita recipient), the Philippines, and Pakistan—is not to reduce poverty but to shore up governments no matter how brutally they deny basic human rights, as long as they at least claim to be on "our side" in the contest of economic ideology.

Under the Reagan administration our foreign aid is also explicitly conditioned on recipient countries removing any restrictions on the market and promoting exports. "Free enterprise development" has replaced "basic needs" as the favored rhetoric of aid officials.

If not foreign aid, what is our responsibility to the hungry? To answer means, first, that we admit the tragic failure in meeting human needs on the part of both capitalism as we know it and statism as we fear it. Can this failure be denied when deaths from hunger equal the toll of a Hiroshima bomb every three days?

Second, accepting failure means accepting the need for fundamental change.

Third, we must understand that pressure for such change is inevitable. We do not have to create it. People will not watch indefinitely while their children die needlessly of hunger. At first they protest peaceably, for few of us risk our lives if we can avoid it. But if our peaceful demands are met with violence, we will risk our lives. I recall a Central American peasant telling me why he had ultimately taken up arms: "For years I watched the owner of the plantation call in a doctor to treat his sick dogs, while my own children, weakened by hunger, died of simple childhood diseases."

Since pressure for change is inevitable, our choice is between whether we block it—shoring up governments that stand in the way of the hungry—or, if we truly understand the roots of hunger, whether we get out of the way. I believe we must remove the obstacles to change. We must give change a chance.

Many Americans might agree, I hope, with much of what I have said, but when faced with the implications, they hesitate. "But, but, but . . . We can't do that! If we remove our support and 'give change a chance,' the Soviet Union will fill in the void, imposing its own version of economic dogma. Nothing new will be allowed to emerge, only another Cuba."

I have thought about this fear long and hard. I understand it. Let us look at what our choices are.

On the one hand, we can let our government continue on its present course. Where will that lead? One country has come to symbolize for me the horror of that course: Guatemala. There we have consistently blocked change; our military and economic aid since the 1950s has abetted the overthrow of a democratic government carrying out genuine land reform and has strengthened the hand of the elite-controlled succession of governments that have murdered tens of thousands of Indians, all opposition leaders (even moderate Christian Democrats), and hundreds of churchpeople. Guatemala has the worst human rights record in Latin America. Poor Guatemalans are living in a state of siege—Guatemala is "a nation of prisoners" according to Amnesty International. March Mejia said, "Even to ask about the disappeared is a subversive act."

Guatemala represents one choice. I have rejected it. The other choice, as I have said, is to give change a chance. Primarily, this would mean simply forcing our government to obey the law—both U.S. law and international treaties that forbid much of our government's policy of shoring up repressive governments and attempting to overthrow those like Nicaragua's that do not fit our dogma.

I think there are two possible outcomes. Yes, one could be the emergence of another Cuba. Then we have to ask, What harm has been caused us by Cuba? One could make the case that it has been more a drain on our adversary, the Soviet Union, than a threat to us. Indeed, if we had been trading with Cuba as we now are with China, Cuba's development would have been a boon to our economy.

I would argue that there is another possibility, however. Having studied underdevelopment all these years, I feel certain of one thing. Every country emerging from decades, even centuries, of domination by an elite beholden to outsiders will, above all else, want to chart its own path. Such peoples will want to do it their way. The last thing they will want is to become a puppet of another superpower.

My close-up observation of Nicaragua during the past six years has strongly confirmed my hunch. Determined not to be a satellite of the Soviet Union, yet knowing that it needs support from abroad, Nicaragua has worked for what it calls "mixed dependency." Until the mid-1980s only 20 percent of its aid and trade has been with the Soviet bloc. Eighty percent of agriculture and 60 percent of the economy as a whole are in private hands.

Perhaps Nicaragua's most dramatic break with past revolutions was one of the first acts of the interim Sandinista government that took over when Somoza fled. It abolished the death penalty, and it gave every captured National Guardsman the benefit of a trial.

Thus, if we can escape the spell cast by Washington, we can view Nicaragua as an example, not of a new model of development but a lesson in the possibility of real change. Just because something is not like us, it does not have to be our enemy. Emerging peoples do want to break free of both rigid dogmas—capitalism and statism.

To appreciate this truth requires a great deal of us but not more than we should be capable of. After all, of all people on earth we should be open to the possibility of something new. Remember that when our nation was born, its very principles were considered madness. Of the Declaration of Independence a high-ranking British officer wrote, "A more false and atrocious document was never fabricated by the hands of man." Shortly before his death James Madison said of our new-born nation, "America has proved what before was believed to be impossible."

Thus, our birthright as Americans should be the belief that something new is indeed possible under the sun. That conviction should allow us to give change a chance in the Third World.

Concluding Remarks

You may think that I have come a long way from the topic of hunger and its roots. I have not. I believe that the causes of hunger are located in belief systems that rob us of our power and teach us to abdicate before the false gods of economic dogma both our moral responsibility and our innate human sympathies for one another's well being.

I have challenged us to break loose. This is a risk, but there is no change without risk. We must risk challenging deeply ingrained ways of thinking, for only if we can experience ourselves changing will we believe the world can change. This challenge takes place in every aspect of our lives, in making choices about where we eat, sleep, study, and save, to name a few.

Willing to change ourselves, we are ready to understand "a politics of hope." We are not putting blind faith in models but putting forth honest hope gleaned from looking at real examples of human courage and innovation from Nebraska to Nicaragua, from Spain to Kerala, India.

My father turned seventy not too long ago. He is a thoughtful man, and he found himself mulling over the fact that if you just multiply his age by one hundred, you have virtually the span of civilization as we know it. I was taken aback. I tend to think of human civilization as being so old that surely we should have given up our folly by now. But if my father's lifetime represents one hundredth of our history, maybe we are merely in our adolescence! Maybe our tragic stubbornness, as we cling to forms that sacrifice life to human-made economic law, is just a sign of the fanaticism of youth. Maybe we are on the brink of a new confidence born of greater maturity, which brings with it the courage to put our values first.

We were warned against the consequences of following false gods, but we mortals did not listen. Now the ante has been upped. Now the stakes are ultimate—the survival of life on earth. Perhaps with this realization we will be jolted into the higher stage of maturity now required of us.

Thus, "a politics of hope" lies in our courage to challenge unflinchingly the false gods of economic dogma. A politics of hope lies in garnering the confidence to trust in our deepest moral sensibilities, our deepest emotional intuitions about our connectedness to others' well-being. Only on this basis can we challenge all dogma, demanding that it serve our values instead of continuing to contort them—while our fellow human beings starve in the midst of plenty.

—1985

Part II   Toward Decentralism and
Community Revitalization

Hazel Henderson
Development Beyond Economism:
Local Paths to Sustainable Development

Futurist, development policy analyst, and much-sought-after consultant, Hazel Henderson is a major player in the global arena. Like
her late mentor and friend E. F. Schumacher, she balances an extraordinary intelligence with an equal measure of down-to-earth common
sense, which she brings to bear with great effectiveness on the present
critical state of human affairs. She is the author of *Creating Alternative
Futures*, *Politics of the Solar Age*, *Paradigms in Progress: Life Beyond Economics*,
and *Building a Win-Win World* and co-edited *The United Nations at Fifty:
Policy and Financing Alternatives*, a special issue of *Futures* (London),
March 1995.

Henderson has been known to refer irreverently to the dominant
economic paradigm—economism—as a form of brain damage. She
contends here that "industrialism's ideology, crystallized in economic
theory, is too narrow a framework for policy formation and the management of the total productivity of societies." Having delineated the
practices by which economism impoverishes regional economies, she
moves on to discuss better criteria, such as new social and economic
indicators of progress, by which to judge the health of an economic
system.

Cautioning against the view that capitalism has won out over com-
munism, Henderson sees the need for a "positive vision of a safer,
multipolar world that is now poised to enter a new Age of Interdependence in which global cooperation with all our rising industrial

partners and the developing countries of the South leads to mutual sustainable development." She asserts that the indicators she advocates will provide the feedback we need to guide structural changes in policy. As she points out, because we have the capacity to create interlinked, interactive electronic communications systems, we can harness this capability to share relevant information on a global scale yet apply it to the population, resources, and ecology of a region. Henderson takes Schumacher's call for economics as if people mattered a step further. As she puts it herself, "My goal is to weigh in on the side of life in human evolution—that's all."

Excerpts from this lecture appear in *Paradigms in Progress*.

The effects of forces of industrialism and economism, foretold so well by E. F. Schumacher, have led to a revolution of unprecedented scope. Technologies and human activities now effect major changes in the global atmosphere and ozone layer, create deserts, pollute ground waters and oceans, and accumulate garbage in space. Industrialism's powerful promise of development, modernization, and economic growth has fueled rising expectations worldwide. However, today's guiding philosophy of macro-economic management has locked most governments and policy analysts into a narrow range of options generated by the conflict between socialist and capitalist models. These sterile Left-Right debates about whether to nationalize or privatize and whether to regulate or deregulate are based on outdated concepts of national sovereignty, the immobility of capital, and material definitions of commodities that, ignoring information and service flows, focus on trade (now swamped by financial flows) and overaggregation of statistical data that exclude nonmoney sectors such as social costs and social benefits.

Meanwhile, seven globalization processes are driving the restructuring occurring in all countries: the globalization of (1) technology and production; (2) employment, work, and migration; (3) trade, finance, debt, and information; (4) the arms race and militarization; (5) pollution and resource depletion; (6) consumption patterns and the emergence of a "global culture"; and (7) the multiple restructuring within and between countries driven by all the foregoing. These processes are circular, interactive, accelerating, and irreversible. A new dialogue to redefine "development" is inevitable, and thanks to global mass communications, this general debate between countries and between different cultures and academic disciplines can accelerate social learning worldwide. Indeed, some countries are even redefining their goals by enhancing problem-solving capabilities and accelerating learning.

The Failure of Economism

Economic reform efforts can be viewed as attempts to clarify the basic values and rules underlying all economies, exemplified by *perestroika* in the Soviet Union, the unplanned "hollowing" of the U.S. economy, the consolidation of the European Community, and the shifts toward "democracy" in Eastern Europe, the Philippines, and Korea, as well as the struggle to end apartheid in South Africa. As the two superpowers further realign out of economic necessity, the rest of the world is breaking out of the ideological prison of the Cold War, which has crippled the United Nations and still preempts massive resources that must be released for true development.

As the new game of mutual development emerges, it is clear that the winner of the Cold War was Japan. With 25 percent of the world's capital, Japan's role as a fulcrum for the shift to the new worldgame is crucial, and many globally minded and influential Japanese politicians, scientists, and business leaders are stressing ever more the transfer of yen surpluses to development projects in the South. Japan's growing internationalism and its key fulcrum role emphasize the importance of redefining development in an Age of Interdependence. "Developing countries" (a misnomer, because all countries are developing) are reassessing their own plans, which have been dashed by crushing debt service and the politically impossible "adjustment" demands of the International Monetary Fund. This new realism is endorsed by the United States in the Bush Administration's Brady Plan, which moves beyond the Baker Plan by accepting the necessity of actual write-offs, as U.S. banks have been doing for some time and as Canada has proposed doing by forgiving debts of the most-pressed African countries. Secondary markets for "country junk bonds" are growing.

Many developing countries almost gave up on the North-South dialogue of the past decade when their proposals for a new international order were consistently ignored. The South Commission, co-chaired by Julius Nyerere, former president of Tanzania, and President Carlos Andres Perez of Venezuela, charted a new course at its meeting in Kuala Lumpur in 1987. The Commission resolved to redefine "sustainable, equitable, people-centered development" without the help of traditional, Eurocentric industrial development theorists. The term "sustainable development" has gained wide acceptance since it was advocated in *Our Common Future* (1987) by the World Commission on Environment and Development, chaired by Dr. Gro Harlem Brundtlandt of Norway. Much imaginative work to operationalize these new concepts of development is underway, such as that in Venezuela and Costa Rica as well as the new Human Development Index of the United Nations Development Program. These new indicators prove that per-capita-averaged income and current national accounts conventions are by no means the best indicators of overall welfare or progress. Many countries moving away from central planning, however, still follow in the spirit of Karl Marx's original invention of the word socialism—that is, a more inclusive, systemic view *beyond* economics.

Indeed, a broader view of development is essential because planet Earth is now providing warning signals from which all humans must learn and to which they must react. The breakdown of the ozone layer and similar large-scale effects, as well as the seven globalization processes mentioned earlier, are teaching us the overriding need for cooperation: our planet is a vast programmed learning environment, with positive and negative feedbacks allowing basic understanding of its functioning. Out of this context comes the criterion of sustainability—that is, providing equitably for the needs of the present generation without jeopardizing the needs of future generations. Thus, the redefinition of development must include more than the avoidance of the boom-bust cycles of market-dominated capitalistic economies and the rigidities of Soviet, Stalinist-style central planning. It must also include measures to curb the excessive pollution and depletion of the earth's resources that both these development models cause. In the United States and other mature industrial societies the excesses of mass consumption have brought major problems, including eroding ethical standards and widespread, increasing levels of drug abuse, crime, illiteracy, homelessness, hunger, and splintered families and communities as well as an increasing gap between rich and swelling poor populations. Short-term industrial values are clearly at a crisis point.

The basic problem is that industrialism's ideology, crystallized in economic theory, is too narrow a framework for policy formation and the management of the total productivity of societies. This economism is an inappropriate basis for sustainable, equitable development. An expanded framework is needed, based on broader interdisciplinary theory, thermodynamics, physics, engineering, ecology, and the life sciences as well as advanced approaches to human motivation, psychology, and anthropology. Similarly, more dynamic models, such as those emerging from chaos theory, are needed to capture the rapid changes and restructuring now occurring.

Once we transcend the box of economism and its false universalism, we can see how many intractable debates between economists, such as those about planning versus markets and competition versus cooperation, can be overcome. The key issue is less whether countries have market-system or centrally planned economies, as many economists still believe, than the extent to which they incorporate at every decision level the necessary feedback loops from those people and resource systems affected by the original decisions. There are two major types of the latter (cybernetic) systems: homeostatic (governed by predominantly negative feedback loops) and morphogenetic (governed primarily by positive feedback loops). Of course, prices are a ubiquitous and useful element of feedback systems. Unfortunately, as technology becomes more complex and as social and technological interlinkages increase, managerial scale and scope must also increase in order to control this complexity. Each order of magnitude of technological and managerial scale in the market sector calls forth an equivalent order of magnitude of government regulation—particularly in democracies where citizens demand it politically.

From a general systems theory perspective, all economic systems are sets of rules devised to fit the specific culture, values, and goals of each society. Thus, even so-called free-market or laissez-faire economies are designed by humans and legislated into existence, while prices and wages reflect the values of each society and its state of knowledge of its real situation in the physical world. In fact, the notion of "objectively set free-market prices" is revealed as a myth (albeit a politically useful one), because all markets are in one way or another created by human rather than "invisible" hand. Resource allocation methods—whether planning, price regulation, rationing, barter, or reciprocity—are only as good as the state of human knowledge.

This decision-theory view of economic systems as "games" with human rules, as management systems employing many feedbacks and strategies, allows an overview of planning, market, and other tools of policy, and it is clear that both competition and cooperation are equally useful strategies that must be continually balanced in all societies. A former economist, Herbert Simon, who won a Nobel Memorial Prize in economics, noted in his acceptance speech that he no longer used the economic method but rather decision theory and other systemic models. Systems scientist Stafford Beer has been designing such models for many years and is applying them under a United Nations grant in Uruguay.

Ecological and natural-resource decisions as well as prices are only as good as human scientific knowledge and must be based on sound science in systematic, dynamic models. Many systems theorists, including Fritjof Capra, Leonard Duhl, M.D., and the World Health Organization itself have proposed substituting health as a basic criterion for development, that is, healthy land and water, healthy cities, healthy public policy—all with healthy people as the goal. Introducing a new set of "green" sin taxes on pollution, depletion, obsolescence, and waste is increasingly popular in Europe, where eighty-five of these kinds of fees are levied. Natural resources can be conserved by correcting prices to include social costs incurred. Full-cost pricing will provide a more accurate market-allocation method, albeit that additional accounting for environmental and social cost will produce an "inflation effect," because most countries have been *overstating* productivity for decades.

New Indicators and Methods

Social indicators have been a theme of discussion in many mature industrial societies for years, but their application has been thwarted by bureaucratic resistance, by intellectual vested interests in methods, textbooks, etc., and by cultural biases (for example, against accounting for the work of women in parenting, housekeeping, and subsistence agriculture). No one correct method will emerge, because multiple models and indicators will be closely fitted to local situations and the different "cultural DNA" of diverse societies. It is clear that guiding societies by overaggregated indices of the late twentieth century is like trying to fly a Boeing 747 with a single oil pressure gauge! The social-indicators debate is about disaggregation, revealing overlooked detail both locally and sectorally and adding a whole row of additional gauges to societies' "instrument panels," so as to plug feedback into appropriate decision levels with more precision and timeliness. For example, an in-house audit evaluated one thousand World Bank projects and found that none of them had met their projected goals, not even in traditional economic terms. Thus, no easy formulas are available for addressing the new needs of developing countries for culturally sensitive, egalitarian, and sustainable development. The first order of business for development officers is to be able to "decode" the cultural DNA of a recipient country and determine what values and goals they are optimizing, which may not be goals they are qualified to assist toward.

Traditional indicators such as Gross National Product (GNP) and Gross Domestic Product (GDP) were developed for military mobilization purposes in Britain and the United States. Their materialistic view of progress cannot guide humanity beyond consumerism toward moral growth and sustainable development. I have traced the evolution of economic indicators and reviewed the many post-economic indicators and policy tools available to decision-makers. The crucial role of the "informal economy" and the unpaid productive work of subsistence agriculture is now the subject of a small but growing body of literature. In addition, there is the inability of national accounts to distinguish between "goods" and "bads" (that is, "wealth" and "illth"), because liquor, tobacco, auto accidents, cleaning up pollution, and the multibillion dollar "stress industry" are all included as progress. It might well be that in the United States these growing social and environmental costs as well as the increasing monetarization of cooking, child care, and other formerly unpaid work are the main growth sectors of

GNP. Yet efforts to add sin taxes to harmful products are fiercely resisted by industry lobbies.

Whereas national accounts and indicators are important, it is local indicators that provide the balance to correct overaggregation at the national level. Local indicators and methods now being tried in the United States are useful; key experiments during the 1980s have been in response to the economic and social problems experienced by many localities as a result of the budget priorities and laissez-faire policies of the Reagan Administration. The local search for alternative models in many states has stemmed partly from the failure of national policies based on overaggregated statistical illusions, including GNP, inflation, unemployment, interest rates, and all the other paraphernalia of the Bush Administration's macro-economic management, which is now lying in ruins. Due to its failure at the local levels, where conditions vary widely across the U.S., macro-economic management has become almost as discredited in the U.S. as the centralized Stalinist policies have in the U.S.S.R. Both became bureaucratic, out of touch with regional needs, and—in the U.S.—relied more on "single bullet" monetary and fiscal policies with very different lead and lag times, which were often in conflict with each other. As the seven great globalizations took hold during the 1980s, it became clear that domestic economic policies were a thing of the past, for such policies were swamped each day by the billions of footloose "hot" dollars sloshing around the planet seeking interest rate advantage. No one knows how to measure these global monetary aggregates, which also began swamping bilateral trade statistics, obscuring policy options, and leading to faulty trade legislation and unnecessary conflict with trading partners.

In all this national confusion, with rising domestic deficits and the rapid shift of the United States from the world's largest creditor nation to the world's largest debtor nation, states and localities were left to fend for themselves. Many of them, suffering severe economic disruption, unemployment, deficits, crime, and other social and environmental problems, turned to experiments and activist policies in spite of their ideological commitments to the free market and nonintervention. Formerly, they had relied on accounting firms to assist in trying to reach their economic development goals. This economic approach offered assessments of a state's business climate in such terms as what conditions a firm's corporate clients were looking for in plant location. Predictably, states were told that a good business climate was one of low taxes, tax credits, cheap land, and cheap labor—what is sometimes called "the plantation model," where the state's people and resources are basically offered on the auction block to lure outside investors. This has always been a risky game, since in the United States approximately twenty-five thousand localities are offering these lures, while only some five hundred location or relocation decisions are made by major companies each year. Such a strategy in the new global economy puts the state in direct competition with countries with much lower wages and resources, countries which in addition offer tax holidays. Indeed, most multinational corporations have financial models that can tell them, on a daily basis, which nation is foolish enough to be offering them an annual 35 percent return on their investment. State officials can never win at such games employing global electronic funds-transfer, where no allegiance exists to any locality.

Another approach in the desperate search for local development is the use of the Quality of Life indicators developed by such firms as the Midwest Research Institute of Kansas City, which rates various cities for their livability and attractiveness to top managers, their schools, research facilities, cultural opportunities, climate, etc., as well as the more traditional factors such as few unions, availability of capital and trained workers, friendly politicians, and weak environmental laws. Another new approach is that of Ameritrust/SRI (formerly called the Stanford Research Institute) of Menlo Park, California. Its revised Indicators of Economic Capacity (1986) are based on a somewhat arbitrary regional grouping of states, one still geared toward traditional economic growth; however, these indicators move away from the plantation model and identify "quality inputs" such as an educated work force, level of state investment, and commitment to academic excellence, research facilities, quality of faculty at universities, numbers of Ph.D.s graduated, levels of state investment in infrastructure, civic services, and other quality-of-life factors as well as capital availability.

Capital availability is the crux and symptom of the globalization process and the largely unanticipated restructuring of the U.S. economy. The capital markets have been deregulated in the U.S., and capital has become highly centralized as a result of this and of the global search for advantageous interest rates and currency differentials in the now tightly interlinked financial markets of the world. This situation leaves most localities short of the liquidity they need as their locally generated money is "vacuumed out" of their communities by branches of the major money-center banks. The chronic lack of liquidity for local circulation and trading as well as investment in local enterprises, dependency on outside investment, and other factors are the basic cause of their vulnerability. Not surprisingly, new forms of computer-assisted barter, skills exchanges, and limited-purpose local currencies are in use in thousands of U.S. and Canadian localities. Meanwhile, capital is available only to those large-scale, successful enterprises that localities are forced to try to lure in head-on competition with other cities and states as well as with other countries.

At the same time, on Wall Street billions of dollars of hot hungry money competes fiercely for fewer and fewer opportunities in higher and higher technology start-ups in the hopes of at least one big winner. No enterprise that does not offer at least 30 percent annual returns on investment is even considered, while millions of modest, useful, small-scale enterprises meeting local and regional market needs are too short of capital to expand, and many are forced into bankruptcy when they experience a temporary cash-flow crunch. Thus, many local needs for locally produced goods and services that would build a stable local economy remain unmet, and many people must drive long distances to shop for vital goods and services at regional outlets of large multinational companies. In this way local initiative is choked off much as it is in a centrally planned economy. This kind of absurdity occurs when market power is allowed to dominate instead of playing a balanced role in resource allocation. Of course, traditional economics recognizes this in monopoly theories, and the U.S. government is supposed to regulate such monopolies. However, political power, campaign contributions, lobbyists, advertising, and the commissioning of research allow many of these monopolistic corporations to control the mass media, large areas of production, and weapons manufacture, thereby capturing enormous government contracts and influencing policies.

Many U.S. localities are waking up to the impossibility of playing the old plantation-model game of economic development, since it leaves them wide open to the new world-trade roller coaster, vulnerable to sudden investment shifts, currency swings, bouncing interest rates, and lower wage competition while their well-laid local plans are disrupted as financial markets open in cities around the world. In Florida a 1987 report, *Florida Sunrise*, shifts emphasis from the plantation model to one in which education and investments in its citizens are seen as new paths to development.

Thus, the old growth view of economic development has hit these kinds of new snags, and most states now developing growth management plans and development impact fees that assess construction and development companies with part of the infrastructure costs their projects will incur. Many large companies that lose out in these new struggles with local governments or refuse to pay impact fees simply move over the Mexican border and set up the now-familiar maquiladoro plants in politically weak Mexican border towns. These towns, too, are becoming overburdened with influxes of migrants looking for jobs, and their tax bases are insufficient to provide infrastructure and basic services. Not surprisingly, many state and city officials and politicians are now learning that development in the world of the late twentieth century requires a dual approach. The first is to learn to play a much smarter role in the global "fast lane" and find export strategies not easily copied by others, that is, genuine niches of true *comparative* (not competitive) advantage, partly based on Adam Smith's concepts. The second is to develop a home-grown economy to provide basic security and meet local needs with local resources in smaller-scale enterprises. One can see common elements in this approach in the United States' and China's contract system, now spreading to the Soviet Union.

As the "luring outside investors" approach to development becomes more risky in today's context, when a whole generation of a technology, from innovation to obsolescence, can have a life cycle of three years or less, more interest is focusing on the "home grown economy" approach. This basic strategy offers a minimal safety net until the world trade roller-coaster is tamed by new global agreements, such as those to which the group of seven leading industrial nations known as the G-7, is already moving (in spite of their free-market preferences). The decline in United States competitiveness is forcing new thinking. The commonality in the new approaches in the United States, one of the world's most mature industrial societies with one of the highest per-capita-averaged income and GNP levels, is the now widely shared insight that sustainable development must begin locally from the grass roots, building on basic agriculture rather than the failed trickle-down model of industrialism. This industrial model saw rural communities and farmers as "backward" people to be mobilized into industrial production. But, as has been discovered in most countries, there are few easy short cuts to sustainable development, which involves steady efforts and prevention of harmful unanticipated consequences often very expensive to remedy or even irreversible.

The U.S. Office of Technology Assessment (OTA)—on whose Advisory Council I served during its start-up years from 1974 until 1980—released a study in June 1988 on grass-roots development in Africa that surveys efforts to enable the poor to participate in the process of development. OTA assessed nineteen countries in which such projects had been funded by a new agency set up by the U.S. Congress, the African Development Foundation (ADF). Between 1984 and 1987 the ADF awarded grants ranging from $700 to $250,000 to 114 projects, two-thirds of which were agricultural activities such as providing potable water, raising vegetable crops for local consumption, improving animal health, renting tractors, helping set up cooperatives, and the like. OTA's study confirmed the validity of ADF's assumptions concerning local participation as key to healthier forms of social and sustainable development. The World Bank is examining other grass-roots development models such as the Grameen Bank in Bangladesh and Women's World Banking, a network headquartered in New York and Amsterdam with local affiliates in fifty-five countries.

A now widely perceived key to sustainable development seems to involve what the World Bank refers to as "capillary lending," where channels are sought to pass through overly large sums of money to many village organizations, bypassing government and political influences in capital cities. It is questionable, however, whether the World Bank can recycle recent Japanese surpluses effectively.

All these new grass-roots lending policies—in mature industrial societies such as the United States, Canada, and many European countries as well as in developing countries—again prove that old "left" or "right" ideologies associated with economic models are obsolete. For example, the World Bank is pressured from both left and right to stop its massive, inappropriate projects, which are geared to host-country governments and infrastructures and which have led to some $30 billion of overinvestment in centralized energy projects on a world basis. Conservatives demand a grassroots private-sector approach in the name of the free market, while liberals want less environmental damage and less maldistribution of income. Still, the blindest spot is hardly addressed: the informal economy of unpaid productive work still uncounted in any national accounting models. These informal sectors still provide the basic safety nets in all societies, even those in the U.S., Canada, and Western Europe. Sociologists (rather than economists) collect this data, using methods of counting productive hours worked, whether paid or unpaid. Most of these studies in France, Sweden, Canada, and the United Kingdom show that approximately 50 percent of all productive work is unpaid, while other studies show that fully 80 percent of all the world's capital formation and investment is not monetarized, a fact that millions of subsistence farmers, rural entrepreneurs, and most of the world's women know only too well. Even industrial value systems are now in conflict.

Accounting for unpaid production is a way to address price inflation and pinpoint specific ways to keep prices in line with true value. A good example of the problem of per-capita-averaged income statistics based on monetarized production alone is evident in Japan. The British journal *The Economist* noted the problem in its June 11, 1988, issue under the headline "Feeling Poor in Japan." The article pointed out that the Japanese per-capita income, GNP-averaged, is the equivalent of $19,200 per year (ahead of the U.S.'s $18,200). But when the Organization for Economic Cooperation and Development worked out what the money actually *buys* in each country, that is, its purchasing-power parity, Japan looked poorer. Each Japanese had only $13,100 compared with the $18,200 for each American. The difference, of course, involves all of the various quality-of-life indicators.

Whereas purchasing-power parities are a very important new social indicator, they still do not get at the full range of social costs and benefits in each country. In Japan, *The Economist* pointed out, extra consumer goodies are consumed in ever more unpleasant surroundings; even New Yorkers have ten times as much green space per capita. True, Japanese life expectancy is in the eighties, but homes are prohibitively expensive due to exorbitant land prices, and standard commutes to work in cities are over an hour each way. In Britain real purchasing-power parity is higher than nominal income, as it is in Italy, whereas in the United States real purchasing power has remained flat for over a decade while nominal wages have increased. One of the key tasks is to examine the unpaid informal sectors and, where necessary, create barter systems locally so that real welfare may be increased without increasing wages and inflation. The interchangeability of money and information is a key to building local "information societies" based on barter and skills-exchanging networks in the same way that barter systems are employed when countries wishing to trade with each other do not have foreign exchange for this purpose as, for example, in the barter trade between China and the Soviet Union and the 25 percent of all world trade now conducted in barter or countertrade systems. Similarly, the Southern Hemisphere countries could set up highly sophisticated computerized countertrading networks to at last create multiple South-South trade systems by bypassing the current financial and trade channels.

The pressing need to overhaul national accounting as currently defined by the United Nations Systems of National Accounts (UNSNA) is underlined by Chilean "barefoot economist" Manfred Max-Neef, winner of the Alternative Nobel Prize, who notes that "nearly half of the world's population and over half the inhabitants of the Third World are statistically invisible in economic terms." Hernando de Soto, in *The Other Path*, documents the extent of Peru's informal sector, which accounts for 38.9 percent of that country's GDP. In *If Women Counted*, Marilyn Waring, a New Zealand legislator, dissects the UNSNA statistics to show how they confirm women's domination by men in most countries. In a speech to the second meeting of the South Commission in March 1988 President Perez of Venezuela offers a new paradigm for development indexes based on his concept of "integral development" and deals fundamentally with quality of life rather than quantities of goods and services produced. Perez's formula includes: (1) satisfaction of basic needs and the need to treat humans as indivisible beings; (2) self-reliance where possible; and (3) sustainability defined as equitable distribution within inbuilt environmental standards. New kinds of indicators are now under widespread discussion.

A comprehensive review of existing economic indicators and their reformulation is under way in the United Kingdom by Victor Anderson. Research in the Netherlands is proceeding along similar lines—based on the work of Roefie Heuting, Wil Albeda, and others—and includes accounting for natural resource stocks and the nonmonetarized work in the informal sector. Many researchers in Europe—including Britain's Gershuny, Robertson, and Shankland, Sweden's Inglestam and Ackerman, Italy's Giarini, Canada's Dyson and Nicholls, and Germany's Huber—have studied the nonmonetarized informal sectors of production, but as of 1989 the results are not included in national accounting in any integral way. Italy's Orio Giarini, author of *Dialogue on Wealth and Welfare* and *The*

Emerging Service Economy, believes it will be necessary to incorporate the nonmone-tarized sectors in order to understand the postindustrial service economies now emerging in Europe and North America. James Robertson reaches the same conclusion and has coined the term "ownwork" for the increasing numbers of part-time or self-employed autonomous workers in Britain and North America. Both Robertson and Giarini agree that this type of work must be accorded much higher status and part-time work must gain more prestige. They also agree that the encouragement of these new services sectors and their expansion will com-plement the formal, monetarized sectors. These views accord with my own work in *The Politics of the Solar Age* and that of Scott Burns in *The Household Economy*.

As industrial societies move further into this "services and amenities" view, beyond production per se, emphasis also shifts from obsolescence/innovation cycles to durability and overall optimization. An underlying problem with all national-income accounting is the focus on *flows* rather than *stocks*, emphasized in what is still the fundamental Western book on overhauling economic models, *The Entropy Law and the Economic Process* by Romanian scientist Nicholas Georgescu-Roegen. Costa Rica already portrays this wider, more systemic view, which au-gurs well for overall performance and efficiency there. The China 2000 studies are a model for other countries, including the United States, where similar studies—such as the Global 2000 Report to President Jimmy Carter—have been ignored. Today the new Solvency School led by Paul Kennedy, author of *The Rise and Fall of the Great Powers*, and including James Chace, Mancur Olsen, David Calleo, and George Kennan, has gained widespread credibility, and the U.S. citizenry is wak-ing up at last to the reality of the *relative* decline of the United States in a new multipolar world of vigorous new trading partners. The importance of the battle over indicators can be seen as essentially political, and the more suspect and meaningless conventional economic indicators of progress become, the more politicians must resort to polls of public opinion.

Savoring Cultural Diversity

A key political task in the United States is to cast its relative decline in positive terms, not to parrot the shallow claims of some commentators that the United States has "won" the Cold War and capitalism has been vindicated but rather to outline the positive vision of a safer, multipolar world that is now poised to enter a new Age of Interdependence, one in which global cooperation with all our rising industrial partners and the developing countries of the South leads to mutual sustainable development. As Soviets and Americans continue to discuss at many levels the newer definitions of security in economic, social, and environ-mental terms, the U.S. debate can adjust to the new global realities, particularly if new social indicators of progress are set up. Two-thirds of Americans agree with the Solvency School that the United States has become weaker compared to other countries, and 45 to 50 percent believe with Stanley Greenberg that America is "slipping dangerously." The Solvency School, while correctly pointing out the mutual economic exhaustion of the superpowers, overlooks the more fundamental set of stresses that their rivalry places on human communities—because, of course, no indicators of these factors are available. When indicators provide scientific, timely feedback on a full range of variables within broader and global parameters, they enable nations to steer with instrument panels of sufficient accuracy and sensitivity to continually course-correct on their paths to sustainable development. Clearly, most of the data are already available—on education, health, literacy, life expectancy, infant mortality, shelter, and water and air quality—to complement per capita income averages. Other categories needing work are human rights and political participation; much data on these subjects are available from Amnesty International and other human rights organizations. More inaccessible but also available are "shadow prices," equivalents for unpaid services and production and for barter and countertrade as well as for much of the productive work of normal systems.

The now widespread interest in global sustainable-development indicators has produced much debate in academic circles, such as that found in two pieces in *Environmental Management*, "The 1987 Forum on Global Sustainability: Toward Definition" and "The 1988 Forum on Global Sustainability: Toward Measurement." The International Institute for Applied Systems Analysis in Austria has produced good work in this area, for example, R. E. Munn's "Environmental Prospects for the Next Century: Implications for Long-Term Policy and Research Strategies (in *Technological Forecasting and Social Change* 33 [1988])," which reviews nonlinear, interdisciplinary models for global change that are adaptive and cross-cultural. Worldwatch's *State of the World Report* for 1988 assesses capital investments required as a "down-payment" on sustainability in six areas: (1) protecting topsoil and cropland, (2) reforesting the earth, (3) slowing population growth, (4) raising energy efficiency, (5) developing renewable energy, and (6) retiring Third World debt. The report then sets out two alternative global security budgets: the first continues the projected $900 billion spent annually on military security; the second defines global security in terms of sustainable development, with the military budget component falling each year from $845 billion in 1990 to $751 billion in the year 2000.

The economic exhaustion of the U.S. and the U.S.S.R. will continue to be the major factor in the development of these new "common security" definitions, but tremendous political activity on the part of the global citizens of every nation will also be needed to force these priorities onto politicians and leaders in business, academia, unions, and other social groups. The more we have improved social and economic indicators to provide better feedback on our current course, the sooner political will can be mustered for the necessary shift in policies. We particularly need to know the social costs of our current course, from the approximately $30 billion the U.S. will have to spend to clean up toxic dumps to the $60 billion per year in medical and absenteeism costs in the U.S. caused by smoking. This information is available, but few governments pay to have it collated and presented as it is by Peter Draper (in the *Royal Society of Health Journal* [U. K.] 1977), who makes it possible to relate which kinds of consumption lead to which kinds of disease—for example, linking sugared cereals to tooth decay—so that the social costs could be charged to manufacturers.

In the last analysis a fundamental understanding of the role of money itself must be propagated. Money is information used to track transactions and to keep score between organizations and individuals—it is not a commodity in itself. The global financial system today, which has reduced money to a series of blips on thousands of computer trading screens, has now made it possible to see that money and information are equivalent and interchangeable. At a time when such money-symbol systems are used to dominate human affairs, as is the case with Third World debts, these new insights are producing demands for reform. Similarly, overusing the price system to accomplish social policy has become a bad habit in the United States, whose free-market economy is actually a crazy quilt of administered prices, tax credits, subsidies, and incentives—all managed by the narrow policy tools of fiscal and monetary string pulling while international economic relationships are giving way to crude, bouncing-dollar policies and interest rates that feed global instabilities. The need for reform of monetary policies as well as of the role of central banks in issuing money and credit is growing urgent in the West.

In the emerging Age of Interdependence the new global, systemic view is allowing serious questioning of traditional disciplines, including economics, and of the proper role of money itself in relation to tangible wealth and human labor, skills, and knowledge. The South Commission, at its meeting in Caracas in August 1989, co-hosted by Venezuelan President Perez, convened experts on national accounts and social indicators from five continents and the United Nations. Perez presented their report, *Toward a New Way to Measure Development*, at the Belgrade meeting of non-aligned nations in September 1989, at which time fifteen countries formed a South Economic Summit.

All development paths are now interlinked and interactive via communication networks, and this will equalize technological means and move the world trade system to one of less "hardware" and more "software" as knowledge diffuses ever more rapidly and widely. Every country will know how to apply basic technologies to its own domestically manufactured goods, and multinational companies will have fewer opportunities to recapture research and development investments. For example, when IBM introduces a new computer line, clones produced by others appear swiftly on the world market. I have termed this the "bursting seed-pod" model of world trade—that is, when mature industrial companies and countries release their knowledge and "scatter" it in spite of patents, since information flows and is not scarce in the same way as are goods but can be shared in a win-win game. Enormous efficiencies can thus be gained and shared, with huge savings in transportation and distribution. As each country is forced to think harder about its true niche, it will seek creative advantage and will become more unique in its exports and less subject to competition. An example is the Netherlands, whose exports of dike technology to keep out sea water have no competitors. Each country's unique gifts can be offered to other countries, and all can savor the growing diversity of the human family, just as today we savor one another's food, art, music, and culture.

—1989

    In addition to the books by Hazel Henderson listed in Further Reading, she has also published articles that have contributed to new directions in economic thought: "Economists versus Ecologists," *Harvard Business Review*, July–Aug. 1973; "The Limits of Traditional Economics," *Financial Analysts Journal*, May–June 1973; "Post-Economic Policies for Post-Industrial Societies," *Revision*, Winter 1984/Spring 1985; "Riding the Tiger of Change," *Futures Research Quarterly* 2 (1), 1986; "Rethinking the Global Vision," *Action Linkage Networker*, Sept. 1987.

Jane Jacobs
The Economy of Regions

Jane Jacobs shares with E. F. Schumacher that lamentably rare attribute
of a remarkable intelligence pragmatically applied. So often brilliant
minds are attracted to the novel, the abstract, or the obscure. However
valuable such preoccupations, the woes of the world are too rarely addressed by them. But Jacobs, like Schumacher, has made the day-to-
day concerns of ordinary people the focus of her refreshingly clear
thinking, to the gratitude of her many readers and listeners.

"Human life, to be fully human, needs the city," wrote Schumacher.
As her influential book *The Death and Life of Great American Cities* makes clear,
Jacobs is a noted proponent of urban life. When the neighborliness and
mutuality she considers among the benefits of city dwelling began to
disintegrate in New York City, she moved her home to Toronto, where
she continues to write. She is also the author of *The Economy of Cities*, *The
Question of Separatism: Quebec and the Struggle over Sovereignty*, *Cities and The Wealth
of Nations*, *Systems of Survival*, and *A Schoolteacher in Old Alaska: The Hannah Breece
Story*.

Critical of simple-minded notions of isolated self-sufficiency, in
her Schumacher lecture Jacobs addresses economics on a regional
scale, defining region in the sense of a specific area that gives rise to
the dynamic exchange of goods, services, and people. She examines
the factors essential to establish a sustainable economic entity. The five
forces that shape and reshape all regional economies—city markets,
city jobs, city technology, transplanted city work, and city capital—
interact to create a pattern of exchange between city and hinterland

that is complex and diverse. The key to an ongoing healthy economy is the ability of the city to replace imports from beyond the region with local production. She explains: "In a natural ecology the more niches that are filled, the more efficiently the ecology uses the energy it has at its disposal and the richer it is in life and means of supporting life. Just so with our own economies." Jacobs's analysis becomes more timely in the light of the alarming decline of many cities. Their problems must be addressed, however, and she is of invaluable assistance in weighing the options.

"The Economy of Regions" became part of her 1984 book *Cities and the Wealth of Nations.*

We tend to take it for granted that nature—being basic to everything—is the place to begin when we try to understand regional economies. The given natural attributes of a region certainly do explain much about subsistence economies: why some people eat seals and caribou while others eat dates and goats; why herders in some places stay put beside their fields while others traipse back and forth between summer and winter pastures; why some people shelter in thatch and mats while others build with stone and timber; why some spin wool, others cotton, and so on. But interesting as such economic travelogues are, they don't go far to explain even subsistence economies. For one thing, they don't explain why these *are* subsistence economies instead of something else.

Nor do regional natural attributes get us anywhere when we try to understand the why of other types of economies. After all, it isn't nature that decrees a tropical island is to be ecologically and socially devastated by the monoculture of sugar cane. Nature merely permits this possibility, along with many others. Even where nature is most limiting, economic life is a grab bag of surprises. Nowadays in the Canadian Arctic, some groups of Eskimos—or Inuit, as they prefer to be called—make their living by selling sculptures and prints to distant museums, publishers, and collectors. They work a regular seven-hour day at it, all year around, fulfilling contracts and meeting deadlines. Other Inuit groups combine hunting, trapping, and fishing with temporary construction jobs for wages. Still others eke out their lives, by all accounts miserably, on welfare. Some others still make their livings much as their ancestors did, except that they make use of guns, steel knives, steel traps, snowmobiles, and radio weather forecasts. All this is unaccountable if we try to explain it in terms of regional attributes or regional ecologies. Our basic information about all the regions of Japan tells us that they are deficient in the proportion of land that is arable, almost lacking in minerals, and that the populations of most parts of the nation are dense. The logical meaning of all this, if we took nothing else into consideration, would seem to be that the Japanese are very hungry people and have no foundation for industry.

I do not suggest that we should ignore given natural attributes, much less that we should neglect the great questions of the uses to which we put what nature has given us. The less we know and the more heedless we are of such matters, the worse for us as well as for nature. Rather, my point is that there are other clues we must recognize if we are to understand regional economies and what we are up against when our symbiosis with nature turns into a mess.

The kinds of clues that help us understand how regional economies are shaped and reshaped are to be found, in microcosm, in the economic history of single settlements—for instance, in a hamlet named Bardou, perched high in the Cevennes Mountains in south central France. This small cluster of stone houses found its way into my Toronto newspaper because it is so charming. It has become, in its small way, a kind of Shangri-la for writers, musicians, artists, and craftsmen fleeing the cities of Europe, the United States, and Canada in search of beauty and a cheap, quiet place in which to work.

Some two thousand years ago, when Gaul became a province of Rome, Bardou was linked into the imperial economy by roads terminating in a collection of iron mines nearby. The iron found there was not shaped at the site into swords, chisels, hinges, wheel rims, and so forth, and where the raw material went for manufacturing is now unknown. A logical guess would be the foundries at Nîmes, an ancient city that had already become a metropolis of this part of Gaul in pre-Roman times. Or it might have been carried to Lugdunum, now Lyons, which also has an ancient tradition as a center of metal working and was the hub of the Roman road system in Gaul. Much less do we know the exact destinations of the finished or semifinished manufactured goods, apart from the fact that they must have been solvent markets. What we do know is that the iron was in sufficient demand to justify mine roads so well engineered and solidly built that they still serve admirably as hikers' trails, although they have gone largely untended and unrepaired for some fifteen centuries. Both the mines and the roads were abandoned when economic life in this part of Gaul disintegrated, probably in the fourth century.

The area then reverted to wilderness—unpopulated, as far as has been discovered, until early in the sixteenth century when landless peasants from the slopes below pushed farther up the mountain and built the houses of the present hamlet. They scratched out little garden plots among the rocks, gathered chestnuts in the forest, and probably caught game there; on their poor and rocky soil they pursued as best they could the subsistence arts they had inherited from economies of the distant past more creative than their own. Lifetime after lifetime nothing changed in this subsistence economy. To romantics this may sound idyllic, but we must suspect that the life here was not only hard but probably also boring and mean. At any rate, tradition has it that people in Bardou were accustomed to stealing one another's garden produce by shifting boundary markers in the night, then interminably squabbling and feuding over the thefts. Such were the excitements for about three and a half centuries.

Then, in the 1870s a radical change abruptly took place. Word penetrated that a better life was to be found in Paris. Perhaps the information percolated in from army recruits or from migrants who had left hamlets on the lower slopes or villages in the valley. Paris had been attracting rural people for untold generations; the word was late in reaching Bardou, yet once a few venturesome souls went there, a slow but almost total exodus followed. By 1900 half the population had left; during the following forty years everybody went except for a single family.

In 1946 when a pair of enterprising German hikers happened by on the old Roman mine roads, the ruins sheltered only one aged man. The hikers bought the hamlet from him and from such descendants of the former inhabitants as they and their Paris lawyers could trace, and when legal title had been established, the new owners moved in and invited kindred spirits to join them and help pay expenses. In this incarnation Bardou has a rotating core of year-round residents who live on savings or by selling their works in distant cities. Holiday renters and campers are welcomed, along with the incomes they bring. Residents and vacationers alike import nearly all their necessities, but they make a virtue of getting along without electricity, telephones, and hot water. A piped cold-water system was financed when a motion picture company rented the hamlet briefly to make location shots and paid well for the privilege.

We could beat our brains out trying to explain Bardou's economic twists and turns in terms of its own attributes, right down to compiling statistics on the amount and quality of iron taken and that remaining, the probable average yield of chestnuts, the tools used there, the man-hours required to build a house, the nature of the soil, the annual rainfall, the delineation of the watershed and Bardou's place in it, and so on. None of this would enlighten us as to why Bardou's economy, at differing times, has taken the wildly differing shapes it has. To understand both the changes that have occurred and why there were periods when nothing changed, we must look to clues that do not define Bardou in any way except as they have acted upon it from a distance.

Bardou is a microcosmic example of a passive economy, meaning an economy that is shaped and reshaped by forces that do not originate within itself but come from outside, specifically from distant cities. Like a toy on a string, time and again Bardou has been jerked by this powerful external economic energy. In ancient times it was exploited for its iron, then abandoned. In modern times it was depopulated by the pull of city jobs and income, then later repopulated by city people. The jerks were never gentle. But when cities and their people had no uses for Bardou and simply let it alone, the place either had no economy whatever, as when it was a wilderness, or else a subsistence economy that remained unchanging. Bardou's history is unique but only in the sense that every place, like every person and every snowflake, is unique. Otherwise, the same kinds of changes and events to be found in Bardou's story are duplicated in principle in many other passive economies and on a large, regional scale.

Five different kinds of forces from distant cities jerked Bardou about, although never all at the same time. Those forces were the power of city markets, the power of city jobs, the power of city technology, the power of city work transplanted out of cities, and the power of city capital. These are the five forces that shape and reshape all regional economies, except those for which cities have no use, as happened in Bardou when it was a place of subsistence life only. These five great forces arise within cities, primarily as a consequence of city import-replacing, a process that shifts and enlarges city markets, rapidly increases city jobs, spurs development and use of rural labor-saving technology, multiplies and diversifies city enterprises, and generates capital—all simultaneously. Because these five forces are consequences of one and the same process, are in fact different facets of the process, within a city they are inextricably intertwined. I am not going to concern myself here with their effects inside a city that repeatedly replaces wide ranges of its imports with its own production but rather with the effects these forces have upon the world outside cities.

In the hinterlands of some cities—beginning just beyond the suburbs—rural, industrial, and commercial workplaces are mingled and mixed together. Such city regions are different from all other regions, having the richest, densest, and most intricate economies to be found, except for those of cities themselves. City regions are not defined by natural boundaries, because they are wholly the artifacts of the cities at their nuclei. The boundaries move outward or halt, only as city economic energy dictates. The largest and densest city region in the world today, and one of the richest, is Tokyo's. As it has grown over the years, it has vaulted over mountains so rugged that to build rail lines and roads across them required spectacular feats of late nineteenth century engineering. The city region of Toronto, the city where I live, is bounded in some directions by the Great Lakes, but in others it simply peters out and halts on gently rolling land presenting no change in natural landscape. Boston's city region extends to the north into the southern part of New Hampshire, a circumstance that exasperates municipal and state officials in New Hampshire who would like to see a nice, even smear of mixed economic activity over the whole state. To that end they contrive special inducements to lure Boston enterprises and people into the northern part of the state, where work is badly needed, and special obstacles to discourage them from moving to the south, which is already prosperous—but to no avail. For the time being, southern New Hampshire and southern Maine are as far north as Boston reaches in its capacity to reshape its own hinterland radically.

Not all cities generate city regions. For example, Glasgow never did so, although in the late nineteenth century and the first decade of this century Glasgow was at the forefront of industry and technology. Marseilles is the most important French seaport and in addition has built up considerable industry alongside its shipping work, but it has no city region to speak of even though it is the metropolis of all southern France. It is closer to Bardou than is Paris, and so are a number of other French cities, such as Montpelier, but that is not where the inhabitants of Bardou went when they sought city jobs. Rome has an amazingly small city region considering its size, and Naples has none, but Milan is at the nucleus of a whole nest of overlapping and abutting city regions. San Francisco and Los Angeles developed significant city regions, but Seattle did not; Philadelphia did but not Pittsburgh, nor did Richmond or Atlanta. Obviously, cities good at working up exports for themselves or at serving as cultural or administrative capitals or transportation and depot centers do not necessarily generate city regions. Something more is required: the practice of repeatedly replacing, from time to time, wide ranges of the city's current imports with local production.

On their own hinterlands, import-replacing cities bring to bear the whole panoply of their forces simultaneously—their markets, jobs, technology, transplants, and capital. That is what makes these regions so complex and diverse. The forces interplay with one another, but let us look at some of their characteristics one at a time.

The markets of an import-replacing city enormously diversify agricultural crops and other rural products produced in a city region. As we shall see, this is not how city markets affect regions other than their own hinterlands. Apples happen to be among the rural cash crops of the Toronto city region. Only a few kinds go to market outside the region: Macintosh, Delicious, Golden Delicious, Spies, Cortlands, Russets, and Spartans. These are of course available in Toronto too, but if it were not for the market of Toronto itself and its several smaller satellite cities, those few varieties would be the only regional apples with solvent markets. However, in the region's own city markets one finds as well, in season, apples that seldom leave the region: Wealthies, Baxters, Snows, Lobos, Melbas, Humes, Wolf Rivers, Ida Reds, Pola Reds, Tydeman Reds, Blenheim Orange, Transparents, Gravensteins, Matsus, Tolmans, Sweets, Duchess, Lady, Delaware, Orange Pippins, Empires, Cravens, and St. Lawrences. The apples merely symbolize the diversity of rural crops and the many kinds of rural work other than agriculture stimulated by the diverse demands an import-replacing city makes upon its own hinterland.

City jobs draw rural young people from a city's own hinterland, and at times when those jobs are rapidly multiplying and diversifying, they draw migrants rapidly and in great numbers. Since that is exactly when a city's markets for rural goods from its hinterland are also increasing and diversifying most rapidly, something has to give. What gives are the traditional ways of doing things out in the countryside. It is no accident that, historically, rural labor-saving equipment has first been put to work in the hinterlands of vigorous, prospering cities and only later—often much later—reaches other types of regions.

The industries dotting a city region are enterprises that have balanced their needs to be close to their suppliers and customers against their conflicting aims of escaping the costs of city space and the congestion or other disadvantages of the city. The balances they strike are reflected in the physical pattern of a city region's industrialization. The transplanted industries typically cluster most thickly just beyond the city and its suburbs, thinning out with distance, here and there forming clots within the region but eventually petering out as the region's current borders are reached. As a city region becomes dotted with industry and commercial establishments, some enterprises can start out in the region itself rather than in the nuclear city or cities, but they too are tethered to the region. So are branch plants drawn by the city and regional markets. In sum, many of the enterprises a city generates can move, but they can't move far. They depend on other nearby producers or customers or both. This is why, in the aggregate, industries and services of city regions produce amply and diversely for the region's own people and producers as well as for others.

Of course, the transplanting of city enterprises out into the region, the rural labor-saving equipment, and the other changes taking place—such as provision of schools, hospitals, transportation, recreation places, and so on—require capital, the fifth great city economic force. It is used, in a city region, in conjunction with other forces.

All these forces come to bear upon a city region not only simultaneously but in roughly reasonable proportion to one another. This is not so when those same forces reach into distant regions, as they always do. It is as if the complete mesh or net of economic ties with which a city binds its own region unravels at the city-region borders. The various strands—markets, jobs, technology, transplants, and capital—separate from the mesh and take off by themselves, each in its own idiosyncratic direction, joining up with similar, not different, strands from other cities. In this fashion, cities shape stunted and bizarre economies in distant regions that lack vigorous city economies of their own.

The most economically important among such regional grotesques are supply regions, which are disproportionately shaped by the markets of distant cities. Not only are supply regions narrowly specialized to supply agricultural or natural resource goods to distant markets, and to do little else, but each one is remarkably narrow in the few kinds of goods it supplies. The markets of cities that stimulate rural diversity in their own hinterlands do quite the opposite when they draw from distant regions. There the markets become highly selective. Furthermore, the markets of many cities agree on what it is they want from distant regions. They are both selective and concerted, and this concerted selectivity is a powerful force when it operates alone, as it were, unmediated by the other city forces.

The concerted selectivity of distant markets is a deep-seated trait, long predating modern transportation and communication as well as modern mass markets. In late medieval times, for instance, Sardinia was exporting cheeses to all the cities of Europe—and nothing but cheeses. Obviously, the distant city markets were both highly selective and highly concerted in what they wanted from Sardinia. The Canary Islands at one time were supplying cane sugar to all the European cities, but nothing else. They were the prototype for the later one-crop sugar islands of the West Indies. At the time when ancient Bardou's iron mines were operating, evidently nothing was wanted of Bardou but iron, and it was wanted badly enough to justify those splendid mountain roads. Supply regions in our own time are so common, in poor countries and in parts of rich countries too, that there is no need to give examples; we are all familiar with them.

Sometimes supply regions are called, in a shorthand we all understand, "colonial" economies. The epithet embodies this piece of truth: imperial powers have typically shaped conquered territories into regions supplying a narrow range of rural or resource goods for distant markets. But the term "colonial" is too optimistic because by reverse implication it suggests that if alien domination of some sort is thrown off, a stunted and narrow economy producing poorly on its own behalf will proceed to become well-rounded and less economically dependent on its narrow specialties. Yet the stultification is not that easily corrected. When Castro disposed of American influence in Cuba, he did not throw off Cuba's servitude to sugar. Indeed, many a supply region, far from being forced into its role, actually courts it or else slides into it for sheer default of alternative ways to make a living.

Supply regions are often poor, and thus the stultification of their economies is often attributed to poverty. But a rich supply region is as stunted and stultified as a poor one. The shortcomings of these regions go deeper than poverty. Although supply regions are warped into their narrow, unbalanced economies by the power of distant city markets, those distant cities are powerless to straighten them out. Only vigorous, innovative, and productive cities of their own, generating city regions instead of supply regions, can perform that service for supply regions.

When the second great city force, the power of its jobs, reaches into a distant region disproportionately, it leads to abandonment of the region—sometimes all but complete abandonment, as happened on a small scale in little Bardou, but more commonly selective abandonment by people of working age or sometimes partial abandonment. Since 1921 Wales has lost a third of its population. The chief settlement, the economically inert little city of Cardiff, offered few opportunities, and so most people seeking city work or better incomes left Wales entirely. Large sections of Sicily and Spain that once were heavily populated now lie almost empty. Even in rich countries one finds abandoned regions. I come from one: a stagnated anthracite-coal supply region in northeastern Pennsylvania. Its two chief settlements, Scranton and Wilkes Barre, are less populated today than when I was a child in the 1920s, and even then people in search of more diverse opportunities and better incomes were beginning to leave, usually for New York and its region if they came from around Scranton, as I did, or for Philadelphia and its region if they were from Wilkes Barre. In Ontario, beyond the city region of Toronto, there are settlements now so deserted by the young and the middle-aged that a recent social services investigation of them summarized the predicament as "the old looking after the old."

The difference between poor stagnant regions losing population and poor stagnant regions in which people stay put is simply that people from places like Scranton, Wales, and the abandoned parts of Ontario can have realistic hopes of doing better some place else and have means to get there, while people in places such as Haiti, where most stay put, lack a way of getting out or a place to go. If people in all poor and stagnant regions were to have ample access to city work, no matter how distant, we may be reasonably sure that virtually all poor, stagnant regions in the world today would be losing populations at a great rate. This is not to say that people in stagnant regions have little attachment to them just because these places are poor and lack economic opportunity or that most people actually like migrating. Emigration to escape poverty, especially if one must leave a familiar culture for a strange one, is a painful and bitter choice of the lesser of two evils.

The most striking and revealing fact about the economies of abandoned regions is that the departures have no effect on these poor economies other than to shrink them. In old Bardou, for almost seventy years people kept trickling away to Paris. As the population diminished, the hamlet's economy shrank but otherwise did not change. Of course, the people who went to Paris totally transformed their own economic lives for better or for worse; we must assume largely for the better, or the exodus would have stopped. But those left behind remained in the same immemorial poverty until they too left. In rural Wales those who stayed on did not change their lot, which is why people kept on leaving. Over the past thirty years tens of millions of workers have left poor villages in stagnant regions of Egypt, Turkey, Italy, Greece, Yugoslavia, Morocco, Algeria, Spain, Portugal, and the Azores to work on contract in cities and city regions of northern Europe.

When unemployment rose in northern Europe and millions of these "guest workers" flooded home, they returned to villages that were no better off than when they left.

In the case of regions people abandon, the disproportionate force from distant cities and their regions, unbalanced by other city forces, is the pull of city jobs. Of course, it doesn't change an abandoned region's economy. That single force can depopulate a distant region, but it can't do anything else to reshape it. Even the remittances that migrants send home to dependents merely alleviate poverty, like any other income-transfer payments. That is beneficial as far as it goes, but it doesn't reshape these pitiable economies themselves.

In the Italian film "Bread and Chocolate," which depicts the loneliness, discrimination, exploitation, dead-end jobs, and painful cultural dislocation foreign workers endure when they migrate to Swiss cities and city regions, one of the characters, instead of blaming Switzerland and the Swiss, exclaims, "Blame Italy, which forces us to emigrate!" He was getting close but not close enough. After all, the north of Italy is not the source of streams of Italian workers abandoning their regions. On the contrary, for generations Milan, Florence, Boulogne, and their great networks of city regions have been accepting an enormous share of southern Italians and Sicilians seeking escape from poverty and have helped them do just that. "Blame southern Italy, which has no import-replacing cities!" would have been a more accurate cry of pain.

When technology to improve rural productivity reaches into regions without vigorous cities of their own, the results usually improve the lot of those still remaining at work on the land but are disastrous for those who become redundant because of new methods and equipment, and so they are cleared from the land and their former livelihoods on it. These clearance regions are mirror images of abandoned regions. In the case of abandoned regions, those who remain stay badly off while those who leave for distant city jobs may improve their lot. But in the case of regions from which workers on the land are cleared, those who remain at work are better off while those no longer needed are worse off unless productive city jobs somewhere or other await them.

These results suggest that to improve everyone's lot—that of those who stay and those who leave—the two kinds of events, abandonments and reception of rural labor-saving equipment and methods, ought to happen in the same places and at the same time. This is precisely what does happen in the rural parts of city regions, where some leave for jobs in the city while technology improves the productivity, yields, and incomes of those remaining at rural work. But in all other types of regions, economic forces seldom, if ever, dovetail that neatly. If people who escaped poverty by leaving Wales, old Bardou, or dying mining towns in Pennsylvania had had to wait until labor-saving equipment uprooted them from their work, they would be waiting still.

The difference between clearances in poor countries that lack alternative productive jobs for redundant rural people and rich countries like the United States is that rich countries can afford doles. But either way, unless productive city jobs do exist, displaced people are left in the lurch. This outcome has been a sequel to the great American agricultural clearances in the south and to the Green Revolution in Third World countries. In some cases the displaced people migrate to cities that have little to offer them but idleness and either doles or poverty without doles. Sometimes they stay on the land, and that can be worse, if anything, should they deforest steep slopes in the struggle to survive or take to cultivating land otherwise unsuitable for cultivation or pasture.

Improved productivity of individual rural workers and increased yields are not the same things, but as a practical matter they might as well be. In agriculture high yields and thrifty use of labor go hand in hand. That is why the most thoroughly rural countries, those in which as many as 80 percent of the people work the land, are paradoxically the hungriest and why countries in which a small proportion of the population devotes itself to growing food are paradoxically the best fed.

In practice it seems that every measure for increasing agricultural yields also reduces need for agricultural labor. So simple a device as a bicycle water pump can appreciably increase yields in fields where irrigation water was formerly drawn by hand. But one single such pump typically eliminates about two thousand work days annually or, to put it another way, takes the place of about seven workers. A single bicycle water pump and a single small mechanical tiller, taken together, make about fourteen workers in Java redundant. Of course, the equipment also increases yields. Yet both the increased productivity of the workers using the new equipment and the increased wealth of the higher yields are largely illusory from a social point of view, and from an economic point of view as well, when fourteen other workers are left less productive and more poor than ever. In India the government sponsored development of a very nice piece of small technology, a bicycle-powered spinning wheel; however, one villager equipped with it can spin as much yarn as twelve workers using traditional wheels. Solvent markets for cloth of hand-spun Indian yarn are not growing at any comparable rate, either in India or anywhere else, so the new wheel has much the same disastrous side effects as the bicycle water pump in Java. Having successfully sponsored the wheel's development, the government of India ironically cannot promote its use.

Like most of you, I assume, I see much hope in the use of the small and intermediate technology Schumacher advocated. Furthermore, small and inter-mediate technology is quite as necessary, valuable, and constructive in the eco-nomic life of cities as it is in rural and village life and in currently rich countries as well as poor ones. The use of large and expensive capital-intensive equipment has become so mindless and rococo that it leads to mechanization poverty, meaning that it actually doesn't pay its way in direct and indirect costs but makes us poorer. Nuclear power plants are an extreme example, but in principle so are many types of equipment now being used for agriculture.

The point I am making here, however, is that in regions where rural workers have no access to alternate productive livelihoods, even the most seemingly appropriate and gentle labor-saving and yield-increasing devices become a curse. On the surface it seems plausible that both the poverty and the hunger of poor rural people can be ameliorated if only their yields can be somewhat increased. What could be more straightforward? But because of the indissoluble links be-tween improved yields and thriftier use of labor, it doesn't work out that way unless alternate, genuinely productive jobs are available at quite other kinds of work.

In the absence of alternate city work, the humane approach seems to be to introduce new rural methods and new devices so gently and with such glacial slowness that few workers become redundant. But then yields also increase only with glacial slowness if at all. The harvest of that policy is frustration and anger that so little material good comes of so much implied promise. The hard truth is that there is simply no decent way of overcoming rural poverty among people who have no access to productive city jobs. To be sure, their economies can be turned into vast charity wards, but that is not my definition of meaningful or hopeful economic development.

The standard diagnosis of the trouble with supply regions, abandoned re-gions, and clearance regions as well as stagnated and declining cities is "not enough industry." To be sure. But the standard prescription for the deficiency is "attract industry." What are these industries that can be lured and hooked? Where do they come from and why?

For the most part they are industries that originally developed in cities or city regions but are no longer tethered there by localized markets or by everyday dependence on multitudes of producers and services close by. Their markets have become far-flung, and they supply so many of their own everyday needs for producers' goods and services internally and have become so practiced at acquir-ing those they must buy from others, whatever the source, that these enterprises have developed great freedom in choosing where to expand or to relocate. They can move to virtually any place providing other special advantages they seek: for instance, exceptionally cheap labor, close proximity to raw materials they use, release from environmental regulations, or the chance to cash in on tax forgiveness and other subsidies commonly offered to enterprises that will move into depressed areas.

The very freedom of location that enables these industries to leave city regions for distant regions means freedom from local markets and freedom from symbiotic nests of other producers. Therefore, their presence does nothing, or little, to stimulate creation of other, symbiotic enterprises. This outcome becomes starkly obvious whenever these transplants pull up stakes and leave for yet a different location, perhaps one with still cheaper labor or still lower electric rates. What they leave behind when they move are merely economic vacuums, very different from what they left behind originally in the cities or city regions of their origin. And as long as they remain in a region with a transplant economy of this sort, they produce only little and only narrowly for the local economy itself. Their markets are distant. In effect, such transplants shape a kind of industrialized supply region incapable of producing amply and diversely for its own people and producers as well as for others.

Furthermore, there aren't anywhere near enough untethered transplants to meet the demand for industry in passive regions. That is why so-called development officials compete so hard for their limited numbers and often enough concede so much in the form of subsidies and the environmental damage they will tolerate from the transplants. Regions with industrial transplant economies are profoundly parasitic, depending as they do for their economic life on what comes to them already developed, already well-established, already ready-made. Yet that is what a region with no creative city of its own must depend upon if it wants industry.

Almost two centuries ago Catherine the Great of Russia said: "Most of our factories are in Moscow, probably the least advantageous spot in all Russia. It is dreadfully overpopulated and the workers become lazy and dissolute. . . . On the other hand, hundreds of small towns are crumbling in ruins. Why not transport a factory to each of them, according to the produce of the district and the quality of the water? The workmen would be more industrious and the towns would flourish."

As was perhaps natural, because Catherine was a monarch, she was thinking of an economy much as if it were an army. If you have a territory and an army, you can deploy the troops where you judge they are needed; never mind if they would rather hang about the glitter and fleshpots of the city.

Because a developing economy is not created in the same way as an army nor supplied and sustained in the same way either, it isn't analogous in the way it can be successfully deployed; nevertheless, this old and simplistic conception of how to defeat stagnation and poverty in the boondocks remains much with us.

Catherine was royally untroubled by the hard questions: What if there are too few factories for all those crumbling towns? What happens when the transplanted factories depart or fail or grow obsolete—what do they leave behind them in their company towns? What if the sources of these factories run dry and new sources don't bubble up?

Capital, the fifth great force generated by cities, can be used in quite as bizarre and unbalanced a fashion as the others when it reaches out into distant regions. The Volta Dam in Ghana, one of the world's great hydroelectric and irrigation projects, was supposed to supply factories with power. But few factories, other than a joint venture of Kaiser Aluminum and Reynolds Metals, materialized to make use of it, even though the power is artificially priced at about one-twentieth of the world's average electric power prices. The dam was supposed to help farmers prosper, but solvent markets for the putative crops didn't materialize either, whether within Ghana or far distant. The eighty thousand people whose traditional village subsistence economies were wiped out to make room for the dam and its reservoir were relocated on land so poor it couldn't support them, and about half have become landless paupers.

Carried away by the power of money to finance great capital undertakings, many people seem to think of such investments as being economic development itself. Build the dam and you have development. But in real life build the dam, and unless you also have proportionately increased solvent city markets and proportionately large numbers of transplanted industries and their jobs, you have nothing. Or even worse than nothing: new and added problems. The economic pointlessness of the Volta Dam is not all that unusual. A United Nations Food and Agriculture Organization official has commented that he could cite about forty dams around the world—major dams—that are completely useless.

In our own time, capital in the form of both loans and gifts for regions lacking cities has been unprecedentedly abundant. Many, many symptoms besides unre-payable loans inform us that these uses of capital are wildly disproportionate and out of balance with other city economic forces reaching distant passive regions; they are also out of balance with the need to maintain existing creative cities or to help new and creative city economies to rise and flourish in regions that lack them and desperately need them.

Many of the processes at work in natural ecologies and in our own economies are amazingly similar. I shall mention only two, although many other similarities are obvious. In a natural ecology the more niches that are filled, the more effi-ciently the ecology uses the energy it has at its disposal and the richer it is in life and means of supporting life. Just so with our own economies. The more fully their various niches are filled, the richer they are in means for supporting life. That is why city regions are so much better off than specialized economies like those of supply, clearance, and transplant regions, to say nothing of abandoned regions and charity-supported regions.

In a natural ecology the more diversity there is, the more stability, too, because of what ecologists call its greater numbers of homeostatic feedback loops, meaning that it includes greater numbers of feedback controls for automatic self-correction. It is the same with our economies, and this is why city regions are economically more resilient and less fragile than other types of regions.

The other, nonhuman animals don't add new kinds of activities to their older kinds in an open-ended way. But we aren't the other animals. It is natural for human beings to build new kinds of work and skills upon earlier kinds because the capacity to do this is naturally built right into us, like the related capacity to understand and use a language in an open-ended way. All normal human beings, starting in infancy, have the capacity to add new kinds of skills to their earlier skills, new kinds of work to earlier work. Without that built-in capacity we might be something else, but we would not be human beings. Collectively, this is also the capacity we use when we develop human economic life. In the process, we create unprecedented problems, but then, if we continue to use our marvelous human gifts creatively and keep our wits about us, we solve the problems in unprecedented ways, at the same time casting up still newer kinds of work as well as more problems. In its very nature, developing economic life has to be open-ended rather than goal-oriented, making itself up as it goes along. The people who developed agriculture couldn't foresee soil depletion; the people who developed automobiles couldn't foresee acid rain.

Cities are the open-ended types of economies in which our human capacities for open-ended economic creation are not only able to establish new and initially tentative little things but also to inject them into everyday life in a practical way. Cities don't work like perpetual motion machines. They require constant new inputs in the form of innovations based on human insights. And if they are to generate city regions, they require repeated, exuberant episodes of import-replacing, which are manifestations of the human ability to make adaptive imitations. Any region without an innovative and import-replacing city of its own right there is bound to be either an unchanging subsistence region or else a stunted and profoundly dependent region, no matter what its given natural attributes or the innate attributes of its people may be.

Any region with an innovative and import-replacing city of its own becomes capable of producing amply and diversely for its own people and producers as well as for others, again no matter what its given natural attributes. Such a city and its city region also automatically become capable of shaping and reshaping the economies of distant regions lacking vigorous cities of their own, shaping them for better or for worse. Too often the shaping is either disappointing or disastrous, but there is no remedy for that other than the emergence of vigorous cities in regions that lack them and need them. Back in 1377 a Tunisian scholar and historian, Ibn Khaldun, explained that the Bedouins of the desert, who sold animal products and grain to urban people, would remain economically weak and dependent "as long as they live in the desert and have not acquired . . . control of the cities." True to a point. But he might have added, "or as long as they do not create a city of their own."

—1983

Robert Swann and Susan Witt

Local Currencies: Catalysts for Sustainable Regional Economies

In 1980 Robert Swann established the E. F. Schumacher Society in
Great Barrington, Massachusetts, in response to Schumacher's earlier
request to start a sister organization to his own Intermediate Technology Development Group. The work of the Society constitutes a direct
link with Schumacher's philosophy and is a tangible embodiment of
his message.

Schumacher chose wisely. Robert Swann brings the pragmatic
skills of a builder to his lifelong commitment to both community and
decentralized economics. Before founding the Society he worked with
Ralph Borsodi to issue a commodity-backed currency on an experimental basis in Exeter, New Hampshire, a forerunner of today's local
currencies. In 1978 he launched the Community Investment Fund,
one of the first investment initiatives with socially responsible criteria,
anticipating a national movement in social investment.

His 1960s civil rights work led to an effort to secure land for
African-American farmers. With Slater King he founded New Communities in Albany, Georgia, using documents modeled on those of
the Jewish National Fund. As founder of the Institute for Community
Economics he helped other groups around the country form similar
community land trusts, which earned him the title of father of the
American land reform movement. Swann continues his innovative
work at the Society, bringing Schumacherian concepts to life.

Susan Witt says that her background in literature gave her invaluable training for promoting community solutions to economic problems. Story-listening and story-telling are tools for sharing new ideas within community. Her talks to groups around the country are sprinkled with examples from her home area in the Berkshires.

She has served as executive director of the Schumacher Society since its inception, leading its national educational programs and at the same time remaining deeply committed to implementing Schumacher's ideas at the local level. She is the founder of the SHARE micro-lending program, administrator of the Community Land Trust in the Southern Berkshires, and a board member of the Great Barrington Land Conservancy and other Berkshire organizations. In 1992 she was elected the first woman president of the Great Barrington Rotary Club.

Her strong local roots have helped her to identify with the plight of indigenous peoples. She is the prime mover of the Society's work with the Buryat people on the western shore of Siberia's Lake Baikal as they struggle to develop a self-sufficient economy in keeping with their traditions.

The E. F. Schumacher Society has legal documents available for people who want to replicate its innovative projects in their own communities.

This essay is based on one of the Eighth Annual E. F. Schumacher Lectures, presented by Robert Swann in 1988.

E. F. Schumacher argued in *Small Is Beautiful: Economics as if People Mattered* that from a truly economic point of view the most rational way to produce is "from local resources, for local needs." Jane Jacobs, one of today's foremost scholars on regional economies, emphasizes Schumacher's point through her analysis of a healthy region as one creating "import-replacing" industries on a continuing basis. A well-developed regional economy that produces for its own needs is possible only when control of its resources and finances lies within the region itself. At present, the ownership of land, natural resources, and industry and the determination of conditions for receiving credit have become increasingly centralized at the national level. Now all but a few large urban areas find that their economic resources are controlled from outside the area.

The banking system is one of the most centralized institutions of our economy and one of the major obstacles to strengthening regional economies and the communities within them. Yet centralized banking in the United States did not get its start until the early twentieth century. The customs of borrowing and lending and money-printing grew up over generations in towns and rural communities to form what we now call our banking systems. These systems were small-scale, regional, and decentralized. Paper money was made standard, or national, in 1863 in order to raise funds for the fight against the Confederate

States, but it was not until 1913 that a central system became formalized with the Federal Reserve Act. Centralized banking and control of money called for large banks and wealthy investors who could assemble huge, unprecedented sums of money. These banks in the money centers, with their industrial customers, could pay a higher interest rate to depositors than could the smaller banks, and these smaller, often rural banks began sending their deposits to the large cities. The national currency made money more fluid and allowed rural dollars to support urban industrial growth. Rural creditors were pleased with this arrangement until the first time a New York bank closed and carried off the savings of a small town or until a local farmer couldn't secure a loan because a Chicago bank was borrowing from his bank at a high rate of interest.

A national currency facilitated the industrialization of the United States, which in turn created many jobs; however, the centralization of the monetary system has served to centralize the benefits of the system as well.

The effect on small farmers and rural economies has been devastating. The ongoing "farm crisis" is a dramatic manifestation of what is really a monetary crisis that began in the deep depression of the 1870s and 1880s and was later codified in the Federal Reserve Act. Credit for small-scale farming and the small rural businesses that are a part of the farm community had dried up long before the Depression of the 1930s, and the United States government had to create the Farmers Home Administration in order to help replace—with tax money—some of the rural capital that had been lost to the large cities.

The "housing crisis" is also in part a monetary crisis. Investors place money in land as a "hedge against inflation," which drives land and housing prices up. The high cost of land is a major factor in the present shortage of affordable housing, and it takes home ownership out of reach for the majority of Americans.

The local and decentralized banking systems of a hundred and fifty years ago had the advantage of diversity. The failure of a local bank—even a New York bank—was still a local failure, and its costs were internalized. But today we are facing the failure of an entire system. Consider the billions of tax dollars spent by the national deposit insurance system to bail out the Savings and Loan industry. And recall that billions were added to the national debt in order to bail out large banks when developing countries defaulted on their loans. These systemic failures are bound to occur if local economic control of banking customs and money supply is compromised by centralization and is sacrificed to serve the heedless demands of growth.

This predicament calls for a reorganization of economic institutions so that they will be responsive to local and regional needs and conditions. These new institutions would decentralize the control of land, natural resources, industry, and financing to serve the people living in an area in an equitable way. We need to create an infrastructure that encourages local production for local needs. Community land trusts, worker-owned and worker-managed businesses, nonprofit local banks, and regional currencies are some of the tools for building strong regional economies.

Because we have all learned to assume that national currencies are the norm, a regional currency is perhaps the least understood of these tools. Jane Jacobs, in her book *Cities and the Wealth of Nations*, views the economy of a region as a living entity in the process of expanding and contracting and a regional currency as the appropriate regulator of this ebbing and flowing life. Just like a nation, a region that does not produce enough of the goods it consumes comes to rely heavily on imports and finds that its currency is devalued. Import costs increase, the exchange of goods is reduced, and the region has to "borrow," which means that it exports its capital—dollars, not goods—and ends up importing nearly everything it needs. But if the region is supplying its own needs, then its currency "hardens" and holds its value relative to other currencies. Imports are cheaper, and trade is more equitable—or even skewed in favor of the self-reliant or "import-replacing" region.

Jacobs describes currencies as "powerful carriers of feedback information . . . and potent triggers of adjustments, but on their own terms. A national currency registers, above all, consolidated information of a nation's international trade." This feedback informs economic policy-makers. But should the industrial Great Lakes region or the farm-belt states adjust their economies in the same manner as the Sunbelt states or the Silicon Valley of the West Coast? A significant part of any region's economy is governed by a monetary and banking system over which members of a community have little or no control. The dependency on national currencies actually deprives regions of a very useful self-regulating tool and allows stagnant economic pockets to go unaided in a seemingly prosperous nation. What we propose instead is the establishment of a system with community accountability.

Regional currencies are not a recent invention—the practice is centuries old. The so-called free banking era of U.S. history, when many currencies circulated, contributed substantially to bringing about Thomas Jefferson's dream of a nation of small, independent, self-reliant farmers who found ready credit with community banks to produce and sell their goods. Even in the early years of this century local banks issued their own currency, which John Kenneth Galbraith says was important for the rapid development of the American economy.

How were these banks different from banks today? Because they were located in small towns, the bankers knew personally the people they were dealing with and could make loans on the basis of "character," not strictly on the basis of how much collateral an individual had to secure the loan. A more striking difference is that each bank could issue a local scrip. Unlike a national currency, which easily leaves the region in which its value is created, the local currency could circulate only in a limited regional area; local currencies and local capital could not travel to the money centers to finance the operations of multinational corporations or interest payments on debt. Credit decisions were made by local bankers with particular personal knowledge not only of the borrowers but also of the needs of the region as a whole.

One of the major objections to "free banking" in the nineteenth and early twentieth centuries has been that some of these local banks failed and some printed money to speculate in land and to make unproductive loans. The argument is that such abuses can be controlled if money is issued centrally. But it was unity—a shared belief in a common responsibility and vigilance—rather than uniformity that was needed. Community development banks like Chicago's South Shore Bank and the Grameen Bank of Bangladesh make up an intellectual diaspora—they are decentralized and unified. The Savings and Loan industry is uniform.

Decentralization and diversity have the benefit of preventing large-scale failure. This is as true in banking as it is in the natural world. Think of seeds. If many different strains of corn are planted by different farmers and a disease hits the crop, some strains will resist and the corn will be harvested. But if all the farmers have shifted to a new hybrid seed and a blight hits the corn, the result can be widespread crop failure and disaster. How do we ensure diversity in banking? As the economist Frederick Hayek has pointed out, to keep banking honest it would be better to return to a banking system that utilizes competing currencies rather than to rely on a central system.

In the 1930s a worldwide deflation encouraged many new forms of exchange that competed with the national currencies. The town of Woergel in Austria created a scrip system that drew international attention. The people in this little town were able to trade in labor and materials, which they did have, rather than in Austrian shillings, which they didn't have, and they managed to pull themselves out of the Depression in a matter of months. Local scrip also sprang up around the United States. A former editor of The Springfield Union in Massachusetts told us the story of a scrip issued by his newspaper. He was just a copyboy at the paper during the bank failures of the 1930s; he remembers that the publisher, Samuel Bowles, paid his newspaper employees in scrip. It could be spent in the stores that advertised in the paper, and the stores would then pay for ads with the scrip, thus closing the circle. The scrip was so popular that customers began to ask for change in scrip—they would see Bowles around town and had more confidence in his local money than in the federal dollars. Newspaper money helped to keep the Springfield economy flowing during a period of bank closures, facilitating commercial transactions that went well beyond the original intent of the issue.

Forty years later in the town of Exeter, New Hampshire, the economist Ralph Borsodi and Robert Swann issued a currency that was based on a standard of value using thirty different commodities in an index similar to the Dow Jones Average. It was called the Constant because, unlike the national currency, it would hold its value over time. The Constant circulated in Exeter for more than a year, proving, as Borsodi had hoped, that people would use currency that was not the familiar greenback. At the time, it received national publicity in Time, Forbes, and other magazines. When asked by a reporter if his currency was legal, Borsodi suggested that the reporter check with the Treasury Department, which the reporter did. He was told, "We don't care if he issues pine cones, as long as it is exchangeable for dollars so that transactions can be recorded for tax purposes."

This is all that the government requires of a local currency, and all that a local currency requires of a community is trust. A currency is only as strong as the confidence that people have in one another to produce something of value. Trust is at the heart of the successes in Springfield and Woergel and Exeter.

Borsodi discontinued his experiment after a year, but he had accomplished his purpose: to demonstrate local acceptance and verify the legality of locally issued, non-governmental currencies.

The Southern Berkshire town of Great Barrington, home of the E. F. Schumacher Society, has made strides toward issuing a Berkshire currency. Our story will make plain the particulars of how local currency works and how it encourages economic self-reliance. In 1982 a discussion group on regional economies led to the incorporation of a nonprofit organization called SHARE (Self-Help Association for a Regional Economy), with open membership and a board elected from its members. The intent was to establish an organizational base for a local currency.

The soundest method for issuing any currency is to extend credit for productive loans. A productive loan differs from a consumer loan in that it provides the recipient with the capability to produce goods for market with a value in excess of the loan, thereby creating new economic wealth in the community. The classic example of a productive loan is one made to the farmer for seeds in the spring, contributing to an abundant harvest of food in the fall. Currency that is created responsibly for productive loans will maintain its value or even strengthen in value as the wealth of the community increases relative to the supply of scrip in circulation.

SHARE's first objective was to gain experience in making productive loans. Its board of directors chose to focus on people who were unable to secure normal bank financing but who had the kind of small, locally owned enterprises that produced quality goods and services for local consumption. Some of these businesses could get bank loans but at rates of 15 or 18 percent; SHARE was determined to make low-cost loans available. SHARE members open savings accounts at the First National Bank of the Berkshires, and these accounts are used by SHARE to collateralize loans. This kind of lending requires that the community separate the functions of banking. The bank makes the loans and handles the accounting, but the lending decisions, based on a set of social, ecological, and financial criteria established by SHARE, are made by the community of depositors.

Sue Sellew of Rawson Brook Farm makes a soft chèvre cheese from the milk of her dairy goats and the herbs she grows on her organic farm. She borrowed $5,000 from SHARE to bring her milking parlor and cheese room up to state standards. This has enabled her to sell the cheese to stores and restaurants.

Jim Golden trained his two draft horses, Spike and Rosie, to haul timber and firewood from forests. Jim can assure his customers that their woods will be treated in an ecologically responsible manner and won't suffer the undue stress caused by heavy equipment. A SHARE loan was made to complete a barn for the team.

Bonnie Nordoff had a poor credit record, but she also had a knitting machine that took bulk-weight yarn, and she had a talent for designing clothes. She knits sweaters, tights, leg-warmers, and scarves in whimsical, colorful designs. Her small SHARE loan bought a bulk supply of wool yarn, which lowered her overall costs and established credit with suppliers. She borrowed again for a second knitting machine when the first loan was repaid. Her business kept on growing, and she applied a third time to buy a machine for an employee. The first two loans had established bank credit for her business, so SHARE sent Bonnie directly to the bank's loan officer, who readily approved a loan.

The payback record on SHARE-collateralized loans has been 100 percent, both because of their scale and because of community support for the loan recipients. SHARE members help maintain this perfect record by recommending these small businesses to their friends.

Most loans have been for start-up businesses requiring no more than $3,000. They are made for equipment or inventory but not for salary or advertising—productive loans, not consumer loans. A piano teacher purchased a piano with loan funds in order to provide lessons in her home, but an application to purchase a piano for private use was sent to the bank's consumer-loan officer.

The SHARE loan-collateralization program is simple to operate and easily copied. Similar programs have started around the United States, using the model created in the Berkshires. It is the "grandmother principal" that has made SHARE a success: When people without credit histories decide to go into business, they frequently turn to a family member, such as a grandmother, for help. Instead of lending directly the grandmother might offer a savings account as collateral for a bank loan. The SHARE program simply extends "the circle of grandmothers," creating a family of place.

SHARE puts a human scale and a human touch back into local economic transactions. A newsletter tells SHARE depositors "what your money is doing tonight"—it is working locally to make cheese or sweaters or to house two very big horses. On weekends SHARE members visit Sue Sellew's farm, where the baby goats nibble at the keys in their pockets. They come by the next weekend with their grandchildren, and on the next weekend they serve Monterey chèvre at their dinner party. Monterey chèvre is not just any cheese; it is a cheese with a story, and SHARE members are a part of that story. They ask for the cheese at local stores. They think of Bonnie's wool sweaters when contemplating a special gift. They root for Spike and Rosie at the draft-horse pulling contest. These local economic relationships encourage social patterns that in turn shape a uniquely local culture.

Frank Tortoriello is the owner of a popular deli on Main Street in Great Barrington. He turned to SHARE when the bank refused him a loan to move his restaurant to a new location. But Frank didn't need SHARE's circle of grand-mothers; he already had a circle of his own in his customers. SHARE suggested that Frank issue Deli Dollars as a self-financing technique. The notes would be purchased during a month of sale and redeemed after the Deli had moved to its new location. A local artist, Martha Shaw, designed the note, which showed a host of people carrying Frank and his staff—all busy cooking—to their new location. The notes were marked "redeemable for meals up to a value of ten dollars." The Deli would not be able to redeem all the notes at once after the move, so SHARE advised Frank to stagger repayment over a year by placing a "valid after" date on each note. To discourage counterfeiting Frank signed every note individually like a check.

We recommended that the notes be sold for ten dollars each, but Frank thought that would be too good a deal for the Deli. With his customers in mind he sold ten-dollar notes for eight dollars and raised $5,000 in thirty days: contractors bought sets of Deli Dollars as Christmas presents for their construction crews; parents of students at nearby Simon's Rock College knew Deli Dollars would make a good gift for their kids; the bankers who turned down the original loan request supported Frank by buying Deli Dollars. The notes even showed up in the collection plate of the First Congregational Church because churchgoers knew the minister ate breakfast at the Deli. Regular customers were pleased to help support what they saw was a sure thing—they knew firsthand how hard Frank worked and believed in his ability to make good on redemption. Frank repaid the loan, not in hard-to-come-by federal notes but in cheese-on-rye sandwiches.

Jennifer Tawczynski worked at the Main Street Deli and carried the idea home to her parents Dan and Martha Tawczynski, who own Taft Farm, one of two farm markets in the area. The Tawczynskis came to SHARE with the idea of issuing "greensbacks" to help them meet the high cost of heating their greenhouses through the winter. Customers would buy the notes in the late fall for redemption in plants and vegetables come spring and summer.

At around the same time the other farm market in town, the Corn Crib, was damaged by fire. Customers of the Corn Crib came to SHARE with the idea of issuing notes to help owners Don and Ruth Zeigler recover from the ravages of the fire. SHARE suggested that the two farms together issue a Berkshire Farm Preserve Note. Martha Shaw designed the note with a head of cabbage in the middle surrounded by a variety of other vegetables. The notes read "In Farms We Trust" and were sold for nine dollars each. The Massachusetts Commissioner of Agriculture traveled from Boston to purchase the first Berkshire Farm Preserve Note, and five national networks showed our farmers using Yankee ingenuity to survive a difficult winter. The Berkshire Women with Infants and Children (WIC) program purchased Berkshire Farm Preserve Notes in order to give them to families, part of a local initiative to supplement the federal food program. The notes do not carry the food-stamp stigma, and the Berkshire agency knows it is supporting local farmers at the same time it is supporting local families.

The notes could be purchased at either farm and were redeemable at either farm. At the end of the redemption period SHARE acted as the clearinghouse for the notes. The farmers received the income (ranging from $3,000 to $5,000 per farm per year) from the sale of the notes, and they found a committed base of customers who would travel out of their way to buy from their local farms rather than purchase the jet-lagged vegetables from supermarket chains.

Deli Dollars started a consumer movement in the Berkshires. The Berkshire Farm Preserve Notes, Monterey General Store Notes, and Kintaro Notes that followed gave Berkshire residents a way to vote for the kind of small independent businesses that help to make a local economy more self-reliant.

The popularity of the scrip inspired the Southern Berkshire Chamber of Commerce to work with the Schumacher Society staff to issue Berk-Shares as a summer promotion. Customers were given one Berk-Share for every ten dollars spent in a participating business over the six-week summer period. During a three-day redemption period customers could spend their Berk-Shares just like dollars in any of the seventy participating stores. The success of the Berk-Share program depended on the energy and cooperation of a small group of merchants and in large part on the sense of community among consumers. Of the seventy-five thousand Berk-Shares handed out (representing three-quarters of a million dollars in Berk-Share trade) twenty-eight thousand were spent during the three-day redemption period! Some families pooled their Berk-Shares for a gift for one member of the family. People who were going away over the redemption week-end were sure to give their Berk-Shares to a neighbor who would use them. A spirit of festivity and excitement filled Main Street that weekend as people chatted about how they planned to use their Berk-Shares.

Although the Berk-Shares and Deli Dollars and Farm Preserve Notes repre-sented a major shift in local attitudes toward an alternative exchange and cap-tured the imagination of both consumers and producers, they were not yet the year-round local currency the organizers had envisioned. A suggestion from several area banks pushed the effort forward to its next stage. The Berk-Share organizing committee proposed that the five local banks participate in a Berk-Shares zero percent loan program during the winter holidays. Spending that would normally flow to catalogue stores and malls would instead go to the locally owned stores that accepted Berk-Shares, helping to secure local jobs and keeping local dollars local. The committee presented the idea at a meeting with the bankers, who in turn proposed that the committee create a year-round Berk-Share that would be a 10 percent discount note. Customers would come to the banks and purchase one hundred Berk-Shares for ninety dollars and redeem them at local stores for one hundred dollars worth of goods and services. The mer-chants would then deposit their Berk-Shares at local banks at ninety cents per share.

But how to clear the Berk-Share accounts among the five banks? The Federal Reserve system transfers dollars from the issuing bank to the receiving bank. This national clearing system is limited to a national currency; a local currency needs a local system. The bankers at the meeting came up with the solution. They said, "Well, we can just walk down the street to one another's banks and make the exchange, the way we used to with checks." It gave these individual bankers, who are caught up in a highly centralized and fast-paced system, great pleasure to imagine recapturing in a small way the early days of banking when transactions had a warmer, more community-spirited tone.

The Schumacher Society and the Main Street Action Association of the South-ern Berkshire Chamber of Commerce are cooperatively seeking funds to staff the first year of issue of a year-round scrip. When the program is in place and local businesses and their customers are familiar with the new Berk-Shares, Main Street Action and the Schumacher Society will work to develop a commodity backing for the scrip. Eventually, loans can be made in Berk-Shares at an interest rate as low as 3 percent—the cost of servicing the loan. Unlike the current SHARE program, which relies on borrowed dollars, a loan in Berk-Shares would carry no interest costs. A 3 percent loan could encourage new business ventures like local food processing that otherwise couldn't compete because investment capital is too expensive. A local scrip can empower Berkshire residents to shape their own economic futures unfettered by high interest rates and credit decisions made in far-away money centers. Each town can be a money center, and local economic problems will have local solutions.

In the summer of 1991 Paul Glover heard a radio interview with Schumacher Society staff about the Deli Dollars and Berkshire Farm Preserve Notes. The story inspired him to issue Ithaca Hours in his hometown of Ithaca, New York, as a way to create more local jobs and more security for Ithacans who are underemployed. Ithaca Hours has grown from its small grass-roots beginning to include over a thousand individuals and stores. The scrip can buy food items, construction work, professional services, health care, and handicrafts. Each Ithaca Hour is worth ten dollars—the average hourly wage in Tompkins County—so the five thousand Ithaca Hours (or $50,000) in circulation have increased local economic transactions by several hundred thousand dollars annually.

Individuals and stores agreeing to accept Ithaca Hours notes are issued two free Hours to begin trading and are listed in the free monthly paper, *Ithaca Money*. This newspaper features articles about the local economy and tells the stories of small home-businesses that have prospered by accepting payment in scrip. Only Ithaca Hour vendors can advertise in *Ithaca Money*, and although the ad will run for two months, it costs only half an Hour (five dollars).

Consumers are led to shop locally because Ithaca Hours can be used only in Ithaca. One market farmer who had difficulty paying bills during the winter was able to secure a loan in Ithaca Hours from a customer who had accumulated more than she could use. She preferred to recirculate them rather than let them lie idle. The farmer's family paid for child care, movie tickets, and other goods and services in Ithaca Hours and then repaid the loan in produce in the summer. The Alternative Credit Union in Ithaca accepts partial repayment of mortgage loans in Hours because its employees have agreed to accept part of their salaries in scrip.

Paul Glover has opened a downtown Hours Bank in order to regulate circulation of the currency, provide visibility, and supply a diverse array of goods for purchase with Ithaca Hours. The organizers work with local businesses by tracking the goods that these businesses buy from outside the region and then connecting them with local producers of the same goods. This is the substance of an import-replacement program that will create sustainable jobs.

A local currency may be dollar-denominated or measured in chickens (as Wendell Berry once suggested for his part of Kentucky) or hours or cordwood, as long as people know they can spend that chicken cash, that cordwood note. Confidence in a currency requires that it be redeemable for some locally available commodity or service. The Schumacher Society recommends the following policies to maintain confidence over the long haul:

- The issuing organization should be incorporated as a nonprofit so the public understands that providing access to credit is a service not linked to private gain. The organization should be democratic, with membership open to all area residents and with a board elected by the members.
- Its policy should be to create new short-term credit for productive purposes. Such credit is normally provided for up to three months for goods or services that have already been produced and are on their way to market—credit for things that pay for themselves in a very short time.
- The regional issuing organization should be free of governmental control—other than inspection—so that investment decisions are independent and reflect community priorities.
- Social and ecological criteria should be introduced into loan-making. (Community investment funds also use a positive set of social criteria particular to their own region. These funds could join with hard-pressed local banks to initiate regional currencies.)
- Loan programs and local currencies should support local production for local needs.

Local currencies can play a vital role in the development of stable, diversified regional economies, giving definition and identity to regions, encouraging face-to-face transactions between neighbors, and helping to revitalize local cultures. A local currency is not simply an economic tool; it is also a cultural tool.

Community groups in Kansas City, Eugene, Boulder, and in little Philmont, New York, are issuing their own currencies, and each is uniquely tailored to the people, culture, and products of the region. Each community has its own tale of how and why people first organized and what they hope to achieve by their efforts. A Schumacher Society member who was visiting Ithaca looked in *Ithaca Money* for a way to spend his scrip before leaving town. He decided on a craft item that a woman made and sold in her home. The daughter who answered the door understood that the visitor was not from Ithaca and asked, "What does your hometown currency look like?"

—1988; revised 1995

For further information about local currencies contact the E. F. Schumacher Society at 140 Jug End Road, Great Barrington, Massachusetts 01230, phone: (413) 528–1737, email: <efssociety@aol.com>, web site: <http://members.aol.com/efssociety>. A handbook of legal documents for starting a SHARE program is also available.

John McClaughry
Bringing Power Back Home:
Recreating Democracy on a Human Scale

John McClaughry brings to the grass-roots citizen activism espoused
by E. F. Schumacher a uniquely Yankee twist. A founding member of
the Schumacher Society and chairman of its board of directors, he is a
former member of the Vermont state House and Senate. Currently he
is president of the Ethan Allen Institute and the Institute for Liberty
and Community in Vermont. He also serves as town moderator for his
hometown of Kirby. In the early 1980s he served as a senior policy advisor in the Reagan White House. In 1989 he co-authored, with Frank
Bryan, *The Vermont Papers: Recreating Democracy on a Human Scale*, from which
this lecture draws.

Adamantly independent and an individualist in his political views,
McClaughry has described himself as a libertarian, agrarian, distribu-
tist, Jeffersonian, Republican decentralist. As such, he believes strongly
in "that peculiarly American spirit of creative self-help, mutual aid
typified by the frontier barn raising."

Much of his thinking is based on his experiences in Kirby with its
population of just under three hundred. There he has learned that
when the seat of power becomes remote, citizens become the subjects
of a central power. This being the case even in independent Vermont,
it is McClaughry's intent to restore participatory democracy.

Because John McClaughry's work is not merely theoretical but
involves committed service in the political life of his town, county,
and state, his career can be singled out as the embodiment of Schumacher's decentralist politics in action.

I am going to begin today on the topic of recreating human-scale democracy by giving you a small-town Vermont fable. This fable, however, is true, and it happened in my town of Kirby, Vermont.

We are a small town, population 285. We have no village store or post office. Our 285 residents have five different postal addresses. We have as public buildings only the town hall, which was donated by a church back in 1934, and the town shed adjacent to the town hall.

The town shed, which is the centerpiece of this little fable, was built in the late 1930s to accommodate a horse-drawn grader that grades dirt roads (I mention its purpose for the benefit of those of you who are from the asphalt jungles). In due course, however, graders got motorized and got longer, and it came to pass that the town owned a grader about four feet longer than the shed. Now, working outside on a piece of equipment in a Vermont winter is no fun, as the road commissioner ruefully reported on several occasions, so we decided we needed a town shed that was insulated, partially heated, and big enough to accommodate the grader and the ten-wheel dump truck.

Acting on that decision, the road commissioner bulldozed down the old shed and had a contractor come and pour a foundation that was about eight feet wider and four feet longer than the old town shed. Then, on the appointed day, after he got some framework up, about twenty-five Kirby residents, men and women, showed up with tools and began to put up the siding, roofing, door frames, and so on, which is the way we do things in a small town where nobody is cash rich and everyone hates to pay higher property taxes to hire people to do things like that.

It was a Saturday morning. We had some staging provided by one of the contractors who lived in town, and we went to work putting up the siding. The older ladies in town who were not able to climb up on scaffolding provided a potluck lunch and supper for the workers. We got most of the work done on Saturday, and Sunday we came back to do the last part of it.

On late Sunday afternoon, when the final stages were being completed, there mysteriously appeared a short, earnest gentleman who announced that he was from one of the state's environmental agencies and had some business with us. The work stopped. He gave a speech, directed toward Buster Wood, our chief selectman and a dairy farmer, which went something like this: "You are building a building on town property without having a permit under 10 V.S.A. Chapter 151, and I am here to inform you that you are going to pay a five hundred dollar fine for your insubordination."

By this time there were twenty-five men standing about watching this action.

Buster pushed his John Deere cap back on his head and said, "Well, actually we aren't building a new town shed." The fellow said, "You're not?" and Buster said, "Nope, we're just renovating the old shed." This bureaucrat looked about and saw twenty-five men standing around with hammers and beat a retreat. I am happy to say he never came back.

There were two schools of thought on how to deal with this. The conservatives defended staring the man into submission. The radicals like myself thought we should have taken him hostage. My side lost that argument to the counterargument that the state of Vermont has hundreds of these guys, and it won't pay to get any of them back.

This little fable illustrates a serious theme: the constant advance of the central power. In William Butler Yeats's poem "The Second Coming" there is the line, "Things fall apart; the center cannot hold." I often think about that when I see the crumbling of the Soviet Empire on television these days and the disarray that Leopold Kohr has spent a lifetime predicting.

Yet even though at the macrostage we see large empires becoming unglued, as prophesied back in the fourteenth century by Ibn Khaldun—a famous Arab historian who ought to be more widely known in the West—when you look at it from the worm's-eye view from, say, the town of Kirby, you see power moving steadily away from the people, away from our little democracies, and toward a state government. To many of you who come from New York or California, the State of Vermont probably resembles a Gilbert and Sullivan government, but to us it looks more like "The Return of Godzilla."

The people of Vermont have steadily been losing power to the state government for twenty or thirty years. I dare say the same thing has been going on in many other states, although perhaps on a different schedule. We are, in our little state, victims of what Frank Bryan and I, in our book *The Vermont Papers: Recreating Democracy on a Human Scale*, call the Systems Axiom. Leopold Kohr has described it wonderfully well in his many works, and I need not embellish his description much. In a nutshell, the Systems Axiom calls for the administration of all public business in accordance with the standards of uniformity, efficiency, and bureaucracy. Uniformity, because we all are entitled to the same whoever we are, wherever we are, and whatever circumstances we are in. Efficiency, because we must be careful not to waste any motion or energy. Bureaucracy, because it takes a trained cadre of professional experts to make sure that all these plans and rules are uniformly obeyed by a submissive population.

Contrast that to the decentralist vision Fritz Schumacher and Leopold Kohr have offered. Diversity, where differences abound and people are not forced into a mold to satisfy someone's picture of how society should operate. Democracy, where people make choices after debate and after considering the effects of their decisions on their community, even though the choices may turn out to be "wrong" later on. Spontaneity, where people deal with problems in a way that suits their culture, their community, and their preferences, rather than obey a plan written for them by somebody who defines the official, orthodox, proper way to do things. It is this decentralist vision, as opposed to the Systems Axiom, which is the central theme of our book. Even though it deals with our own particular state, I think the same contrast can be found in many areas of human life and in many parts of the world.

Our governor, a rather liberal and well-meaning woman who wants good things for everyone, gave a speech last January in which she called for "a new era of planning" for our state. Her words in describing the new era were that it will be "specific in requirements, uniform in standards, and tough on delinquents." It could have been Benito Mussolini in Milan in 1922.

The idea is that planning done by the right people—people who are public spirited, well-educated, dedicated, and honest—will minimize the chances of things going wrong, which they will certainly do if the future is left to the whims and vagaries of ordinary unwashed citizens who don't pay enough attention to how the world ought to work. This is the vision of a lot of people in history, some of them not very savory. It is the kind of thing we have had to face from well-meaning people who decide that everything should be ordered correctly.

The Vermont Papers is a protest against that. It is more than a protest, though; it is an attempt to present a positive vision in the tradition of the others who have preceded me today and in past years at the Schumacher Lectures.

That tradition is called "civic humanism." It came from the time of Machiavellian Florence, not Machiavelli as cynical advisor to the Prince but Machiavelli as expositor of the Florentine ethic that can be traced back to Polybius and forward to James Harrington in England and to many of the founders in this country, notably Thomas Jefferson. This civic humanist ideal held that in order to have a free republic, to protect the republic against the degeneracy that results in despotism on the one hand and mob rule on the other, and to maintain the immortal republic one must have the following circumstances:

1. Because politics is based on a material foundation, there must be widespread distribution of power and thus of property. There should be no great aggregations of wealth and no grinding poverty; the great bulk of the free men and women of a society should have some independent means and not remain beholden to an economic superior.

2. The citizen must be a participant in the civic life and the repository of virtue. This virtue is acquired not by piety and adherence to religious ideals,

although that was all right, but by a devotion to advancing the common good through participation in the public process, obviously a democratic ideal.

3. Every able-bodied male served in the militia to protect the republic against invasion.

The civic humanist ideal required an economic base widely shared, near-universal political participation, and a military obligation to protect the republic from its enemies. This tradition has been very deeply rooted in my little state of Vermont, which, as some of you may know, was an independent republic for fourteen years before it agreed to merge with the United States of America. That decision is not always appreciated in these times, especially when Washington thinks up more and more things for us to do. Most recently, for instance, the Federal government required us to offer special education for disabled children, which it promised substantially to pay for. The Federal government ended up paying only 4 percent, leaving the state of Vermont, and all the other states for that matter, to pick up a very sizable burden or else face very expensive law suits by those who are now entitled to receive the benefits.

This civic humanist ideal means restoring to people their role as citizens instead of subjects. For citizens are people who participate in making the decisions about the way they live their lives. Merely going to the polls every two years or four years to cast a ballot for one or another television personality who happens to be running for office is a pretty cheap version of citizenship. Voting on a state referendum question, as Californians are famous for, is also a fairly insubstantial form of citizenship.

By citizenship we mean active participation in public affairs at a level such as the town or neighborhood where the individual's contribution can be appreciated and can count for something. The small human community, celebrated by Aristotle and Lao Tzu, the place where you belong and where you recognize those who belong and those who are strangers, where the good of everyone is tied together in an interconnected web that is ruptured only at the peril of everyone in the community—that is where citizenship resides.

By contrast, in a society that is planned to be "specific in requirements, uniform in standards, and tough on delinquents," you are no longer a citizen but rather the subject of a central power. Once we become subjects, we lose those sparks of humanity and democracy and freedom that have made this country such a great country in world history.

How then do we regain our role as citizens when we are confronted by this Godzilla bearing down upon us? Well, in Vermont we have 246 towns. A town in Vermont is a box on the map, more or less square, which was laid out by surveyors in the eighteenth century long before people came here to live in any numbers. Town populations range from 40,000 people in Burlington to zero people in a couple of the small towns that never had any people at all at any time in their history. These 246 towns, up until the 1950s, were the primary governing units of Vermont; the state government was something of an afterthought. The state government consisted of such entities as the Supreme Court and the offices of the Secretary of State and the Attorney General, but the business of welfare, of education, and of transportation at the local level was basically conducted by the town.

As you might suspect, however, a tiny little town like mine is hard-pressed to manage that kind of responsibility in the decade of the 1990s. Frankly, we do not have the material or intellectual resources to do it. Our town employment consists of a town clerk, who spends five hours a day two days a week in the office, and the road commissioner, who works a forty-hour week on the roads. Our three selectmen meet every month. They often find it really difficult to deal with complicated programs and rules sent down from Montpelier and Washington.

Now, the conservatives in Vermont have always said they deplore the removal of power from the communities to the distant state government. Their cry has been, first, "Don't take any more away" and, second, "Give it back." On both counts they have lost. The state continues to take power away from the communities and never gives any of it back. So drawing the battle line between the 246 towns and the state is guaranteed to be a losing cause. If one were to ask those conservatives who are urging restoration of power to the towns, "How can your little town possibly carry out the functions that have been taken away in the last thirty years?" chances are that in two-thirds of the towns the answers would be pretty unconvincing, for as the conservatives have defended town government against the state, the state has gobbled up more and more of the functions of town government. We are now heading for a day, not too far away, when all we will do at the local level will be to issue dog licenses.

On the other hand, the liberals as a rule will say, "If we have 246 towns making all these decisions about welfare, education, and transportation, a bunch of them will do it wrong. People in one town would be entitled to higher welfare checks than people in other towns. People in one town will have a different high-school curriculum than people in another town. Obviously this would be unfair to somebody. They've all got to be the same somehow, and even if nine-tenths of the towns are capable of doing it right, which we doubt, the fact that the remaining 10 percent will commit some flagrant offense against the Greater Good requires that the whole process be regulated, coordinated, and ultimately financed by the central government."

The liberals will definitely win this debate. Frank and I have tried to open up a new dimension in the debate by saying that we reaffirm the goals of citizenship, democracy, diversity, of people managing their own lives on a participatory basis and looking after the needs of their own community. To do that, however, we have to create a new kind of local government that is big enough in scope and resources to receive those functions back from the state government and carry them out efficiently.

To that end we propose resurrecting an old Vermont name, "shire." We have a reprint of a map issued about 1780 which shows what is now the entire state of Vermont divided up into three shires, an old English term that meant a local government composed of a number of separate parishes or tuns, the predecessor of the towns or tunscipes, the predecessor of today's townships.

If one looks at a book like George E. Howard's *Local Constitutional History of the United States*, published in 1888, one will find a fascinating analysis of how the local government system emerged from the Teutonic tribes—the Angles and the Saxons—and how it was replaced after the Norman conquest in 1066 by a system of counties subservient to Norman counts.

We do have counties in Vermont. They are judicial districts, not real governmental bodies. At the county level we elect only a sheriff and a state's attorney and a couple of judges. All the judges really do is operate the courthouse for a state court system that is now increasingly centralized. None of this activity is related to a body politic. We have no county council, no county chief executive, or anything like that, although many midwestern states do. We need to create some kind of governmental body to receive the usurped governmental functions back from the state.

The shire is that kind of a body. In population shires would range, by geographical necessity, from about 2500 in the upper reaches of the Connecticut River to about 60,000 in the Burlington area. We believe that within those shires the people can restore the kind of participatory democracy Vermont has always been famous for.

In a way, this is practiced today in the town of Brattleboro, with a population of about twelve thousand. It's too large for a mass assembly of its people. The town is divided into four wards, each of which has a ward assembly meeting, where the issues are debated by all the citizens in person. Then each ward elects a hundred people to take part in the representative town meeting. That is the model we propose to follow in most of our shires. There would be neighborhood assemblies or the present small towns or whatever other divisions the shire cares to make. Their people would send delegates to the shire council to make the rules of the shire community.

To finance the functions that are brought back we propose an innovative adaptation of the Canadian revenue-sharing system. It would assure that there is an equal tax base per capita in all the shires. The shires would then levy taxes on that base to carry out their governmental responsibilities.

Finally, the shire needs to have an identity. Because it, unlike the institutions that grew out of human experience, is being created de novo, we need to create a citizen loyalty to the shire that protects its independence vis-à-vis the higher-level government. To do that, we propose a lot of identity building through things such as the shire flag, the shire anthem, and shire license plates.

The financing plan, the political structure, and the cultural identity of the shire can be pretty much what the citizens want to make of it. Those of you who are familiar with local government know there is something in the law known as the Dillon Rule. Named after a long-dead jurist, it says that local governments are the creatures and subjects of the state. The state can merge, abolish, reform, or reorganize local government at its pleasure, by the majority vote of the legislature and the signature of the governor. That is the rule in all of our states.

We propose reversing that rule. We propose making Vermont a federation of shires, where the state government is a body with certain limited and specific powers. The shire would retain residual power to deal with shire problems. To explain the division of functions between state and shire is an interesting but time-consuming exercise that I won't go through for you today; there are, however, a couple of functions that must be retained on the state level.

Obviously the justice system, the laws against crimes, would have to be uniform throughout the state. The protection of the environment against actions that transcend shire boundaries would be under the jurisdiction of the state government, which would deal with entire watersheds and with air pollution (to the extent possible, given the fact that we are captives of the Midwest with its coal-generated power plants).

Without going into a lot of details, I invite you to think about the idea of recreating local government in a way that makes it possible to finally bring back the power that has slipped away to the central government. Think about recreating in your localities a true democratically governed polity, where the decisions are made by people who care about how they live and how they are governed, where the decisions are made by neighbors, not by distant functionaries whom no one knows.

Unless we do this, even in a small state like mine, we are in real danger of losing the essential ingredient of citizenship: the attachment and the loyalty of the citizen to his or her own little polity, where he or she counts for something. It's a community where citizens know their fellow citizens, greet them at the drugstore or the post office or the landfill, and interact with them through the Cub Scout troop, the church, Alcoholics Anonymous, the Extension Service, or any of the activities that form the foundation of community life. To the extent it is possible, those decisions about how our community operates need to be brought back home.

Just as during the Vietnam War the cry was "Bring the boys back home," we are saying in *The Vermont Papers*, "Bring government back home," "Bring citizenship back home," or perhaps more appropriately, "Allow citizenship to flourish once again in a small, human-scale, democratic society where it has its only chance of being meaningful in the lives of the people and in the life of the republic we live in."

—1989

Wendell Berry

People, Land, and Community

Wendell Berry is one of the best known and most respected writers in America. One reason for this is the extraordinary clarity and honesty of his thinking, which is an inseparable component of his writing, whatever the form: essay, novel, or poem. It is equally integral to how he lives as a writer, farmer, teacher, and family and community member. Berry has farmed a Kentucky hillside in his native Henry County for two decades; his family has lived there for eight generations. He has long been an embattled guardian of rural communities and landscapes.

The title of his lecture (later revised to become a chapter in his book *Standing by Words*, from which it is reprinted here), reflects his intellectual bond with E. F. Schumacher. Two more similar sensibilities would, in fact, be hard to find. The ideas of the two men echo over time, like birds calling back and forth across a valley. Both are passionate critics of the industrial modern/postmodern world structure, yet neither directs his efforts to analysis alone. Both have offered compelling and sustainable alternatives to the giantism that afflicts us, and here too their ideas are similar, leading back again and again to community, commitment, place, sufficiency—all values that Winona LaDuke declares fundamental to the beliefs of her Native American people. The implications of such a belief-system become obvious if one summons up a mental image of the despoiled landscapes of today and contrasts them with those in which Native Americans lived before the arrival of the Europeans.

For growing numbers of people, as they renew their commitment to place and community and to future generations, their views resonate with Wendell Berry's statement of absolute commitment of many years ago: "All my dawns cross the horizon and rise from underfoot. What I stand for is what I stand on."

I would like to speak more precisely than I have before of the connections that join people, land, and community—to describe, for example, the best human use of a problematical hillside farm. In a healthy culture, these connections are complex. The industrial economy breaks them down by oversimplifying them and in the process raises obstacles that make it hard for us to see what the connections are or ought to be. These are mental obstacles, of course, and there appear to be two major ones: the assumption that knowledge (information) can be "sufficient," and the assumption that time and work are short.

These assumptions will be found implicit in a whole set of contemporary beliefs: that the future can be studied and planned for; that limited supplies can be wasted without harm; that good intentions can safeguard the use of nuclear power. A recent newspaper article says, for example, "A congressionally mandated study of the Ogallala Aquifer is finding no great cause for alarm from [sic] its rapidly dropping levels. The director of the . . . study . . . says that even at current rates of pumping, the aquifer can supply the Plains with water for another forty to fifty years. . . . All six states participating in the study . . . are forecasting increased farm yields based on improved technology." Another article speaks of a different technology with the same optimism: "The nation has invested hundreds of billions of dollars in atomic weapons and at the same time has developed the most sophisticated strategies to fine-tune their use to avoid a holocaust. Yet the system that is meant to activate them is the weakest link in the chain. . . . Thus, some have suggested that what may be needed are warning systems for the warning systems."

Always the assumption is that we can first set demons at large, and then, somehow, become smart enough to control them. This is not childishness. It is not even "human weakness." It is a kind of idiocy, but perhaps we will not cope with it and save ourselves until we regain the sense to call it evil.

The trouble, as in our conscious moments we all know, is that we are terrifyingly ignorant. The most learned of us are ignorant. The acquisition of knowledge always involves the revelation of ignorance—almost is the revelation of ignorance. Our knowledge of the world instructs us first of all that the world is greater than our knowledge of it. To those who rejoice in the abundance and intricacy of Creation, this is a source of joy, as it is to those who rejoice in freedom. ("The future comes only by surprise," we say, "—thank God!") To those would-be solvers of "the human problem," who hope for knowledge equal to (capable of controlling) the world, it is a source of unremitting defeat and bewilderment. The evidence is overwhelming that knowledge does not solve "the human problem." Indeed, the evidence overwhelmingly suggests—with Genesis—that knowledge *is* the problem. Or perhaps we should say instead that all our problems tend to gather under two questions about knowledge: Having the ability and desire to know, how and what should we learn? And, having learned, how and for what should we use what we know?

One thing we do know, that we dare not forget, is that better solutions than ours have at times been made by people with much less information than we have. We know too, from the study of agriculture, that the same information, tools, and techniques that in one farmer's hands will ruin land, in another's will save and improve it.

This is not a recommendation of ignorance. To know nothing, after all, is no more possible than to know enough. I am only proposing that knowledge, like everything else, has its place, and that we need urgently now to put it in its place. If we want to know and cannot help knowing, then let us learn as fully and accurately as we decently can. But let us at the same time abandon our superstitious beliefs about knowledge: that it is ever sufficient; that it can of itself solve problems; that it is intrinsically good; that it can be used objectively or disinterestedly. Let us acknowledge that the objective or disinterested researcher is always on the side that pays best. And let us give up our forlorn pursuit of the "informed decision."

The "informed decision," I suggest, is as fantastical a creature as the "disinterested third party" and the "objective observer." Or it is if by "informed" we mean "supported by sufficient information." A great deal of our public life, and certainly the most expensive part of it, rests on the assumed possibility of decisions so informed. Examination of private life, however, affords no comfort whatsoever to that assumption. It is simply true that we do not and cannot know enough to make any important decision.

Of this dilemma we can take marriage as an instance, for as a condition marriage reveals the insufficiency of knowledge, and as an institution it suggests the possibility that decisions can be informed in another way that is sufficient, or approximately so. I take it as an axiom that one cannot know enough to get married, any more than one can predict a surprise. The only people who possess information sufficient to their vows are widows and widowers—who do not know enough to remarry.

What is not so well understood now as perhaps it used to be is that marriage is made in an inescapable condition of loneliness and ignorance, to which it, or something like it, is the only possible answer. Perhaps this is so hard to understand now because now the most noted solutions are mechanical solutions, which are often exactly suited to mechanical problems. But we are humans—which means that we not only *have* problems but *are* problems. Marriage is not as nicely trimmed to its purpose as a bottle-stopper; it is a not entirely possible solution to a not entirely soluble problem. And this is true of the other human connections. We can commit ourselves fully to anything—a place, a discipline, a life's work, a child, a family, a community, a faith, a friend—only in the same poverty of knowledge, the same ignorance of result, the same self-subordination, the same final forsaking of other possibilities. If we must make these so final commitments without sufficient information, than what *can* inform our decisions?

In spite of the obvious dangers of the word, we must say first that love can inform them. This, of course, though probably necessary, is not safe. What parent, faced with a child who is in love and going to get married, has not been filled with mistrust and fear—and justly so. We who were lovers before we were parents know what a fraudulent justifier love can be. We know that people stay married for different reasons than those for which they get married and that the later reasons will have to be discovered. Which, of course, is not to say that the later reasons may not confirm the earlier ones; it is to say only that the earlier ones must wait for confirmation.

But our decisions can also be informed—our loves both limited and strengthened—by those patterns of value and restraint, principle and expectation, memory, familiarity, and understanding that, inwardly, add up to *character* and, outwardly, to *culture*. Because of these patterns, and only because of them, we are not alone in the bewilderments of the human condition and human love, but have the company and the comfort of the best of our kind, living and dead. These patterns constitute a knowledge far different from the kind I have been talking about. It is a kind of knowledge that includes information, but is never the same as information. Indeed, if we study the paramount documents of our culture, we will see that this second kind of knowledge invariably implies, and often explicitly imposes, limits upon the first kind: some possibilities must not be explored; some things must not be learned. If we want to get safely home, there are certain seductive songs we must not turn aside for, some sacred things we must not meddle with:

> Great captain,
> a fair wind and the honey lights of home
> are all you seek. But anguish lies ahead;
> the god who thunders on the land prepares it . . .
>
> .   .   .   .
>
> One narrow strait may take you through his blows:
> denial of yourself, restraint of shipmates.

This theme, of course, is dominant in Biblical tradition, but the theme itself and its modern inversion can be handily understood by a comparison of this speech of Tirêsias to Odysseus in Robert Fitzgerald's Homer with Tennyson's romantic Ulysses who proposes, like a genetic engineer or an atomic scientist,

> To follow knowledge like a sinking star,
> Beyond the utmost bound of human thought.

Obviously unlike Homer's Odysseus, Tennyson's Ulysses is said to come from Dante, and he does resemble Dante's Ulysses pretty exactly—the critical difference being that Dante thought this Ulysses a madman and a fool, and brings down upon his Tennysonian speech to his sailors one of the swiftest anticlimaxes in literature. The real—the human—knowledge is understood as implying and imposing limits, much as marriage does, and these limits are understood to belong necessarily to the definition of a human being.

In all this talk about marriage I have not forgot that I am supposed to be talking about agriculture. I am going to talk directly about agriculture in a minute, but I want to insist that I have been talking about it indirectly all along, for the analogy between marriage making and farm making, marriage keeping and farm keeping, is nearly exact. I have talked about marriage as a way of talking about farming because marriage, as a human artifact, has been more carefully understood than farming. The analogy between them is so close, for one thing, because they join us to time in nearly the same way. In talking about time, I will begin to talk directly about farming, but as I do so, the reader will be aware, I hope, that I am talking indirectly about marriage.

When people speak with confidence of the longevity of diminishing agricultural sources—as when they speak of their good intentions about nuclear power—they are probably not just being gullible or thoughtless; they are likely to be speaking from belief in several tenets of industrial optimism: that life is long, but time and work are short; that every problem will be solved by a "technological breakthrough" before it enlarges to catastrophe; that any problem can be solved in a hurry by large applications of urgent emotion, information, and money. It is regrettable that these assumptions should risk correction by disaster when they could be cheaply and safely overturned by the study of any agriculture that has proved durable.

To the farmer, Emerson said, "The landscape is an armory of powers. . . . ." As he meant it, the statement may be true, but the metaphor is ill-chosen, for the powers of a landscape are available to human use in nothing like so simple a way as are the powers of an armory. Or let us say, anyhow, that the preparations needed for the taking up of agricultural powers are more extensive and complex than those usually thought necessary for the taking up of arms. And let us add that the motives are, or ought to be, significantly different.

Arms are taken up in fear and hate, but it has not been uncharacteristic for a farmer's connection to a farm to begin in love. This has not always been so ignorant a love as it sometimes is now; but always, no matter what one's agricultural experience may have been, one's connection to a newly bought farm will begin in love that is more or less ignorant. One loves the place because present appearances recommend it, and because they suggest possibilities irresistibly imaginable. One's head, like a lover's, grows full of visions. One walks over the premises, saying, "If this were mine, I'd make a permanent pasture here; here is where I'd plant an orchard; here is where I'd dig a pond." These visions are the usual stuff of unfulfilled love and induce wakefulness at night.

When one buys the farm and moves there to live, something different begins. Thoughts begin to be translated into acts. Truth begins to intrude with its matter-of-fact. One's work may be defined in part by one's visions, but it is defined in part too by problems, which the work leads to and reveals. And daily life, work, and problems gradually alter the visions. It invariably turns out, I think, that one's first vision of one's place was to some extent an imposition on it. But if one's sight is clear and if one stays on and works well, one's love gradually responds to the place as it really is, and one's visions gradually image possibilities that are really in it. Vision, possibility, work, and life—all have changed by mutual correction. Correct discipline, given enough time, gradually removes one's self from one's line of sight. One works to better purpose then and makes fewer mistakes, because at last one sees where one is. Two human possibilities of the highest order thus come within reach: what one wants can become the same as what one has, and one's knowledge can cause respect for what one knows.

"Correct discipline" and "enough time" are inseparable notions. Correct discipline cannot be hurried, for it is both the knowledge of what ought to be done, and the willingness to do it—all of it, properly. The good worker will not suppose that good work can be made properly answerable to haste, urgency, or even emergency. But the good worker knows too that after it is done work requires yet more time to prove it's worth. One must stay to experience and study and understand the consequences—must understand them by living with them, and then correct them, if necessary, by longer living and more work. It won't do to correct mistakes made in one place by moving to another place, as has been the common fashion in America, or by adding on another place, as is the fashion in any sort of "growth economy." Seen this way, questions about farming become inseparable from questions about propriety of scale. A farm can be too big for a farmer to husband properly or pay proper attention to. Distraction is inimical to correct discipline, and enough time is beyond the reach of anyone who has too much to do. But we must go farther and see that propriety of scale is invariably associated with propriety of another kind: an understanding and acceptance of the human place in the order of Creation—a proper humility. There are some things the arrogant mind does not see; it is blinded by its vision of what it desires. It does not see what is already there; it never sees the forest that precedes the farm or the farm that precedes the shopping center; it will never understand that America was "discovered" by the Indians. It is the properly humbled mind in its proper place that sees truly, because—to give only one reason—it sees details.

And the good farmer understands that further limits are imposed upon haste by nature which, except for an occasional storm or earthquake, is in no hurry either. In the processes of most concern to agriculture—the building and preserving of fertility—nature is never in a hurry. During the last seventeen years, for example, I have been working at the restoration of a once exhausted hillside. Its scars are now healed over, though still visible, and this year it has provided abundant pasture, more than in any year since we have owned it. But to make it as good as it is now has taken seventeen years. If I had been a millionaire or if my family had been starving, it would still have taken seventeen years. It can be better than it now is, but that will take longer. For it to live fully in its own possibility, as it did before bad use ran it down, may take hundreds of years.

But to think of the human use of a piece of land as continuing through hundreds of years, we must greatly complicate our understanding of agriculture. Let us start a job of farming on a given place—say an initially fertile hillside in the Kentucky River Valley—and construe it through time:

1. To begin using this hillside for agricultural production—pasture or crop—is a matter of a year's work. This is work in the present tense, adequately comprehended by conscious intention and by the first sort of knowledge I talked about—information available to the farmer's memory and built into his methods, tools, and crop and livestock species. Understood in its present tense, the work does not reveal its value except insofar as the superficial marks of craftsmanship may be seen and judged. But excellent workmanship, as with a breaking plow, may prove as damaging as bad workmanship. The work has not revealed its connections to the place or to the worker. These connections are revealed in time.

2. To live on the hillside and use it for a lifetime gives the annual job of work a past and a future. To live on the hillside and use it without diminishing its fertility or wasting it by erosion still requires conscious intention and information, but now we must say *good* intention and *good* (that is, correct) information, resulting in *good* work. And to these we must now add *character*: the sort of knowledge that might properly be called familiarity, and the affections, habits, values, and virtues (conscious and unconscious) that would preserve good care and good work through hard times.

3. For human life to continue on the hillside through successive generations requires good use, good work, all along. For in any agricultural place that will waste or erode—and all will—bad work does not permit "muddling through"; sooner or later it ends human life. Human continuity is virtually synonymous with good farming, and good farming obviously must outlast the life of any good farmer. For it to do this, in addition to the preceding requirements, we must have community. Without community, the good work of a single farmer or a single family will not mean much or last long. For good farming to last, it must occur in a good farming community—that is, a neighborhood of people who know each other, who understand their mutual dependences, and who place a proper value on good farming. In its cultural aspect, the community is an order of memories preserved consciously in instructions, songs, and stories, and both consciously and unconsciously in ways. A healthy culture holds preserving knowledge in place for a long time. That is, the essential wisdom accumulates in the community much as fertility builds in the soil. In both, death becomes potentiality.

People are joined to the land by work. Land, work, people, and community are all comprehended in the idea of culture. These connections cannot be understood or described by information—so many resources to be transformed by so many workers into so many products for so many consumers—because they are not quantitative. We can understand them only after we acknowledge that they should be harmonious—that a culture must be either shapely and saving or shapeless and destructive. To presume to describe land, work, people, and community by information, by quantities, seems invariably to throw them into competition with one another. Work is then understood to exploit the land, the people to exploit their work, the community to exploit its people. And then instead of land, work, people, and community, we have the industrial categories of resources, labor, management, consumers, and government. We have exchanged harmony for an interminable fuss, and the work of culture for the timed and harried labor of an industrial economy.

But let me bring these notions to the trial of a more particular example.

Wes Jackson and Marty Bender of The Land Institute have recently worked out a comparison between the energy economy of a farm using draft horses for most of its field work and that of an identical farm using tractors. This is a project a generation overdue, of the greatest interest and importance—in short, necessary. And the results will be shocking to those who assume a direct proportion between fossil fuel combustion and human happiness.

These results, however, have not fully explained one fact that Jackson and Bender had before them at the start of their analysis and that was still running ahead of them at the end: that in the last twenty-five or thirty years, the Old Order

Amish, who use horses for farmwork, doubled their population and stayed in farming, whereas in the same period millions of mechanized farmers were driven out. The reason that this is not adequately explained by analysis of the two energy economies, I believe, is that the problem is by its nature beyond the reach of analysis of any kind. The real or whole reason must be impossibly complicated, having to do with nature, culture, religion, family and community life, as well as with agricultural methodology and economics. What I think we are up against is an unresolvable difference between thought and action, thought and life.

What works *poorly* in agriculture—monoculture, for instance, or annual accounting—can be pretty fully explained, because what works poorly is invariably some oversimplifying *thought* that subjugates nature, people, and culture. What works well ultimately defies explanation because it involves an order which in both magnitude and complexity is ultimately incomprehensible.

Here, then, is a prime example of the futility of a dependence on information. We cannot contain what contains us or comprehend what comprehends us. Yeats said that "Man can embody truth but he cannot know it." The part, that is, cannot comprehend the whole, though it can stand for it (and by it). Synecdoche is possible, and its possibility implies the possibility of harmony between part and whole. If we cannot work on the basis of sufficient information, then we have to work on the basis of an understanding of harmony. That, I take it, is what Sir Albert Howard and Wes Jackson mean when they tell us that we must study and emulate on our farms the natural integrities that precede and support agriculture.

The study of Amish agriculture, like the study of *any* durable agriculture, suggests that we live in sequences of patterns that are formally analogous. These sequences are probably hierarchical, at least in the sense that some patterns are more comprehensive than others; they tend to arrange themselves like internesting bowls—though any attempt to represent their order visually will oversimplify it.

And so we must suspect that Amish horse-powered farms work well, not because—or not just because—horses are energy-efficient, but because they are living creatures, and therefore fit harmoniously into a pattern of relationships that are necessarily biological, and that rhyme analogically from ecosystem to crop, from field to farmer. In other words, ecosystem, farm, field, crop, horse, farmer, family, and community are in certain critical ways *like* each other. They are, for instance, all related to health and fertility or reproductivity in about the same way. The health and fertility of each involves and is involved in the health and fertility of all.

It goes without saying that tools can be introduced into this agricultural and ecological order without jeopardizing it—but only up to a certain kind, scale, and power. To introduce a tractor into it, as the historical record now seems virtually to prove, is to begin its destruction. The tractor has been so destructive, I think, because it is *unlike* anything else in the agricultural order, and so it breaks the essential harmony. And with the tractor comes dependence on an energy supply that lies not only off the farm but outside agriculture and outside biological cycles and integrities. With the tractor, both farm and farmer become "resources" of the industrial economy, which always exploits its resources.

We would be wrong, of course, to say that anyone who farms with a tractor is a bad farmer. That is not true. What we must say, however, is that once a tractor is introduced into the pattern of a farm, certain necessary restraints and practices, once implicit in technology, must now reside in the character and consciousness of the farmer—at the same time that the economic pressure to cast off restraint and good practice has been greatly increased.

In a society addicted to facts and figures, anyone trying to speak for agricultural *harmony* is inviting trouble. The first trouble is in trying to say what harmony is. It cannot be reduced to facts and figures—though the lack of it can. It is not very visibly a function. Perhaps we can only say what it may be like. It may, for instance, be like sympathetic vibration: "The A string of a violin . . . is designed to vibrate most readily at about 440 vibrations per second: the note A. If that same note is played loudly not on the violin but near it, the violin A string may hum in sympathy." This may have a practical exemplification in the craft of the mud daubers which, as they trowel mud into their nest walls, hum to it, or at it, communicating a vibration that makes it easier to work, thus mastering their material by a kind of a song. Perhaps the hum of the mud dauber only activates that anciently perceived likeness between all creatures and the earth of which they are made. For as common wisdom holds, like *speaks to* like. And harmony always involves such specificities of form as in the mud dauber's song and its nest, whereas information accumulates indiscriminately, like noise.

Of course, in the order of creatures, humanity is a special case. Humans, unlike mud daubers, are not naturally involved in harmony. For humans, harmony is always a human product, an artifact, and if they do not know how to make it and choose to make it, then they do not have it. And so I suggest that, for humans, the harmony I am talking about may bear an inescapable likeness to what we know as moral law—or that, for humans, moral law is a significant part of the notation of ecological and agricultural harmony. A great many people seem to have voted for information as a safe substitute for virtue, but this ignores—among much else—the need to prepare humans to live short lives in the face of long work and long time.

Perhaps it is only when we focus our minds on our machines that time seems short. Time is always running out for machines. They shorten our work, in a sense popularly approved, by simplifying it and speeding it up, but our work perishes quickly in them too as they wear out and are discarded. For the living

Creation, on the other hand, time is always coming. It is running out for the farm built on the industrial pattern; the industrial farm burns fertility as it burns fuel. For the farm built into the pattern of living things, as an analogue of forest or prairie, time is a bringer of gifts. These gifts may be welcomed and cared for. To some extent they may be expected. Only within strict limits are they the result of human intention and knowledge. They cannot in the usual sense be made. Only in the short term of industrial accounting can they be thought simply earnable. Over the real length of human time, to be earned they must be deserved.

From this rather wandering excursion I arrive at two conclusions.

The first is that the modern stereotype of an intelligent person is probably wrong. The prototypical modern intelligence seems to be that of the Quiz Kid—a human shape barely discernable in fluff of facts. It is understood that everything must be justified by facts, and facts are offered in justification of everything. If it is a fact that soil erosion is now a critical problem in American agriculture, then more facts will indicate that it is not as bad as it could be and that Iowa will continue to have topsoil for as long as seventy more years. If facts show that some people are undernourished in America, further facts reveal that we should all be glad we do not live in India. This, of course, is machine thought.

To think better, to think like the best humans, we are probably going to have to learn again to judge a person's intelligence, not by the ability to recite facts, but by the good order or harmoniousness of his or her surroundings. We must suspect that any statistical justification of ugliness and violence is a revelation of stupidity. As an earlier student of agriculture put it: "The intelligent man, however unlearned, may be known by his surroundings, and by the care of his horse, if he is fortunate enough to own one."

My second conclusion is that any public program to preserve land or produce food is hopeless if it does not tend to right the balance between numbers of people and acres of land, and to encourage long-term, stable connections between families and small farms. It could be argued that our nation has never made an effort in this direction that was knowledgeable enough or serious enough. It is certain that no such effort, here, has ever succeeded. The typical American farm is probably sold and remade—often as part of a larger farm—at least every generation. Farms that have been passed to the second generation of the same family are unusual. Farms that have passed to the third generation are rare.

But our crying need is for an agriculture in which the typical farm would be farmed by the third generation of the same family. It would be wrong to try to say exactly what kind of agriculture that would be, but it may be allowable to suggest that certain good possibilities would be enhanced.

The most important of those possibilities would be the lengthening of memory. Previous mistakes, failures, and successes would be remembered. The land would not have to pay the cost of a trial-and-error education for every new owner. A half century or more of the farm's history would be living memory, and its present state of health could be measured against its own past—something exceedingly difficult *outside* of living memory.

A second possibility is that the land would not be overworked to pay for itself at full value with every new owner.

A third possibility would be that, having some confidence in family continuity in place, present owners would have future owners not only in supposition but in *sight* and so would take good care of the land, not for the sake of something so abstract as "the future" or "posterity," but out of particular love for living children and grandchildren.

A fourth possibility is that having the past so immediately in memory, and the future so tangibly in prospect, the human establishment on the land would grow more permanent by the practice of better carpentry and masonry. People who remembered long and well would see the folly of rebuilding their barns every generation or two, and of building new fences every twenty years.

A fifth possibility would be the development of the concept of *enough*. Only long memory can answer, for a given farm or locality, How much land is enough? How much work is enough? How much livestock and crop production is enough? How much power is enough?

A sixth possibility is that of local culture. Who could say what that would be? As members of a society based on the exploitation of its own temporariness, we probably should not venture a guess. But we can perhaps speak with a little competence of how it would begin. It would not be imported from critically approved cultures elsewhere. It would not come from watching certified classics on television. It would begin in work and love. People at work in communities three generations old would know that their bodies renewed, time and again, the movements of other bodies, living and dead, known and loved, remembered and loved, in the same shops, houses, and fields. That, of course, is a description of a kind of community dance. And such a dance is perhaps the best way we have to describe harmony.

—1981

Wes Jackson
Becoming Native to This Place

Although Wes Jackson is best known for his leadership in defining and promoting ecologically based agriculture, that is not his only initiative in the revival of rural life. Here he explores how we (the immigrants of the past five hundred years) might at last become truly native to the continent we inhabit and, in doing so, evolve an appropriate awareness of place. He begins with a series of vignettes about the settlement of this country, showing how we never truly tried to become native and yet, as is confirmed by Winona LaDuke, drove out those who were.

It is Jackson's thesis that the land cries out for resettling—but not only the land. Paraphrasing Wendell Berry, he says, "The unsettling of . . . towns and communities with the migration of people to urban areas has led to an unsettling of the culture at large with its rising crime rate, increasing national debt, increase in soil erosion, and increase in chemical contamination of the countryside." Deploring what Schumacher called footlooseness, Jackson contends—like Schumacher, who called for policies to reconstruct rural culture—that what we now need is a new generation of settlers. To that end he describes how, as part of a Land Institute project in ecological community accounting, he has bought some of the buildings in the small, almost deserted Kansas town of Matfield Green. Spending time there, he is discovering traces of the day-to-day rural wisdom evolved by Matfield Green's residents earlier in the century. Through learning about them, he feels, we may inch a little further along in the necessary journey of becoming native to our place. Jackson's ultimate vision is one of homecoming.

This lecture became a chapter in Jackson's book *Becoming Native to This Place*.

In March of 1977 Fritz Schumacher came to Salina, Kansas (he died in August of that year). We had just started The Land Institute the previous September. Six weeks later our building burned down with all of our books and tools. I had resigned my position in California to begin this work, and what little retirement money we had, had been put into that building. There was really no reason to keep going except that we had some ideas. When Schumacher came in March, we were rebuilding, mostly with scrap materials. While he was there we arranged for him to give a public lecture, during which he told a story about traveling across the United States with some German friends at the height of the Great Depression, around 1935 or 1936, I imagine. They stopped for gas in a small town near Salina and asked a fellow there, "How are things?"

And he said, "They're all right."

"What do you do?" Schumacher asked him.

He answered, "Well, I work on that farm right over there."

"Oh, that's interesting—you're a farmer."

"Yes. In fact, I work for the man who used to work for me. I didn't have any money to pay him, so I paid him in land. And now he owns my farm and I work for him."

Schumacher said, "That's a very sad story."

And the man replied, "Oh, no; he doesn't have any money either, so he's paying me back in land!"

I like that story because at a very basic level this farmer shows us that in a certain sense all we have to do is figure out a way to stay amused while we live out our lives as inexpensively as possible within the life support system. It's what I call "the Mill-Around Theory of Civilization": if we can simply mill around and not expend too many resources, then we won't do much harm to ourselves or the planet. The problem is, how do we learn to quit *doing* in a manner that uses up the earth's capital? Or stated otherwise, how do we make our vessel so small that it doesn't take much to fill it? Should not this be our journey?

I have a friend, Leland, who gets by on five hundred dollars a year. He lives in a six-by-sixteen-foot shack. He began his journey some twenty-five years ago. In some respects, he's more important to me than Thoreau, for Thoreau's tenure at Walden was brief. Leland's idea is that once we start seeking pleasure, we start doing violence to people and to the landscape. He says there's nothing wrong with the experience of pleasure, but when you start *seeking* pleasure, violence happens. He believes that my intellectual pursuits, for example, are a form of pleasure-seeking, that they create a kind of violence. He even quit growing his beautiful garden because he thought that too was a form of pleasure-seeking; now he just harvests the greens that grow wild in the yard and lives mostly on wheat. Leland took out Social Security because his wife, who lives in the house—he lives in the shack, less than a hundred yards away—felt she needed three hundred dollars a month to live on. He could get four hundred dollars from Social Security, and she could get two hundred, and because she needed only three hundred, he had three hundred dollars a month piling up in the bank. This money was making him have "creative thoughts," which he thought might start causing violence. So he scratched his name off Social Security and says he's a free man again. There are other ways to think about living in the world, but Leland is important to me because he's the most bottom-line person I know. He is very careful not to be judgmental of others. Seeing his example has made me pretty impatient with people who say, "We just can't make it."

But that's not what I came here to talk about. I came to talk about becoming native to this place—meaning, first of all, this continent. I can do no better than to quote Wendell Berry in *The Unsettling of America*. He said that we came to this country with vision but not with sight: "We came with visions of former places but not the sight to see where we are." Later, in a letter he wrote to me, he said that as we came across the continent, cutting the forests and plowing prairies, we never knew what we were doing because we have never known what we were undoing. Dan Luten, the now retired geographer at Berkeley, in a paper maybe twenty-five years ago, said that we came poor people to a seemingly empty land that was rich in resources. And based on that *perception* of reality—"poor people," "seemingly empty land," "rich"—we built our political, educational, economic, and religious institutions. Now we've become rich people in an increasingly poor land that's filling up, and the old institutions don't hold. So here we are. We patch things up, give them a lick and a promise, and things don't quite work.

Well, that's all true as far as our settlement is concerned. But there is more, for standing behind settlement is conquest, which has left its legacy. Kirk Sale said it well in *The Conquest of Paradise*, a book I greatly enjoyed. After 1492 there was the gold of Mexico and the gold of Peru. So the conquerors thought (being Christians, believing things had to come in threes) there must be gold somewhere else. (It was a holy idea.) One of these men was Coronado, who was hanging out down in Mexico, married to the daughter of the bastard son of the king of Spain. He and a bunch of second sons out of Spain were mighty interested in finding that third hunk of gold. So in 1540 Coronado started northward from Compostela to Culiacán and followed the Culiacán Valley northward and finally reached an area straddling parts of present-day New Mexico and Arizona. They were looking for the Seven Cities of Cíbola. No matter that there were only six (seven, like three, is a sacred number, maybe because there are seven holes in our head). These cities reportedly had lots of gold. Coronado's party consisted of a thousand horses and mules, a large flock of sheep, Indians to carry the baggage, and some three hundred Spaniards on horseback. It must have been quite a contingent. They made it all the way up there only to find poor—to their eyes—Pueblo Indians. No gold! But the Pueblo Indians had a slave whom the Spaniards called El Turko because they thought he looked like a Turk. I don't know where he had been captured, but his people were the Herahay, who lived in the area of southeastern Nebraska and northeastern Kansas. Now, this poor slave was homesick and wanted to go home. I'll make a long story short. "There isn't any gold here," he said, "but there's gold up in Quivira." The easternmost part of the former kingdom of Quivira is about eighteen miles south of where I live in Kansas. The story gets a little complicated here. For some reason Coronado and his men left Pueblo country and lit out toward the plains of Texas. They got over to the Llano Estacado and ran into that dropping off point there. The Turk was telling them, "No, no; north and east of here, that's where the gold is." So they thought, "Well, what the heck" (that's a loose translation), and Coronado picked some thirty men to go with him and sent the others back to Pueblo country. (Most of these young adventurers, by the way, were from some of the finest families in Europe. Most were in their twenties. The discoverers of the Grand Canyon were in their early to mid-twenties. Coronado himself was barely thirty.)

Convinced by the Turk, he and his thirty men took off "northward by the needle" of the compass—which is to say, north and east. Finally they came into the kingdom of Quivira—what is now part of Kansas—and what did they find? Houses made of sticks and straw, tall people, some of whom measured six feet eight inches high. Chief Tatarrax, summoned by Coronado, arrived. The only metal he had was a copper ring around his neck. There they were, frustrated and out of sorts, and Coronado was eventually convinced by his men that they should strangle the Turk. They put a rope around his neck, put a stick through the loop, and garroted him. Thus, the first murder by a European of a native of the central part of this continent happened somewhere between eighteen and fifty miles south and west of where I live.

At that particular time, de Soto was over on the Mississippi building barges. Had those two men and their armies marched toward one another for seven days, they would have met. De Soto died near the terminus of his expedition. Some three hundred years passed before that area would be filled in by settlers!

It is time to draw a long breath here and reflect on this history: the first

Europeans in this country came as conquerors of the natives. The settlers who followed, as Wendell Berry says in *The Unsettling of America*, designated them "redskins" and treated them as surplus people. That designation and attitude are sins for which we have never atoned. Furthermore, by ignoring the wisdom and example of the native peoples, the settlers ensured that their own great-great-grandsons and -granddaughters would one day become redskins, surplus people. The loss of people from the land, the small towns and rural communities that have dried up—to the point that now only 1.9 percent of the United States population lives on farms—means that the new redskins have nearly been exterminated. So few farmers are there now that the U.S. Census Bureau has quit counting them as a category. And now we are all candidates for "redskinhood" because we never really came to terms with the attitude and the institutions—the system of laws, the justifications—that made the extermination of both the natives and the farmers possible.

Another long breath, another take on our history—1776. Here is Thomas Jefferson carrying in his mind the image of the Virgilian pastoral landscape. Jefferson had the idea that nature combined with farming carried with it a certain sort of virtue, an inherent virtue that would inform this new chance on earth. This was the Jeffersonian ideal of the small town and the rural community, informed as it was by his Enlightenment worldview, Jefferson believing in rationalism and giving us the grid and the system of laws. But here we have the reality of a highly diverse continent, an ecological mosaic, the product of a time long before any "Enlightenment mind" would appear. Part of this land would accommodate the Jeffersonian ideal; most of it would not.

To illustrate this difference, I would like to contrast an experience I had on a ranch in South Dakota one summer with the experience I had growing up on a farm in the Kansas River Valley. It was the summer I turned sixteen that I abandoned myself to the prairies of South Dakota to work on a ranch belonging to my mother's eccentric and childless first cousin and her Swedish immigrant husband, Andrew. Ina was Andrew's second wife; his first had been her sister Bertha. Andrew and Bertha homesteaded one half-section of land and Ina another. When Bertha died, Andrew and Ina married and joined their holdings. This was near the Rosebud Indian reservation, and on Sundays I sometimes rode with half-breed kids over those prairies, hearing stories of how their Indian grandfather had trapped eagles on this hill or that. Andrew, Ina, and I would go to White River on Saturday afternoon. These Rosebud Sioux would lie in the shade of the stores, and as the sun moved, they would pick up their belongings and move to the shade of the other side. Out on the ranch I would hear Andrew cuss and swear about how the Indians never did anything with the land. In town the very Indian from whom Andrew and Ina were leasing Indian land had once again charged groceries to their account. Andrew always paid, for to fail to meant that a neighboring rancher would be only too willing to lease the same land next year, perhaps forgetting that he too would be trapped into buying a bottle of whiskey at the liquor store, that he too would have to tolerate coming upon what was left of one of his steers butchered by the same redskins.

I fell in love that summer at a Saturday night dance. She was a beautiful white girl, her magic so overwhelming that I failed to sleep the entire night after I met her. Thirty-five years later when I saw her again, she was seriously overweight, had lost most of her teeth, her slip was showing, and she neither recognized nor remembered me as she lugged one of her grandchildren into the bar. I think it was the same bar where, as a teenager, I learned more interesting content at low tuition than at any time before or since. For it was there that I scrutinized, with the civilized eye of a Kansas River Valley Methodist, drunk cowboys—married or not—hugging and smooching young natives and from time to time disappearing with them into the shadows of the dusty back streets of White River.

The land was mostly unplowed and still is. The horse was central to that way of life then, less so now. Out on the ranch, besides the moon and stars the only lights were from Murdo and Okaton across the river twelve to fifteen miles distant. It was a summer of branding and castrating cattle, fixing fences, discovering dens of rattlesnakes, and catching pond bass. Many evenings on the ranch I'd drive out on "the Point" in a Cadillac coup or the pickup to shoot prairie dogs or to see the hundred head of horses in the bottoms or out on the range; "junk horses" Ina called them, for in the dry '30s she would pump water for hours for the cattle, only to have fifty to a hundred head of Andrew's horses show up, run the cattle away, and drink all the water. Andrew justified keeping these mostly wild creatures around by insisting it was horse trading that had made it possible for him to be so solidly positioned. But think of the slack Andrew and Ina enjoyed to be able to afford those hundred head of mostly unbroken horses.

I lived in a small wooden hillside shack set on steel wheels, a shack Andrew had bought from Millette County, which used it to house the county road crew. It had been pulled by horses, perhaps the same horses used to pull the grader blade. Andrew and Ina lived in a small two-room house with a large attic, whose floor bowed from the weight of such old magazines as *Life*, *The Saturday Evening Post*, and Ina's *True Stories* about romance. Some evenings Andrew and I would sit on his steps, which overlooked the White River a half mile away. Andrew would cuss Roosevelt, the Yalta Conference, Indians, neighbors—everybody but Ike, who happened to be Ina's first cousin (otherwise I suspect President Eisenhower would have caught it too).

There was no electricity and only cistern water, which was used at least twice, the last time always to water a small backyard garden or the chickens.

With Ina on her buckskin, Dickey, and me on Bonnie or Violet (the names of two girls back home), we rode the range from one dam to the other, where poles were kept with lures so we could catch some bass on the way home. Or we might go to the abandoned school on the school section for some cottonseed cake to distribute as cattle feed somewhere across the nearly four thousand acres of paradise. I didn't want to go home, and had it not been for high school football in September, I might have stayed. The place became my American dream, and looking back, even though Jefferson's and Lewis and Clark's Missouri was only fifty miles away, I now see that little of Jefferson's vision was there beyond the section lines and the system of laws. His vision of the family farm must have been predicated upon thirty inches or more of moisture per year. Here, the land determined; no yeoman farmer existed. Even so, I loved everything about it. The Indians, the rodeos, the Danish and Swedish immigrants delighted with their land holdings, the rattlesnakes, even the colorful prejudice and how the natives got a little bit even through the butchered steer, the grocery bill, and the whiskey.

In the Kansas River Valley it had been another story. We were farmers there. Hoeing was endless during the summer, what with watermelons, sweet potatoes, cantaloupes, strawberries, peonies, phlox, sweet corn, potatoes, tomatoes, rhubarb, asparagus, etc. It was a relief to put up alfalfa hay or to harvest wheat, rye, or corn (my dad won the county corn-growing contest at least three years). Our farm and market were along U.S. 24 and 40, a two-lane road called the Pacific Highway, a subconscious naming, I suppose, because the nation looked westward. Six children were born to my parents. I was the last in 1936, a sister the first, born in 1914. Dad was fifty the year I was born, my mother forty-two. They were agrarians—fiercely so, I see now—Jeffersonians. They were also Methodist and Congregationalist: don't waste time, motion, or steps. Don't drink pop, alcohol in any form, or eat out. The contrast between that Kansas truck farm and the South Dakota ranch was striking. The row crops required cultivating and hoeing. Sweat of the brow, good manners, and quotable scripture went together. And from what I learned in that market, with people stopping in on their trips from coast to coast, I now sense that we were countrymen then in a way that we are not now. There were no bad jokes about either California or New Jersey then; we all inquired into one another's well-being.

Here was agriculture—row crop variety, of course—which I knew and, I will even say, loved in a certain restricted sense, but it did not compare to the life of the range with the juxtaposition of natives and grassland, ranchers and rodeos. I made up my mind that I would have that South Dakota ranch one day or one like it. But Andrew died of prostate cancer and Ina died of injuries sustained in the pickup she was driving. The ranch was sold and the money willed to one of Ina's nephews, who within a year paid it all out in a lawsuit due to being at fault in a car wreck.

Football and love kept me in college in what must have been one of the most misspent youths in history. And what smoldered in me were two experiences with the land: that of the sodbusting, Jeffersonian agrarian and that of the cattleman. I preferred the latter.

My great-grandfather entered Kansas the first day it was legal: May 30, 1854, the day the Kansas-Nebraska Act was ratified. Twenty-six years old, he had already been to San Francisco by way of Panama. Fifty miles into Kansas he broke tallgrass prairie sod and set right out to farming his 160 acres, Jefferson style, interrupting normal life to fight against pro-slavery forces with John Brown at Black Jack Creek on the Santa Fe Trail in 1857. Later a man who was to become his son-in-law, one of my grandfathers, arrived in Kansas in 1877 with three hundred dollars the day before turning twenty-two. He felt lucky not to have put his money in the bank, for it closed the next day. He thus preserved his grubstake and threw himself onto the Flint Hills grassland of Kansas to run cattle on more or less free grass. By the end of ten years he had enough to go half-and-half with a partner and purchase half of 160 acres of sandy loam in the Kansas River Valley on the second bench, an alluvial terrace high enough above the river bed to assure no more than a flood or two per century. In five years he had bought his partner out.

I was born on that farm, love those soils, love to plow them, love to smell them. Even so, I have wondered why that grandfather, when the grass had been so good to him, would give up his cattle to farm. I think I know the reason: he had come from the Shenandoah Valley of Virginia. A Virginian! An agrarian! He was likely an unconscious Jeffersonian. He played the role and he played it well, for he was a well-off man when he died in 1925. As a school board member, he convinced his neighbors that the new school should be completely paid for in the year it was built. It was a fine, well-built school with two rooms for eight grades. My mother and I both went there. How could the community pay such a debt so quickly? I can't speak for the neighbors, but I know that times were good on the farm generally and that Granddad never expanded his income by expanding his acreage. It was said of him that no matter what he did, things turned out right. I suspect that the reason this was true comes from something revealed in an offhand statement my mother made once. She said he would lean on his scoop shovel for a half hour or more, watching his hogs eat the ear corn, or lean against the barn door to watch them eat the soaked oats or boiled potatoes or whatever he had raised. As much as it was due to the times, his good fortune was due to a combination of joy, sympathy, art, and love rolled into one and tuned to the demands of his place. Here was the Jeffersonian dream, as imperfect as it was, at its high-water mark. The actuality or reality of that dream has been compromised and in decline at dazzling speed ever since World War II.

In fact, it seems now as if the Jeffersonian dream is, as Aldo Leopold said about conservation, like "a bird which flies faster than the shot we aim at it." I won't offer further evidence that the Jeffersonian ideal is receding—that is simply a matter of going through the checklist of environmental and social problems and the irrational patterns of current settlement. Nor will I even attempt an analysis as to why.

The question is, how do we reconcile these two situations on the prairies, scarcely one day's drive apart at the modern speed limit? What is the lesson to be learned from them about future land use? We need to recognize the reality of the ecological mosaic across the country and to realize that there are some places, such as that South Dakota ranch, where the land determines and other places where human beings can actually make useful and appropriate changes in a favorable environment.

From Oklahoma to Saskatchewan, from east of Denver deep into the Midwest, thousands of small towns and rural communities are dying. Thousands of them! Schools are being closed, churches are being closed. This decline is the consequence of what Wendell Berry talked about in The Unsettling of America, explaining the title's double meaning: the unsettling of these towns and communities with the migration of people to urban areas has led to an unsettling of the culture at large with its rising crime rate, increasing national debt, increase in soil erosion, and increase in chemical contamination of the countryside. We have to face it: the reward for destroying communion is power: power over nature, power over the indigenous, power over the constantly newly emerging redskins. Rather than looking to Washington, we must start thinking that small is beautiful. One way to effect this change would be to introduce a second major into our universities and colleges. Right now there's only one major: upward mobility. It's the major which accommodates the original set of assumptions we settled the continent with, the mind-set that fuels the extractive economy. The new major would be "homecoming." It would educate people to go back to a place and dig in. We need a new generation of settlers, people who could go into these places with a fundamentally different mind-set, with the skills for what we might call "ecological community accounting." They would start at the beginning, asking such questions as, "How does one set up the books for this accounting?" We have examples to follow, technological possibilities and idealistic notions from people like John and Nancy Todd, co-founders of the New Alchemy Institute and Ocean Arks International, who inspired our work at The Land Institute years ago; Amory and Hunter Lovins at the Rocky Mountain Institute; David Orr of the Meadowcreek Project, and so on. It is frightening how terribly underfunded these organizations are, making them so fragile that it doesn't take much more than a blip before some of them are extinguished.

The Land Institute is running a project in a little town called Matfield Green, located in Chase County, Kansas. We say we are "setting up the books" for ecological community accounting as a necessary step for understanding how to proceed when it comes to revitalizing community anywhere. The book *Prairie Earth*, written by William Least-Heat Moon, is about Chase County. It has a population of three thousand. Eighty-five percent of the county has never been plowed. It has one traffic light (a yellow blinker, so you don't have to slow down *too* much). And in this county is a little town called Matfield Green, with a population of fifty. Matfield Green doesn't have a lifestyle; it's too small. It just has the basics: a half-time post office, a church that hosts eight to fifteen people per Sunday, and a beer joint. That's all there is left.

I've managed to purchase several buildings in Matfield Green. Five of us went in together and bought the school, a ten-thousand-square-foot building, for five thousand dollars and gave it to The Land Institute. For the gym we paid four thousand dollars. My nephew bought the bank for five hundred dollars. I bought the lumber yard for a thousand. The hardware store we bought for six thousand and renovated; now some of our interns live there. I bought several houses for a thousand dollars or less, one of them for three hundred and fifty. Now, these are not what you call "top of the line" houses, but we've started renovations, putting new roofs on the buildings and replacing rotting stud walls. I'm living part-time in a house there that has cost me—including the purchase price and the purchase of the refrigerator, stove, and everything else in it—less than seventeen thousand dollars. The average homeless person now in New York City costs the city seventeen thousand dollars a year. There are people in that town living on *seven* thousand dollars a year.

At work on my houses in Matfield Green, I've had great fun tearing off the porches and cleaning up the yards. But it has been sad as well, going through the abandoned belongings of families who lived out their lives in this beautiful, well-watered, fertile setting (Matfield Green averages thirty-three inches of rainfall a year). In an upstairs bedroom of Mrs. Florence Johnson's former home, I came across a dusty but beautiful blue padded box labeled "Old Programs—New Century Club." Most of the programs from 1923 to 1964 were there. Each listed the officers, the club flower (sweet pea), the club colors (pink and white), and the club motto ("Just Be Glad"). The programs for each year were gathered under one cover and nearly always dedicated to a local woman who was special in some way.

Each month the women were to comment on such subjects as canning, jokes, memory gems, a magazine article, poems, flower culture, misused words, birds, and so on. The May 1936 program was a debate: "Resolved that movies are detrimental to the young generation." The August 1936 program was dedicated to coping with the heat: roll call was "Hot Weather Drinks"; next came "Suggestions for Hot Weather Lunches"; a Mrs. Rogler offered "Ways of Keeping Cool." The June roll call in 1929 was "The Disease I Fear Most." That was eleven years after the great flu epidemic. Children were still dying in those days of diphtheria, whooping cough, scarlet fever, and pneumonia. On August 20 the roll call question was, "What Do You Consider the Most Essential to Good Citizenship?" In September that year it was "Birds of Our County"; the program was on the mourning dove.

What became of it all?

From 1923 through 1930 the program covers are beautiful, done at a print shop. From 1930 until 1937 the effects of the Depression are apparent; programs are either typed or mimeographed and have no cover. The programs for the next two years are missing. In 1940 the covers reappear, this time typed on construction paper. The printshop printing never reappears. The last program in the box is from 1964. I don't know the last year Mrs. Florence Johnson attended the club. I do know that she and her husband Turk celebrated their fiftieth wedding anniversary, for in the same box are some beautiful white fiftieth-anniversary napkins with golden bells and with "1920 Florence and Turk 1970" printed on them. A neighbor told me that Mrs. Johnson died in 1981. The high school had closed in 1967. The lumber yard and hardware store closed about the same time, but no one knows when for sure. The last gas station went after that.

But back to those programs. The motto never changed. The sweet pea kept its standing. So did the pink and white club colors. The club collect which follows persisted month after month, year after year:

A Collect for Club Women

Keep us, O God, from pettiness;
Let us be large in thought, in word, in deed.
Let us be done with fault-finding and leave off self-seeking.
May we put away all pretense and meet each other face to face, without self-
    pity and without prejudice.
May we never be hasty in judgment and always generous.
Let us take time for all things; make us grow calm, serene, gentle.
Teach us to put into action our better impulses,
straightforward and unafraid.
Grant that we may realize it is the little things that create differences; that in the
    big things of life we are as one.
And may we strive to touch and to know the great common woman's heart of
    us all, and oh, Lord God, let us not forget to be kind.

—Mary Stewart

By modern standards, these people were poor. There was a kind of naiveté among these relatively unschooled women. Some of their poetry in those programs was not good. Some of their ideas about the way the world works seem silly. Some of their club programs don't sound very interesting; some sound tedious. But the monthly agendas of these women were filled with decency, with efforts to learn about everything from the birds to our government and to cope with their problems, the weather, and diseases. Here is the irony: they were living up to a far broader spectrum of their potential than most of us do today!

I am not suggesting that we go back to 1923 or even to 1964. But I will say that those people in that particular generation, in places like Matfield Green, were further along in the necessary journey to become native to their places, even as they were losing ground, than we are today.

Why was their way of life so vulnerable to the industrial economy? What can we do to protect such attempts to be good and decent, to live out our modest lives responsibly? I don't know. But we need to engage in this discussion, for it is particularly problematic. Even most intellectuals who have come out of such places as Matfield Green have not felt that their early lives prepared them adequately for the "official" formal culture.

I will quote from two writers to illustrate this discomfort with the reality of rural culture. The first is Paul Gruchow (in Townships, edited by Michael Marome [Iowa State University Press, 1991]), who grew up on a farm in southern Minnesota:

> I was born at mid-century. My parents, who were poor and rural, had never amounted to anything, and never would, and never expected to. They were rather glad for the inconsequence of their lives. They got up with the sun and retired with it. Their routines were dictated by the seasons. In summer they tended; in fall they harvested; in winter they repaired; in spring they planted. It had always been so; so it would always be.
>
> The farmstead we occupied was on a hilltop overlooking a marshy river bottom that stretched from horizon to horizon. It was half a mile from any road and an eternity from any connection with the rest of the culture. There were no books there; there was no music; there was no television; for a long time, no telephone. Only on the rarest of occasions—a time or two a year—was there a social visitor other than the pastor. There was no conversation in that house.

Similarly, Wallace Stegner, the great historian and novelist, confesses to his feeling of inadequacy coming from a small prairie town in the Cypress Hills of Saskatchewan. In Wolf Willow he writes:

> Once, in a self-pitying frame of mind, I was comparing my background with that of an English novelist friend. Where he had been brought up in London, taken from the age of four onward to the Tate and the National Gallery, sent traveling on the Continent in every

school holiday, taught French and German and Italian, given access to bookstores, libraries, and British Museums, made familiar from infancy on with the conversation of the eloquent and the great, I had grown up in this dung-heeled sagebrush town on the disappearing edge of nowhere, utterly without painting, without sculpture, without architecture, almost without music or theater, without conversation or languages or travel or stimulating instruction, without libraries or museums or bookstores, almost without books. I was charged with getting in a single lifetime, from scratch, what some people inherit as naturally as they breathe air.

How, I asked this Englishman, could anyone from so deprived a background ever catch up? How was one expected to compete, as a cultivated man, with people like himself? He looked at me and said dryly, "Perhaps you got something else in place of all that."

He meant, I suppose, that there are certain advantages to growing up a sensuous little savage, and to tell the truth I am not sure I would trade my childhood of freedom and the outdoors and the senses for a childhood of being led by the hand past all the Turners in the National Gallery. And also, he may have meant that anyone starting from deprivation is spared getting bored. You may not get a good start, but you may get up a considerable head of steam.

Countless writers and artists have been vulnerable to the "official" culture, as vulnerable as the people of Matfield Green. Stegner comments:

I am reminded of Willa Cather, that bright girl from Nebraska, memorizing long passages from the *Aeneid* and spurning the dust of Red Cloud and Lincoln with her culture-bound feet. She tried, and her education encouraged her, to be a good European. Nevertheless she was a first-rate novelist only when she dealt with what she knew from Red Cloud and the things she had "in place of all that." Nebraska was what she was born to write; the rest of it was got up. Eventually, when education had won and nurture had conquered nature and she had recognized Red Cloud as a vulgar little hold, she embraced the foreign tradition totally and ended by being neither quite a good American nor quite a true European nor quite a whole artist.

It seems that we still blunt ourselves by learning long passages from the *Aeneid* while wanting to shake from us the dust of Red Cloud or Matfield Green. The extractive economy cares for neither Virgil nor Mary Stewart. It lures just about all of us to its shopping centers on the edge of Lincoln or Wichita, Louisville or Lexington. And yet, for us the *Aeneid* is part of our story. It is embedded in our thought processes as part of Western civilization. Therefore, it is as essential to becoming native to towns like Matfield Green as the bow and arrow were to the paleolithic Asians who walked here across the Bering land bridge of the Pleistocene.

Our task is to build cultural fortresses to protect our emerging nativeness. They must be strong enough to hold at bay the powers of consumerism, the powers of greed and envy and pride. We have to call the shopping malls and Wal-Marts what they are: the modern cathedrals of secular materialism. One of the most effective ways for this to come about would be for our universities to assume the awesome responsibility of both validating and educating those who want to be homecomers—not to return, necessarily, to their original home, but to go someplace and dig in and begin the long journey to becoming native.

Then we can hope for the resurrection of the likes of Florence Johnson and her women friends, who took their collect seriously. Unless we can affirm and promote the sorts of attitudes and efforts that the New Century Club exhibited, we are doomed. An entire club program devoted to coping with the heat of August is indicative of its members being native to a place. That club was more than a support group; it was cultural information-in-the-making, keyed to place. The alternative, of course, is air-conditioning, not only yielding greenhouse gases but contributing to global warming and the ozone hole as well—and, if fueled by nuclear power, to future Chernobyls. As I see it, we can make technology our leading edge or we can make rich cultural information our leading edge. If we choose programs devoted to coping with the heat, we have a chance. But if we choose exercises in human cleverness in the technological realm as our primary focus, then we've had it. Becoming native to one's place means making everything from our domestic livestock to our domesticated plants native too. And this is a very long process.

Finally, I come back to 1542, to Coronado and those thirty or so avarice-driven adventurers who made the side trip "northward by the needle" from the plains of Texas to the land of Quivira, the land that would one day become Kansas. When their guide, a native Indian slave, admitted that there was no gold, Coronado allowed this native of the land to be strangled with a rope twisted about a stick. What was his offense? He had told a series of lies to men made gullible by greed. He was no fool, and he must have known the risk, but he did it anyway, and he did it for one reason: he was homesick. Because he was a slave, the lure of gold was his ticket home. He thought he could outwit them in the end, but he failed. He was not cunning enough to overcome the power of conquest. The homecomer of today still confronts that power.

—1993

John L. McKnight
John Deere and the Bereavement Counselor

As a professor of community studies and urban affairs at Northwestern University and co-director of its Asset Based Community Development Institute, John McKnight has an informed perspective on how social programs, however well-intended, too often disempower the individuals and communities they were intended to serve. As a lecturer and consultant he has shared his expertise in social-service delivery systems, health policy, community organization, neighborhood policy, and institutional racism, working with community groups, government agencies, foreign governments, and international agencies. He has, as a result, observed the social system of this country from the street as well as from academe.

Like E. F. Schumacher and Kirkpatrick Sale, McKnight bases his thinking concerning social phenomena on the certain knowledge that throughout human experience there have been cultures that lived in timeless and intimate balance within an inherited ecosystem. Such cultures evolved a social and spiritual community, larger than the sum of the parts, in which the needs—physical, emotional, and spiritual— of its members were addressed. He begins his lecture with a recollection of such a community in Sauk County, Wisconsin, before the arrival of European settlers.

McKnight brings his honed social awareness to the Schumacher study of community, both rural and urban. Here he uses the professionalization of bringing comfort to the bereaved as an example of the dehumanizing effects of abnegating to experts those acts of concern and empathy normally filled by family, friends, and neighbors. In the case of bereavement, he points out, the community of mourners disappears and the fabric of the community as a whole is rendered less complete, as is the richness of individual experience. Comparable critiques can be made, he maintains, in the areas of medicine, education, and crime prevention.

John McKnight has not, however, lost faith in human resilience. He characterizes here those social forms still resistant to appropriation by service technologies as being "uncommodified, unmanaged, and uncurricularized." The success of the La Leche League and the early Hospice movement can be attributed, he claims, to being similarly unfettered. With the present wave of renewed commitment to community and to place gathering momentum, these authentic social forms are beginning to be rekindled.

This lecture appears in McKnight's book *The Careless Society* (1995).

In 1973, only eleven years ago, E. F. Schumacher startled Western societies with a revolutionary economic analysis that proclaimed "small is beautiful." His book of that title concluded with these words: "The guidance we need . . . cannot be found in science or technology, the value of which utterly depends on the ends they serve; but it can still be found in the traditional wisdom of mankind." Because traditional wisdom is passed on through stories rather than studies, it seems appropriate that this lecture should take the form of a story.

The story begins as European pioneers crossed the Alleghenies and started to settle the Midwest. The land they found was covered with forests. With incredible effort they felled the trees, pulled the stumps, and planted their crops in the rich, loamy soil.

When they reached the western edge of the place we now call Indiana, the forests stopped; ahead lay a thousand miles of the great grass prairie. The Europeans were puzzled by this new environment. Some even called it "the Great Desert." It seemed untillable. The earth was often very wet, and it was covered with centuries of tangled and matted grasses.

With their cast iron plows, the settlers found that the prairie sod could not be cut and the wet earth stuck to their plowshares. Even a team of the best oxen bogged down after a few yards of tugging. The iron plow was a useless tool to farm the prairie soil. The pioneers were stymied for nearly two decades. Their western march was halted and they filled in the eastern regions of the Midwest.

In 1837 a blacksmith in the town of Grand Detour, Illinois, invented a new tool. His name was John Deere, and the tool was a plow made of steel. It was sharp enough to cut through matted grasses and smooth enough to cast off the mud. It was a simple tool, the "sodbuster," which opened the great prairies to agricultural development.

Sauk County, Wisconsin, is the part of that prairie where I have a home. It is named after the Sauk Indians. In 1673 Father Marquette became the first European to lay eyes upon their land. He found a village laid out in regular patterns on a plain beside the Wisconsin River. He called the place Prairie du Sac. The village was surrounded by fields that had provided maize, beans, and squash for the Sauk people for generations reaching back into unrecorded time.

When the European settlers arrived at the Sauk prairie in 1837, the government forced the native Sauk people west of the Mississippi River. The settlers came with John Deere's new invention and used the tool to open the area to a new kind of agriculture. They ignored the traditional ways of the Sauk Indians and used their sodbusting tool for planting wheat.

Initially the soil was generous and the farmers thrived. Each year, however, the soil lost more of its nurturing power. Within thirty years after the Europeans arrived with their new technology, the land was depleted. Wheat farming became uneconomic, and tens of thousands of farmers left Wisconsin seeking new land with sod to bust.

It took the Europeans and their new technology just one generation to make their homeland into a desert. The Sauk Indians, who knew how to sustain themselves on the Sauk prairieland, were banished to another kind of desert called a reservation. And even they forgot about the techniques and tools that had sustained them on the prairie for generations unrecorded.

And that is how it was that three deserts were created—Wisconsin, the reservation, and the memories of the people.

A century later, the land of the Sauks is now populated by the children of a second wave of European farmers who learned to replenish the soil through the regenerative powers of dairying, ground cover crops, and animal manures. These third and fourth generation farmers and townspeople do not realize, however, that a new settler is coming soon with an invention as powerful as John Deere's plow.

The new technology is called "bereavement counseling." It is a tool forged at the great state university, an innovative technique to meet the needs of those experiencing the death of a loved one, a tool that can "process" the grief of the people who now live on the prairie of the Sauk.

As one can imagine the final days of the village of the Sauk Indians before the arrival of the settlers with John Deere's plow, one can also imagine these final days before the arrival of the first bereavement counselor at Prairie du Sac: the farmers and the townspeople mourn at the death of a mother, brother, son, or friend. The bereaved are joined by neighbors and kin. They meet grief together in lamentation, prayer, and song. They call upon the words of the clergy and surround themselves in community.

It is in these ways that they grieve and then go on with life. Through their mourning they are assured of the bonds among them and are renewed in the knowledge that this death is a part of the past and the future of the people on the prairie of the Sauk. Their grief is common property, an anguish from which the community draws strength and gives the bereaved the courage to move ahead.

It is into this prairie community that the bereavement counselor comes with the new grief technology. The counselor calls the invention a service and assures the prairie folk of its effectiveness and superiority by invoking the name of the great university while displaying a diploma and certificate.

We can imagine that at first the local people will be puzzled by the bereavement counselor's claims; however, the counselor will tell a few of them that the new technique is meant merely to assist the bereaveds' community at the time of death. To some other prairie folk who are isolated or forgotten the counselor will offer help in grief processing. These lonely souls will accept the intervention, mistaking the counselor for a friend.

For those who are penniless, the counselor will approach the County Board and advocate the right to treatment for these unfortunate souls. This right will be guaranteed by the Board's decision to reimburse those too poor to pay for counseling services.

There will be others, schooled to believe in the innovative new tools certified by universities and medical centers, who will seek out the bereavement counselor by force of habit. And one of these people will tell a bereaved neighbor who is unschooled that unless his grief is processed by a counselor, he will probably have major psychological problems later in life.

Several people will begin to use the bereavement counselor because the County Board now taxes them to insure access to the technology, and they will feel that to fail to be counseled is to waste their money and to be denied a benefit or even a right.

Finally, one day the aged father of a Prairie du Sac woman will die. And next-door neighbors will not drop by because they don't want to interrupt the bereavement counselor. The woman's kin will stay home because they will have learned that only the bereavement counselor knows how to process grief in the proper way. The local clergy will seek technical assistance from the bereavement counselor to learn the correct form of service to deal with guilt and grief. And the grieving daughter will know that it is the bereavement counselor who really cares for her because only the bereavement counselor comes when death visits this family on the prairie.

It will be only one generation between the time the bereavement counselor arrives and the community of mourners disappears. The counselor's new tool will cut through the social fabric, throwing aside kinship, care, neighborly obligations, and community ways of coming together and going on. Like John Deere's plow, the tools of bereavement counseling will create a desert where a community once flourished.

Eventually even the bereavement counselor will see the impossibility of restoring hope in clients once they are genuinely alone with nothing but a service for consolation. With the inevitable failure of the service, the bereavement counselor will find the desert even in herself.

There are those who would say that neither John Deere nor the bereavement counselor has created a desert. Rather, they would argue that these new tools have great benefits and that we have focused unduly on a few negative side effects. Indeed, they might agree with Eli Lilly, whose motto was, "A drug without side effects is no drug at all."

To those with this perspective, the critical issue is the amelioration of the negative side effects. In Eli Lilly's idiom, they can conceive of a new drowsiness-creating pill designed to overcome the nausea caused by an anti-cancer drug. They envision a prairie scattered with pyramids of new technologies and techniques, each designed to correct the error of its predecessor but none without its own error to be corrected. In building these pyramids they will also recognize the unlimited opportunities for research, development, and badly needed employment. Many will even name this pyramiding process "progress" and will note its positive effect upon the gross national product.

The countervailing view holds that these pyramiding service technologies are now counterproductive constructions, essentially impediments rather than monuments.

Schumacher helped clarify for many of us the nature of those physical tools that are so counterproductive that they become impediments. There is growing recognition of the waste and devastation created by these new physical tools, from nuclear generators to supersonic transports. They are the sons and daughters of the sodbuster.

It is much less obvious to many that the bereavement counselor is also the sodbuster's heir. It is more difficult for us to see how service technology creates deserts. In fact, there are even those who argue that a good society should scrap its nuclear generators in order to recast them into plowshares of service. They would replace the counterproductive goods technology with the service technology of modern medical centers, universities, correctional systems, and nursing homes. It is essential, therefore, that we have new measures of service technologies that will allow us to distinguish those that are impediments from those that are monuments.

We can assess the degree of impediment incorporated in modern service technologies by weighing four basic elements. The first is the monetary cost. At what point does the economics of a service technology consume enough of the commonwealth that all of the society becomes eccentric and distorted?

Schumacher helped us recognize the radical social, political, and environmental distortions created by huge investments in covering our land with concrete in the name of transportation. Similarly, we are now investing 12 percent of our national wealth in "health-care technology" that blankets most of our communities with a medicalized understanding of well-being. As a result we now imagine that there are mutant human beings called health consumers. We create costly "health-making" environments that are usually large windowless rooms filled with immobile adult bicycles and dreadfully heavy objects purported to benefit one if they are lifted.

The second element to be weighed has been identified by Ivan Illich as "specific counterproductivity." Beyond the negative side effects is the possibility that a service technology can produce the specific inverse of its stated purpose. Thus, one can imagine sickening medicine, stupid-making schools, and crime-making correctional systems.

The evidence grows that some service technologies are now so counterproductive that their abolition would be the most productive means to achieve the goal for which they were initially established. Take, for example, the experiments in Massachusetts, where, under the leadership of Dr. Jerome Miller, the juvenile correctional institutions were closed. As the most recent evaluation studies indicate, the Massachusetts recidivism rate has declined while comparable states with increasing institutionalized populations see an increase in youthful criminality. There is also the unmentionable fact that during doctors' strikes in Israel, Canada, and the United States, the death rate took an unprecedented plunge.

Perhaps the most telling example of specifically counterproductive service technologies is the Medicaid program that provides "health care for the poor." In most states the amount expended for medical care for the poor is now greater than the cash welfare income provided that same population. Thus, a low-income mother is given one dollar in income and a dollar fifty in medical care. It is perfectly clear that the single greatest cause of her ill health is her low income. Nonetheless, the response to her sickening poverty is an ever-growing investment in medical technology—an investment that now consumes her income.

The third element to be weighed is loss of knowledge. Many of the settlers who came to Wisconsin with John Deere's sodbuster had been peasant farmers in

Europe. There they had tilled the land for centuries, using methods that re-plenished its nourishing capacity; however, once the land seemed unlimited and John Deere's technology came to dominate, they forgot the tools and methods that had sustained them for centuries in the old land and created a new desert.

The same process is at work with the modern service technologies and the professions that use them. One of the most vivid examples involves the methods of a new breed of technologists called pediatricians and obstetricians. During the first half of the century these technocrats believed that the preferred method of feeding babies was with a manufactured formula rather than breast milk. Acting as agents for the new lactation technology, these professionals persuaded a gener-ation of women to abjure breast-feeding in favor of their more "healthful" way.

In the 1950s in a Chicago suburb, there was one woman who still remem-bered that babies could be fed by breast as well as by can, but she could find no professional who would advise her to feed by breast. Therefore, she began a search throughout the area for someone who might still remember something about the process of breast-feeding. Fortunately, she found one woman whose memory included the information necessary to begin the flow of milk. From that faint memory breast-feeding began its long struggle toward restoration in our society. These women started a club that multiplied into thousands of small communities and became an international association of women dedicated to breast-feeding: La Leche League. This incredible movement reversed the tech-nological imperative in only one generation and has established breast-feeding as a norm in spite of the countervailing views of the service technologists.

Indeed, it was just a few years ago that the American Academy of Pediatrics finally took the official position that breast-feeding is preferable to nurturing infants from canned products. It was as though the Sauk Indians had recovered the Wisconsin prairie and allowed it once again to nourish a people with popular tools.

The fourth element to be weighed is the "hidden curriculum" of the service technologies. As they are implemented through professional techniques, the invisible message of the interaction between professional and client is, "You will be better because I know better," and as these techniques proliferate across the social landscape, they represent a new praxis, an ever-growing pedagogy that teaches this basic message of the service technologies. Through the propagation of belief in authoritative expertise, the professionals cut through the social fabric of community and sow clienthood where citizenship once grew.

It is clear, therefore, that to assess the purported benefits of service tech-nologies they must be weighed against the sum of the socially distorting mone-tary costs to the commonwealth, the inverse effects of the interventions, the loss of knowledge regarding the natural tools and skills of community, and the anti-democratic consciousness created by a nation of clients. If we weigh these factors, we can begin to recognize how often the techniques of professionalized service make social deserts where communities once bloomed.

Unfortunately, the bereavement counselor is but one of many new profession-alized servicers that plow over our communities like John Deere's sodbusting settlers. These new technologists have now occupied much of the community's space and represent a powerful force for colonizing what remains of social relations. Nonetheless, resistance against this invasion can still be seen in local community struggles against the designs of planners, in parents' unions demand-ing control over their children's education, women's groups struggling to reclaim their medicalized bodies, and community efforts to settle disputes and conflicts by stealing the property claimed by lawyers.

Frequently, as in the case of La Leche League, this decolonization effort is successful. Often, however, the resistance fails and the new service technologies transform communities into social deserts grown over with the scrub of clients and consumers.

This process is reminiscent of the final British conquest of Scotland after the Battle of Culloden. The British were convinced by a history of repeated uprisings that the Scottish tribes would never be subdued. Therefore, after the battle, the British killed many of the clansmen and forced the rest from their small crofts into coastal towns where there was no choice but to emigrate. Great Britain was freed of the tribal threat. The clans were decimated and their lands given to the English lords who grazed sheep where communities once flourished. My Scots' ancestors said of this final solution of the Anglo-Saxons, "They created a desert and called it freedom."

Our modern experience with service technologies tells us that it is difficult to recapture professionally occupied space. We have also learned that whenever such space is liberated, it is even more difficult to construct a new social order that will not be quickly co-opted again. A vivid example is the unfortunate trend develop-ing within the hospice movement. Those who initiated the movement in the United States were attempting to detechnologize dying—to wrest death from the hospital and return it to the family. Only a decade after the movement began, we can see the rapid growth of "hospital-based hospices" and new legislation reim-bursing those hospices that will formally tie themselves to hospitals and employ physicians as central "care givers."

The professional co-option of community efforts to invent appropriate tech-niques for citizens to care in community has been pervasive. Therefore, we need to identify the characteristics of those social forms that are resistant to colonization by service technologies while enabling communities to cultivate and care. These authentic social forms are characterized by three basic dimensions: they tend to be *uncommodified*, *unmanaged*, and *uncurricularized*.

The tools of the bereavement counselor make grief into a *commodity* rather than an opportunity for community. Service technologies convert conditions into commodities and care into service.

The tools of the *manager* convert communality into hierarchy, replacing consent with control. Where once there was a commons, the manager creates a corporation.

The tools of the *pedagogue* create monopolies in the place of cultures. By making a school of every-day life, community definitions and citizen action are degraded and finally expelled.

It is this hard-working team—the service professional, the manager, and the pedagogue—that pulls the tools of "community busting" through the modern social landscape. If we are to recultivate community, we will need to return this team to the stable, abjuring their use.

How will we learn to cultivate community again? It was E. F. Schumacher who concluded that the "guidance we need . . . can still be found in the traditional wisdom." Therefore, we can return to those who understand how to allow the Sauk prairie to bloom and sustain a people. One of the leaders of the Sauk was a chief named Blackhawk. After his people were exiled to the land west of the Mississippi and his resistance movement was broken at the Battle of Bad Axe, Blackhawk said of his Sauk prairie home:

There, we always had plenty; our children never cried from hunger, neither were our people in want. The rapids of our river furnished us with an abundance of excellent fish, and the land, being fertile, never failed to produce good crops of corn, beans, pumpkins and squashes. Here our village stood for more than a hundred years. Our village was healthy and there was no place in the country possessing such advantages, nor hunting grounds better than ours. If a prophet had come to our village in those days and told us that the things were to take place which have since come to pass, none of our people would have believed in the prophecy.

But the settlers came with their new tools and the prophecy was fulfilled. One of Blackhawk's Wintu sisters described the result:

The white people never cared for land or deer or bear. When we kill meat, we eat it all. When we dig roots, we make little holes. When we build houses, we make little holes. When we burn grass for grasshoppers, we don't ruin things. We shake down acorns and pine nuts. We don't chop down trees. We only use dead wood. But the white people plow up the ground, pull down the trees, kill everything. The tree says, "Don't. I am sore. Don't hurt me!" But they chop it down and cut it up.

The spirit of the land hates them. They blast out trees and stir it up to its depths. They saw up the trees. That hurts them. . . . They blast rocks and scatter them on the ground. The rock says, "Don't. You are hurting me!" But the white people pay no attention. When [we] use rocks, we take only little round ones for cooking . . . .

How can the spirit of the earth like the white man? Everywhere they have touched the earth it is sore.

Blackhawk and his Wintu sister tell us that the land has a Spirit. Their community on the prairie was a people guided by that Spirit. When John Deere's people came to the Sauk prairie, they exorcised the prairie Spirit in the name of a new God, technology. Because it was a God of their making, they believed they were Gods themselves. And they made a desert.

There are incredible possibilities if we are willing to fail to be Gods.

—1984

Cathrine Sneed
The Garden Project:
Creating Urban Communities

Cathrine Sneed is the founder and director of the San Francisco Garden
Project, one of the most courageous and unlikely of the many innovative endeavors described by the contributors to this volume. U. S. News
and World Report once headlined an article on her work as "A Garden
That Grows People." The New York Times titled its report on the project
"A Jail's Garden Harvest: Hope and Redemption." Neither exaggerates
the synergistic benefits of Sneed's having brought together her love of
gardens and her deep empathy for people who, because of the hope-
lessness of their lives, are in or recently released from jail.

Sneed works at the crux of human and environmental problems.
She has shown that bringing people out of the jails and off the streets
into the garden can be transformative in both the human and natural
realms. One former inmate has stated that the Garden Project saved his
life. Sneed's work also corroborates several theories many of us want
to believe, but until learning of the Garden Project, we lacked sufficient evidence to do so.

Like Schumacher and many of the voices represented here, Sneed
is living proof that a single individual with, in the words of poet
Adrienne Rich, "no extraordinary power," can make a profound difference through nurturing life and the lives around her. Her own recovery from a life-threatening illness bears witness to the healing
potential of the kind of faith Schumacher had in relevant work and
stewardship. Sneed's project has demonstrated that, given a chance,
even the most disadvantaged of us can change our lives. Those who

hold to the restorative and healing potential of the natural world are vindicated by the results of the Garden Project, as are those who believe that gardening has the potential to bridge the prevailing human alienation from nature.

The English word "garden" is derived from the ancient Persian word for paradise. Sneed thinks that what this world needs is "gardens everywhere." To eyes dulled by the harshness of urban and institutional environments, her Garden Project gives a glimpse of how paradisiacal can seem a patch of tended earth sprouting flowers and vegetables.

I'm honored to have been invited to talk to you about what we are doing at the Garden Project in San Francisco, and I'd like to emphasize the "we." People tend to give me the credit for it, but I'm just one of many who have made our project happen. The project involves people coming together to give what they can give, to do what they can do, to get something done. We don't see enough of that in our society. When I first started the project, I asked myself, What are we trying to do here? I realized that our goal should be "gardens everywhere," because that's what I think this world needs.

I am glad to be speaking in a lecture series named after E. F. Schumacher. I hadn't known anything about him, but before I was invited to speak here, someone sent me his book *Small Is Beautiful*. Ever since I started the project, I have been getting letters, books, notes, little crocheted booties, and all sorts of things from people all over the country. I also get a lot of checks—for two dollars and thirty cents all the way up to a thousand dollars and more. It's just amazing. Whoever sent me the book scribbled a name in it that I couldn't read, so I don't know to this day who gave it to me, but I read it. Even the title itself relates to our Garden Project. We are told that it takes a lot to bring about change: a lot of money and a lot of people. I don't think that's the case, because I've seen a small garden move people in a big way. I've seen our small garden affect what to me is one of the biggest disgraces in this country: our criminal justice system. I've seen the system move a little bit. And I haven't seen that happen anywhere else.

The project has three parts, the first of which I started in San Francisco's county jail. Because I was working within the penal system, the first part of the program was the hardest to get going. Yesterday I was in Vermont; some people from the state Department of Corrections had asked me to come and talk to them about what I'm doing. They told me they have a thousand people in their entire system. I tried to keep from laughing, because there are a thousand people in just one of our jails.

When I first started working for the Sheriff's Department as a law student, I was twenty-one years old. I had just had a baby, and in order to go to law school I started a child-care co-op of mothers, mostly single, like me. I grew up wanting to be a lawyer, yet I was shocked that I was accepted into law school.

One of my professors ran a legal services program for prisoners, and I became infected with his enthusiasm and concern. He struck me as different from the other lawyers I had met. He seemed to care and to understand that the people in our county jail needed the kind of help that people with money routinely got. He also seemed to understand that some of the people in our jail were there just because they were poor.

His outlook impressed me deeply, and it made me change from wanting to be a hot-shot lawyer to realizing that although criminal lawyers can sometimes get people out of jail, they don't deal with why they are there in the first place, and they're usually not allowed to help them stay out of jail; that's not their role. But by working with Michael Hennessey, my professor, I realized that keeping people out of jail was what I wanted to do.

Michael decided that he wanted to be the sheriff of San Francisco County. He ran in the election and won. When he took office, he asked me to come and work with him as a counselor. He felt that was what I had already been doing in the jail as his law student—a few divorces here and there and a few restraining orders but counseling as well. When Michael offered me the job, it not only gave me the chance to do something I wanted to do, it also represented an income for me, a single mother, a way to be able to take care of that little baby and her brother.

So I jumped at the opportunity to work in the jail. But when I had been there long enough to see the same people returning to jail, I began to question whether I was really making a difference. It wasn't enough just to get people a new set of clothes when they left the jail; it wasn't enough just to find out for a mother jailed for prostitution where the police had taken her children. Perhaps because I grew up in the sixties, I thought that's what we were supposed to do: make a difference. Actually, I was motivated more by the example of someone who had dedicated her life to making a difference: my mother, who worked in the civil rights movement.

I began to look at why people were there, to deal with causes, but I didn't have many tools to work with. The women—and I was working primarily with women then—tended to have zero work experience, at least of the legally acceptable kind, and very little educational background. They also had a big "C" for Criminal tattooed on their foreheads. Once released, if they tried to get a job, the application would almost always include the question, "Have you ever been arrested?" If they wrote "yes," they didn't get the job. Who wants to hire someone who's been arrested?

That rejection reinforced their belief that they weren't worth anything, that they didn't belong, that they deserved to be the outlaws everyone said they were. So they would go back to selling drugs, to selling themselves, in order to feed themselves, in order to survive. That was hard on me, physically and emotionally, because I identified with the people in the jail. Most of them looked like me and my family: African-American men and women, men and women of color.

It bothered me that I wasn't able to keep a woman out of jail just by giving her my old suit and talking one of my lawyer friends into hiring her. Even though she had on my suit and even though she had a job, she still had to go home to a hotel where someone was either getting arrested or shooting up. She didn't have a supportive environment; often she didn't even have the resources to get to work. The little things I was doing weren't making a real difference.

At that point I ended up in a hospital, going in and out for almost a year. At the end of the year, after chemotherapy, I wasn't getting better. My doctor, who was very nice, very earnest, and, I had thought, very caring, said, "Well, you're not responding, so you can go home and die, or you can stay here and die." Fortunately for me, the very day he said that, Michael Hennessey and Ray, another friend, came to visit me. Ray brought me a book I had never read, Steinbeck's *The Grapes of Wrath*. In it I found a powerful message: hope lies in the land, and if people who are feeling hopeless can connect to the land, and stay connected, then they will be okay.

I had been noticing the way visitors—my family, my friends—would look at me. I could tell they were thinking, "Oh, she's dying. It's so sad." When Michael and Ray came back to see me, they had that look on their faces. But I was feeling excited and inspired by *The Grapes of Wrath*, and I hopped out of bed. "What's up, Cath?" they asked in surprise. I said, "I want to get out, and I want to start a program to garden with the prisoners at the jail." Michael said, "Okay, if you get better, we'll do that. But you've got to get better first." I said, "I am better! And I'm going home."

I did go home. And I did go to the prisoners, but I could barely walk. This time I noticed something I hadn't really seen when I worked with the prisoners before: I saw that they cared. They were concerned because I looked horrible, and they cared enough to practically carry me, every day through the winter, to the old farm where our garden was to be. As they prepared the area for the garden, they really began to work together, to feel that they were making a difference. They would go back to the jail all excited, and the guards—the "deputies," as we call them—would say, "Shut up and get in your cell." It always amazed me that these prisoners would go back to their cages for the night and then in the morning would be up and ready to go with their new enthusiasm: "Let's go. Let's make a difference." I had never seen that in a prisoner before. I think what I was witnessing was that they had found an activity that meant something to them. Sitting in a cell, they had not had that. I also think they had not had a sense of purpose in their lives either, a feeling that they could do something that mattered.

Very slowly—mostly because I couldn't walk—we began to work: to tear down the old buildings on the farm and to clean up areas where we could plant a few things. Within a couple of years, my illness was in remission.

By then I had a waiting list of prisoners, men and women, who wanted to get into the program. The deputies were scratching their heads and saying, "What are you doing with them out there? They come back all tired out, and they're so cooperative. What's going on?" A few years later, one of my friends who was a deputy (it's a miracle that I became friends with a deputy) told me that earlier the staff had had a betting pool, and everyone put money down on whether or not the prisoners would hurt me—beat me up or rape me. That demonstrates the lack of support we had. The deputies were completely and totally against the Garden Project, and as far as they were concerned, the sheriff was a communist anyway, who hired people who looked like me and was making them do this. They just didn't get it. They thought the prisoners were taking us for a ride.

It's been thirteen years now. Gradually the deputies started to come around. They saw that the prisoners were excited about working together and accomplishing something. Then, when we started bringing in the vegetables we were growing and giving them to the deputies, they got excited too. They finally became supportive. I think they stopped betting on what would happen to me.

For ten years we concentrated on making the program at the jail work. During that period I saw tremendous change in thousands of prisoners. But what began to worry me was what happened to them when they got out of jail. Where did they go? They went back to wherever it was they had been when they were arrested, because they had nowhere else to go. They went back to the hotel rooms that no one should live in. They went back to the housing projects that no one should live in. And then, of course, they began to come back to jail. But the difference was that this time they were excited about being in the program again. That was disconcerting to see. It was very clear to me that the gardening experience had meant so much to them that going back to jail didn't seem so bad. This is an indication of an outrage that America needs to deal with: for many people jail is better than home.

I began to consider what we could do about this situation. Even though I was raising my children, I still managed to make time to look for projects that released prisoners could participate in, because I couldn't bear to see them back in jail.

That was the beginning of the second part of our work. One of the projects was tree-planting. Every Saturday for years, ex-prisoners would come, many of them in their jail clothes because they didn't have anything else, and they would plant trees with me for a volunteer citizen organization in San Francisco called

Friends of the Urban Forest. They would come and work very hard. Planting those trees gave them an opportunity to do something, to make a contribution. During the week a lot of them would bring family members and proudly show the trees they had planted.

The problem was that many of them couldn't get to the tree-planting site because they didn't have money for the bus. At that time all we were able to give each prisoner released from our jail was eighty-five cents for one bus fare. That was all most of them had when they left. While they were in jail, they lost whatever possessions they might have had because if they had been living in a hotel, their belongings would be thrown away as soon as they didn't come back. Or if they had been living with their families, whatever resources the prisoners had accumulated were used up because their families were poor.

It became clear to me that we had to find a way to help them get to the planting sites. I started off by giving out eighty-five cents for bus fare from my own pocket, but that didn't last long, because I didn't have enough in my pocket. Then, with the help of some business people and again with the help of Michael Hennessey, we were able to give out bus money on a regular basis, expanding the tree-planting program so we could plant on Saturdays too, and then we would get together and weed on Tuesdays, just as a way to connect with one another.

The expanded program was very successful, but it still wasn't enough. So again I spent a lot of time talking about what we had done at the jail, trying to win financial support for our work. One person I spoke with knew of a vacant lot behind his bakery. We rushed over there to look at it, and sure enough, there was a lot filled with about six dump-truck loads of garbage. Because I felt we had enough to do already, I wasn't immediately excited by this garbage dump. Instead, I persisted in asking this very nice man just to write us a check so we could have more bus money to expand the program. But he kept pointing to the empty garbage lot, and he said something I wouldn't have thought of: "Grow something, and I'll buy what you grow. Then you'll have the money." I said, "Okay, I can do that."

So we cleaned up the garbage dump. We started planting. And now we have a beautiful garden that's about half an acre (remember, small is beautiful). Last summer we had 125 people working on this half acre, which meant that every head of lettuce got stroked quite a bit. There were absolutely no weeds there, mountains of compost, and probably the most pampered spinach you've ever met.

I remember the way we hurried around trying to make sure Alice Waters got the spinach she had ordered for her restaurant on the day and at the time she specified. And I remember talking to a young man one day who was picking spinach, and he was patting the spinach and arranging it in a box. Finally I said,

"Look, we have to get this to the restaurant right away. Just put it in the box. Let's go!" But at the same time I realized what was happening: this young man was producing something that was valuable, that looked beautiful, that was going somewhere, and some money was going to come back in return. He knew he was beginning to make himself and his family self-sufficient. He also knew that before patting the spinach he had been selling crack. Now he had made a conscious choice: he was going to sell spinach, not crack.

I believe small is not only beautiful; small is necessary. I suggest a sequel with that title to Schumacher's book. I say that because I don't have thousands and thousands of dollars, I don't have a fancy office in a fancy place and a great big area to grow enough produce to supply the restaurants that want our food. What I do have is a lot of people who say, "I want a chance. I want to work. I want to work *here*." I know this is true not only in San Francisco but in other places as well. The idea that there isn't enough work for people or the other idea that the kind of people I'm working with don't want to work is just plain wrong.

It's stupid for us to spend the amount of money we do to keep young people in jail instead of preparing them for a job. They certainly didn't have the opportunity to get an education at a place like Yale right here in New Haven. I am sure we are spending more to keep them in jail than parents pay to send their children to Yale. It's stupid.

I said that small is necessary. I think it's necessary for us to begin with what we have: a lot of land and a lot of people. Seems to me we can put them together. The lunch we were served earlier as part of today's program, made from locally and organically grown food, is what schoolkids as well as grownups should eat all the time. That kind of meal should be the rule rather than the exception. I had to take a plane here; the food that was served was unrecognizable. You look at that lettuce and you have to ask, "What is this?" The people working with me—some of them with sixty-five, seventy criminal convictions—can do better than that. They can grow beautiful spinach and lettuce; they can feed people. And while they're feeding people, they're feeding themselves, and they're creating new communities. They're creating hope in their communities.

Now about the second part of our program. The year after we started the garden for released prisoners, the next step was signing a contract with the city; now our people are paid for planting trees. They graduate from the garden to the Tree Corps.

In the four years since we started the garden, we've hired six hundred people—I have files for them all. It's the first time many of them have ever had a Social Security card. One young man was actually very upset when he got his first check; he wanted to know who FICA was and why his money was being taken. The shame of it is that this means he had never seen a paycheck before. It means that the only money he knew was probably his mother's welfare check and her mother's welfare check. How do we break the cycle?

My file drawer represents six hundred people I have to worry about. I have to wonder where they are. I have ten on salary now, and those ten are paid for only two hours of work a day. The funds come from a block grant of around four thousand dollars that has to be divided ten ways. That's not a lot of money. The drug dealers can offer a lot more. It's just a matter of time before they'll go back to those dealers. They have to eat, the same way you do. That's not to condone drug dealing; it's to be realistic.

Yesterday I read an article in *The New York Times* about the Million Man March. A young African-American man was quoted as saying, "What we need is guidance." So often what we read about African-American men is negative, but in this case it was something positive. It's interesting he said that, because last Thursday one of the people in my program said to me, "I just need guidance." He is a sweet boy who at eighteen has been in jail maybe twelve times. Do you know how much money that costs? I don't. No one really does. We only know how much it costs to feed him, clothe him, and keep him in jail, but figures aren't available for how much it costs for the cop to arrest him or for the court proceedings or for the probation officer to write a report, so we don't really know how much it's costing us.

This boy said, "I need guidance." It broke my heart, because I thought to myself, "You're right. You do need guidance. And how are you going to get it here in jail? Who's going to give it to you when you leave the jail? I can't because I have only two hours' pay for you. In two hours you're not going to get the guidance you need to say to that drug dealer, 'No, man, I don't need to sell that stuff.' "

I have to think of all those like him who are regarded by America as criminal. America wants to believe they are somehow "other." There have been many movements throughout history when people have thought of a person or a group as something so terrible that they felt justified in saying, "We'll have to get rid of them." That's what's happening, I think, to all the men and women who are sent to our jails. We're saying, "They are other, they are criminal, they are not us." It's three strikes and they're out.

Small is necessary. I think we have to start with that. When I go home, I have to start with Rashawn. On Monday I'll say, "Rashawn, even though I can't pay you for more than two hours' work a day, at least you know that if you come to the garden, that's two hours when you're not out on the street. And you could volunteer for us for the other six hours, because when you go home you're not getting more money anyway. You might as well just come and stay with us for the day." Unfortunately, if he does, he'll get hungry at some point, and then I'll have to feed him, but the good thing about having a garden is that you can have salad, and we do eat a lot of that. I routinely make lunch for my workers, and they say, "Oh-oh, we're having salad again," and they say, "Oh, vegetarian," but they also say, "Hey, I didn't know vegetables tasted like this. I didn't know you could eat that."

People always ask me why our garden doesn't get vandalized. I know it's because the young people in the neighborhood where the garden is don't know that vegetables are growing there. They don't know this is spinach, they don't know that is lettuce, because they think food comes from the Safeway supermarket in California.

I believe people use drugs because their body is craving something they aren't getting from anywhere else. Over the seventeen years I've worked for the Sheriff's Department, I always ask people, "What were you eating when you were arrested?" Then they look at me as though I'm crazy, and they say, "What do you mean, what was I eating? I was eating whatever I could get." Or, "I was eating at McDonald's." What they can get at McDonald's is a lot of potato chips, a lot of HoHos. I believe I know why people go on crack and from crack to heroin: it's because the poor of America, the poor in urban communities, don't eat real food, and so they crave, and they try to ease that craving. Of course, some people might say, "Then why are there rich kids on drugs?" I don't know. Although again I would guess that it's a similar craving, except that it's not physical; it's in here, in the heart. There's something missing in here.

Educating is the third part of the Garden Project, first of all teaching nutrition and exposing people to good food. I talk to people about what they've been eating. I say, "Think about it. You were eating junk food when you got on crack; now you're not eating that mess, and you feel better. We have good food here. Try it." I think I'm respected enough now so that they do try it. In the old days they just laughed. That was probably because I don't have a fancy classroom where I can teach nutrition and parenting.

Most of the people in our program are young parents. They learn how to be parents from what they observed as children. Unfortunately, if they weren't well cared for, if they were mistreated, then the pattern is likely to repeat itself when they become parents. So instead of saying to them, "Oh, don't hit those kids, it's bad for their development," I say, "Bring your kids with you to the garden. Someone here will take care of them, and then you won't feel so frustrated. You won't feel as though you can't do anything. Here's a way to get a break from your kids." So another parent takes care of their kids for a little while. That cooperation is important, because I can't hire a child-care worker.

We're also trying to use the garden to teach people how to live. And I really believe it's working. But I often think it's not enough. Until others start doing what we're doing, we're in big trouble. It's not only the environment that's in trouble, as we hear all the time; it's also people who are in trouble. And it's not just poor black people, because if poor black people are in trouble, then we all are in trouble, because despite what some may think, we're all connected, and we are a community in more ways than one.

I believe we must say, "We're going to start very small, and we're going to do what we can with what we've got." We start by using examples that are good and building on them.

People want to know where I get my inspiration. They ask how I know that what I'm doing works. The answer is because I am recreating what I have known all my life: family love. I come from a family of fourteen children. And with those thirteen sisters and brothers, I always know somebody cares about me. I really believe that what draws my students to our garden is not the little bit of money they make but the feeling they get working there that they're part of a family where everybody is lifting them up instead of putting them down.

My two children also keep me going. Whenever I look at them, I feel inspired. There are times when I wonder why I am doing this work. Deep down, I know I'm doing it because my son and daughter are just like the sons and daughters in our jail. The only difference is, my children have me, and those kids don't have anybody. What we're trying to do, what I know I have to do, is help them to see that they can have a supportive family, and it doesn't have to be blood relatives.

Schumacher says in his book that the greatest resource is education. What we're really doing in the Garden Project is educating. Showing people that there's another way, showing them that we are family. They can come to the garden and feel safe, feel encouraged, because someone is there for them.

Small is beautiful.

—1995

Part III   Toward a New Era in Human-Earth Relations

Thomas Berry
The Ecozoic Era

The following section of the book takes its title from Thomas Berry's call for "a new era in human-Earth relations." As is clear from the preceding lectures, such a renewal is the key not only to human survival but to the attainment of wisdom. In this area there is no more valued teacher than Thomas Berry.

It has been his inestimable contribution to build a philosophical bridge between ecology and traditional Christianity. He has written: "The natural world is the maternal source of our physical, emotional, aesthetic, moral, and religious existence. The natural world is the larger sacred community to which we belong." In thus initiating a greater understanding of the longstanding breach between people of European extraction—the originators of the industrial revolution and therefore the perpetrators of the ecological crisis—and the natural world, Berry points toward a path that can lead to mutual healing.

A historian of culture, a philosopher and scholar, a Passionist priest and self-designated "geologian," Thomas Berry is the founder of the Riverside Center for Religious Research. He played an integral role in the creation of physicist Brian Swimme's epic video *The Universe Story: A Celebration of the Unfolding of the Cosmos*, which has proved invaluable in teaching the unique and sacred unfolding of life on the planet and in the universe beyond. His 1988 book *The Dream of the Earth* brings the same message to its readers.

He begins his Schumacher lecture with a geological overview of planetary evolution leading up to the present. It is his thesis that in our postmodern period we have lost the sense of our beginnings, of our mythos, of who we are in this life and in the Cosmos. "Whenever we forget this Great Story," he says, "we become confused." As he travels the world with his message, there is no more effective interpreter of the Great Story than Thomas Berry.

It is indeed a high honor to be with you today and to discuss the significance of these terminal decades of the century, which are also the terminal decades of the millennium. Far beyond any of these in its significance is the terminal phase of the Cenozoic Era of Earth history, in which we presently find ourselves. In these fateful years we are terminating sixty-five million years in the biological history of the planet. It is most important that we appreciate the order of magnitude of what is happening in our times.

Lewis Mumford has been mentioned here today in commemoration of his career as our foremost cultural historian in the twentieth century. He extended the horizons of our vision to include a vast range of human cultural development. In doing this he was extremely sensitive to the rootedness of human affairs in the geological and biological systems of the planet. This perception we now need to extend beyond anything that he could envisage in his day.

The changes presently taking place in human and earthly affairs are beyond any parallel with historical change or cultural modification as these have occurred in the past. This is not like the transition from the classical period to the medieval period or from the medieval to the modern period. These changes reach far beyond the civilizational process, beyond even the human process, into the biosystems and even the geological structures of the Earth itself.

There are only two other moments in the history of this planet that offer us some sense of what is happening. These two moments are the end of the Paleozoic Era 220 million years ago, when some 90 percent of all species living at that time were extinguished, and the terminal phase of the Mesozoic Era sixty-five million years ago, when there was also very extensive extinction.

Then, in the emerging Cenozoic Era the story of life on this planet flowed over into what could be called the lyric period of Earth history. The trees had come before this, the mammals already existed in a rudimentary form, the flowers had appeared perhaps thirty million years earlier. But in the Cenozoic Era, there was wave upon wave of life development, with the flowers, the birds, the trees, and the mammalian species particularly all leading to that luxuriant display of life upon Earth such as we have known it.

In more recent times, during the past million years this region of New England went through its different phases of glaciation, also its various phases of life development. New England's trees especially developed a unique grandeur. Possibly no other place on Earth has such color in its fall foliage as this region. It was all worked out during these past sixty-five million years. The songbirds we hear also came about in this long period.

Then we, the human inhabitants of the Earth, came into this region with all the ambivalences we bring with us. Not only here but throughout the planet we have become a profoundly disturbing presence. In this region and to the north in southern Quebec, the native maple trees are dying out in great numbers due to pollutants we have put into the atmosphere, the soil, and the water.

Their demise is largely a result of the carbon compounds we have loosed into the atmosphere through the use of fossil fuels, especially of petroleum, for our fuel and energy. Carbon is, as you know, the magical element. The whole life structure of the planet is based upon the element carbon. So long as the life process is guided by its natural patterns, the integral functioning of the Earth takes place. The wonderful variety expressed in marine life and land life, the splendor of the flowers and the birds and animals—all these could expand in their gorgeous coloration, in their fantastic forms, in their dancing movements, and in their songs and calls that echo over the world.

To accomplish all this, however, nature must find a way of storing immense quantities of carbon in the petroleum and coal deposits, also in the great forests. This process was worked out over some hundreds of millions of years. A balance was achieved, and the life systems of the planet were secure in the interaction of the air and the water and the soil with the inflowing energy from the sun.

But then we discovered that petroleum could produce such wonderful effects. It can be made into fertilizer to nourish crops; it can be spun into fabrics; it can fuel our internal combustion engines for transportation over the vast highway system we have built; it can produce an unlimited variety of plastic implements: it can run gigantic generators and produce power for lighting and heating of our buildings.

It was all so simple. We had no awareness of the deadly consequences that would result from the residue from our use of petroleum for all these purposes. Nor did we know how profoundly we would affect the organisms in the soil with our insistence that the patterns of plant growth be governed by artificial human demands met by petroleum-based fertilizers rather than by the spontaneous rhythms within the living world. Nor did we understand that biological systems are not that adaptable to the mechanistic processes we imposed upon them.

I do not wish to dwell on the devastation we have brought upon the Earth but only to make sure we understand the nature and the extent of what is happening. While we seem to be achieving magnificent things at the microphase level of our functioning, we are devastating the entire range of living beings at the macrophase level. The natural world is more sensitive than we have realized. Unaware of what we have done or its order of magnitude, we have thought our achievements to be of enormous benefit for the human process, but we now find that by disturbing the biosystems of the planet at the most basic level of their functioning, we have endangered all that makes the planet Earth a suitable place for the integral development of human life itself.

Our problems are primarily problems of macrophase biology. Macrophase biology, the integral functioning of the entire complex of biosystems of the planet, is something biologists have given almost no attention. Only with James Lovelock and some other more recent scientists have we even begun to think about this larger scale of life functioning. The delay is not surprising, for we are caught in the microphase dimensions of every phase of our human endeavor. This is true in law and medicine and in the other professions as well as in biology.

Macrophase biology is concerned with five basic spheres: land, water, air, life—and how these interact with one another to enable the planet Earth to be what it is—and a very powerful sphere: the human mind. Consciousness is certainly not limited to humans. Every living being has its own mode of consciousness. We must be aware, however, that consciousness is an analogous concept. It is qualitatively different in its various modes of expression. Consciousness can be regarded as the capacity for intimate presence of things to one another through knowledge and sensitive identity. But obviously the consciousness of a plant and the consciousness of an animal are qualitatively different, as are the consciousness of insects and the consciousness of birds or fish. Similarly, there is a difference in consciousness between fish and human: for the purposes of the fish, human modes of consciousness would be more a defect than an advantage. So too, tiger consciousness would be inappropriate for the bird.

It is also clear that the human mode of consciousness is capable of unique intrusion into the larger functioning of the planetary life systems. So powerful is this intrusion that the human has established an additional sphere that might be referred to as a technosphere, a way of controlling the functioning of the planet for the benefit of the human at the expense of the other modes of being. We might even consider that the technosphere in its subservience to industrial-commercial uses has become incompatible with the other spheres that constitute the basic functional context of the planet.

The biggest single question before us in the 1990s is the extent to which this technological-industrial-commercial context of human functioning can be made compatible with the integral functioning of the other life systems of the planet. We are reluctant to think of our activities as inherently incompatible with the integral functioning of the various components of the planetary systems. It is not simply a matter of altering our ways of acting on a minor scale by recycling (which presupposes a cycling that is devastating in its original form), by mitigating pollution, reducing our energy consumption, limiting our use of the automobile, or by fewer development projects. Our efforts will be in vain if our purpose is to make the present industrial system acceptable. These steps must be taken, but according to my definition of the Ecozoic Era there must be more: there must also be a new era in human-Earth relations.

Our present system, based on the plundering of the Earth's resources, is certainly coming to an end. It cannot continue. The industrial world on a global scale, as it functions presently, can be considered definitively bankrupt. There is no way out of the present recession within the context of our existing commercial-industrial processes. This recession is not only a financial recession or a human recession even. It is a recession of the planet itself. The Earth cannot sustain such an industrial system or its devastating technologies. In the future the industrial system will have its moments of apparent recovery, but these will be minor and momentary. The larger movement is toward dissolution. The impact of our present technologies is beyond what the Earth can endure.

Nature has its own technologies. The entire hydrological cycle can even be regarded as a huge engineering project, a project vastly greater than anything humans could devise with such beneficent consequences throughout the life systems of the planet. We can differentiate between an acceptable human technology and an unacceptable human technology quite simply: an acceptable one is compatible with the integral functioning of the technologies governing the natural systems; an unacceptable one is incompatible with the technologies of the natural world.

The error has been to think that we could distort the natural processes for some immediate human benefit without incurring immense penalties, penalties that might eventually endanger the well-being of the human as well as that of most other life forms. This is what has happened in the twentieth-century petroleum economy we have developed.

The petroleum at the base of our present industrial establishment might at its present rate of use last another fifty years—probably less, possibly more. But a severe depletion will occur within the lifetime of young people living today. The major part of the petroleum will be gone. Our youngest children may see the end of it. They will likely see also the tragic climax of the population expansion. And with the number of automobiles on the planet estimated at six hundred million in the year 2000, we will be approaching another saturation level in the technological intrusion into the planetary process.

It is awesome to consider how quickly events of such catastrophic proportions are happening. When I was born in 1914, there were only one and a half billion people in the world. Children of the present will likely live to see ten billion. The petrochemical age had hardly begun in my early decades. Now the planet is saturated with the residue from spent oil products. There were fewer than a million automobiles in the world when I was born. In my childhood the tropical rain forests were substantially intact; now they are devastated on an immense scale. The biological diversity of life forms was not yet threatened on an extensive scale. The ozone layer was still intact.

In evaluating our present situation, I submit that we have already terminated the Cenozoic Era of the geo-biological systems of the planet. Sixty-five million years of life development are terminated. Extinction is taking place throughout the life systems on a scale unequaled since the terminal phase of the Mesozoic Era.

A renewal of life in some creative context requires that a new biological period come into being, a period when humans would dwell upon the Earth in a mutually enhancing manner. This new mode of being of the planet I describe as the Ecozoic Era, the fourth in the succession of life eras thus far identified as the Paleozoic, the Mesozoic, and the Cenozoic. But when we propose that an Ecozoic Era is succeeding the Cenozoic, we must define the unique character of this emergent era.

I suggest the name "Ecozoic" as a better designation than "Ecological." Eco-logos refers to an understanding of the interaction of things. Eco-zoic is a more biological term that can be used to indicate the integral functioning of life systems in their mutually enhancing relations.

The Ecozoic Era can be brought into being only by the integral life community itself. If other periods have been designated by such names as "Reptilian" or "Mammalian," this Ecozoic period must be identified as the Era of the Integral Life Community. For this to emerge, there are special conditions required on the part of the human, for although this era cannot be an anthropocentric life period, it can come into being only under certain conditions that dominantly concern human understanding, choice, and action.

When we consider the conditions required of humans for the emergence of such an Ecozoic Era in Earth history, we might list these as follows:

The first condition is to understand that the universe is a communion of subjects, not a collection of objects. Every being has its own inner form, its own spontaneity, its own voice, its ability to declare itself and to be present to other components of the universe in a subject-to-subject relationship. Whereas this is true of every being in the universe, it is especially true of each component member of the Earth community. Each component of the Earth is integral with every other component. This is also true of the living beings of the Earth in their relations with one another.

The termination of the Cenozoic Era of Earth history has been brought about by the incapacity of humans in the industrial cultures to be present to the Earth and its various modes of being in some intimate fashion. Ever since the time of Descartes in the first half of the seventeenth century, Western humans, in their dominant life attitudes, have been autistic in relation to the non-human components of the planet. Whatever the abuse of the natural world by humans prior to that time, the living world was recognized until then in its proper biological functioning as having an "anima," a soul. Every living being was by definition an ensouled being, with a voice that spoke to the depths of the human of wondrous and divine mysteries, a voice that was heard quite clearly by the poets and musicians and scientists and philosophers and mystics of the world, a voice heard also with special sensitivity by the children.

Descartes, we might say, killed the Earth and all its living beings. For him, the natural world was mechanism. There was no possibility of entering into a communion relationship. Western humans became autistic in relation to the surrounding world. There could be no communion with the birds or animals or plants, because these were all mechanical contrivances. The real value of things was reduced to their economic value. A destructive anthropocentrism came into being.

This situation can be remedied only by a new mode of mutual presence between the human and the natural world, with its plants and animals of both the sea and the land. If we do not get that straight, then we cannot expect any significant remedy for the present distress experienced throughout the Earth. This capacity for intimate rapport also needs to be extended to the atmospheric phenomena and the geological structures and their functioning.

Because of this autism, my generation never heard the voices of that vast multitude of inhabitants of the planet. They had no communion with the non-human world. They would go to the seashore or to the mountains for some recreation, a moment of aesthetic joy. But this was too superficial to establish any true reverence or intimate rapport. No sensitivity was shown to the powers inherent in the various phenomena of the natural world, no depth of awe that would have restrained their assault on the natural world in order to extract from it some human advantage—even if this meant tearing to pieces the entire fabric of the planet.

The second condition for entering the Ecozoic Era is a realization that the Earth exists, and can survive, only in its integral functioning. It cannot survive in fragments any more than any organism can survive in fragments. Yet the earth is not a global sameness. It is a differentiated unity and must be sustained in the integrity and interrelations of its many bioregional contexts. This inner coherence of natural systems requires an immediacy of any human settlement with the life dynamics of the region. Within this region the human right to habitat must respect the right to habitat possessed by the other members of the life community. Only the full complex of life expression can sustain the vigor of any bioregion.

A third condition for entering the Ecozoic Era is recognition that the Earth is a one-time endowment. We do not know the quantum of energy contained in the Earth, its possibilities or its limitations. We must reasonably suppose that the Earth is subject to irreversible damage in the major patterns of its functioning and even to distortions in its possibilities of development. Although there was survival and further development after the great extinctions at the end of the Paleozoic and the Mesozoic Eras, life was not so highly developed as it is now. Nor were the very conditions of life at those times negated by such changes as we have wrought through our toxification of the planet.

Life on Earth will surely survive the present decline of the Cenozoic, but we do not know at what level of its development. The single-cell life forms, the insects, the rodents, the plants, and a host of other forms of life found throughout the planet—these will surely survive. But the severity of the damage to the rain forests, to the fertility of the soils, to species diversity, and to the chances for survival of the more developed animals; the consequences throughout the animal world of the diminishment of the ozone shield; the extension of deserts; the pollution of the great freshwater lakes; the chemical imbalance of the atmosphere—all are signs of disturbance on a scale that might make restoration to their earlier grandeur impossible, certainly within any time frame that is conceivable to human modes of thinking or planning. Almost certainly we have witnessed in these past centuries a grand climax in the florescence of the Earth.

A fourth condition for entering the Ecozoic Era is a realization that the Earth is primary and humans are derivative. The present distorted view is that humans are primary and the Earth and its integral functioning only a secondary consideration—thus the pathology manifest in our various human institutions. The only acceptable way for humans to function effectively is by giving first consideration to the Earth community and then dealing with humans as integral members of that community. The Earth must become the primary concern of every human institution, profession, program, and activity, including economics. In economics the first consideration cannot be the human economy, because the human economy does not even exist prior to the Earth economy. Only if the Earth economy is functioning in some integral manner can the human economy be in any way effective. The Earth economy can survive the loss of its human component, but there is no way for the human economy to survive or prosper apart from the Earth economy. The absurdity has been to seek a rising Gross National Product in the face of a declining Gross Earth Product.

This primacy of the Earth community applies also to medicine and law and all the other activities of humans. It should be especially clear in medicine that we cannot have well humans on a sick planet. Medicine must first turn its attention to protecting the health and well-being of the Earth before there can be any effective human health. So in jurisprudence, to poise the entire administration of justice on the rights of humans and their limitless freedom to exploit the natural world is to open the natural world to the worst predatory instincts of humans. The prior rights of the entire Earth community need to be assured first; then the rights and freedoms of humans can have their field of expression.

A fifth condition for the rise of the Ecozoic Era is to realize that there is a single Earth community. There is no such thing as a human community in any manner separate from the Earth community. The human community and the natural world will go into the future as a single integral community, or we will both experience disaster on the way. However differentiated in its modes of expression, there is only one Earth community—one economic order, one health system, one moral order, one world of the sacred.

As I present this outline of an emerging Ecozoic Era, I am quite aware that such a conception of the future, when humans would be present to the Earth in a mutually enhancing manner, is mythic in its form, just as such conceptions as the Paleozoic, Mesozoic, and Cenozoic are mythic modes of understanding a continuing process, even though this continuing process is marked by an indefinite number of discontinuities amid the continuity of the process itself.

My effort here is to articulate the outlines of a new mythic form that would evoke a creative entrancement to succeed the destructive entrancement that has taken possession of the Western soul in recent centuries. We can counter one entrancement only with another, a counterentrancement. Only thus can we evoke the vision as well as the psychic energies needed to enable the Earth community to enter successfully upon its next great creative phase. The grandeur of the possibilities ahead of us must in some manner be experienced in anticipation. Otherwise we will not have the psychic energy to endure the pain of the required transformation.

Once we are sufficiently clear as to where we are headed and once we experience the urgency and the adventure of what we are about, we can get on with our historic task. We can accept and even ignore the difficulties to be resolved and the pain to be endured, for we are involved in a great work. In creating such a great work, the incidentals fall away. We can accept the pathos of our times, the sorrow that we will necessarily go through. We can, I think, assist the next generation as they take up this creative effort, mainly by indicating just where they can receive their instructions. It is the role of elders at the present time to assist them in fulfilling their role in this moment of transformation. Elders. We have a lot of older people but few elders. Tribal people, for their part, depend on elders for their instructions. I was privileged to see this process at work some years ago when I was invited to participate at a meeting of indigenous Indian peoples—mostly Ojibwa, Cree, and Six Nations—on Cape Croker along Georgian Bay in northwest Ontario. The purpose of the meeting was to consider the future and the direction their lives should take.

I hope we will be able to guide and inspire our next generation as they attempt to shape the future. Otherwise they will simply survive with all their resentments amid the destroyed infrastructures of the industrial world and the ruins of the natural world itself. The challenge itself is already predetermined. There is no way for the new generation to escape this confrontation. The task to which they are called and the destiny that is before them are, however, not simply theirs alone. The human is linked to every earthly being, to the entire planet. The whole universe is involved. The successful emergence of the Ecozoic Era can presently be considered the great creative task of the universe itself.

This destiny can be understood, however, only in the context of the Great Story of the universe. All peoples derive their understanding of themselves from their account of how the universe originally came into being, how it came to be as it is, and the role of the human in the story. We in our Euro-American traditions have in recent centuries, through our observational studies, created a new story of the universe. The difficulty is that this story was presented in the context of the mechanistic way of thinking about the world and so has been devoid of meaning. Supposedly, everything has happened in a random, meaning-less process.

It is little wonder, then, that we have lost our Great Story. Our earlier Genesis story long ago lost its power over our historical cultural development. Our new scientific story has never carried any depth of meaning. We have lost our rever-ence for the universe and the entire range of natural phenomena.

Our scientific story of the universe has no connection with the natural world as we experience it in the wind and the rain and the clouds, in the birds, the animals, and the insects we observe around us. For the first time in human history the sun and moon and stars, the fields and mountains and streams and woodlands fail to evoke a sense of reverence before the deep mystery of things. These wondrous components of the natural world are somehow not seen with any depth of appreciation. Perhaps that is why our presence has become so deadly.

But now all this is suddenly being altered. Shocked by the devastation we have caused, we are awakening to the wonder of a universe never before seen in quite the same manner. No one ever could tell in such lyric language as we can now the story of the primordial flaring forth of the universe at the beginning, the shaping of the immense number of stars gathered into galaxies, the collapse of the first generation of stars to create the ninety-some elements, the gravitational gather-ing of scattered stardust into our solar system with its nine planets, the formation of the Earth with its seas and atmosphere and the continents crashing and rifting as they move over the asthenosphere, and the awakening of life.

Such a marvel is this fifteen billion year process: such infinite numbers of stars in the heavens and living beings on Earth, such limitless variety of flowering species and forms of animal life, such tropical luxuriance, such magnificent scenery in the mountains, and such springtime wonders as occur each year. Now we are experiencing the pathos of witnessing the desecration of this sublimity.

We now need to tell this story, meditate on it, and listen to it as it is told by every breeze that blows, by every cloud in the sky, by every mountain and river and woodland, and by the song of every cricket. We have lost contact with our story. Yet we can come together, all the peoples of Earth and all the various members of the great Earth community, only in this Great Story, the story of the universe. For there is no human community without the human community story, no Earth community without the Earth story, and no universe community without the universe story. These three constitute the Great Story. Without it, the various forces of the planet become mutually destructive rather than mutually coherent.

We need to listen to one another's way of telling the Great Story. But first we in the West, with our newly developed capacity to observe the universe through our vast telescopes and to hear its sounds as these come to us from the beginning of time and over some billions of years, need really to listen to this story as our own special way of understanding and participating in the Great Story.

Whenever we forget our story we became confused. But the winds and the rivers and the mountains never become confused. We must go to them constantly to be reminded of it, for every being in the universe is what it is only through its participation in the story. We are resensitized whenever we listen to what they are telling us. Long ago they told us that we must be guided by a reverence and a restraint in our relations with the larger community of life, that we must respect the powers of the surrounding universe, that only through a sensitive insertion of ourselves into the great celebration of the Earth community can we expect the support of the Earth community. If we violate the integrity of this community, we will die.

The natural world is vast and its lessons fearsome. One of the most ominous expressions of the natural world has to do with nuclear energy. When we go deep into the natural world and penetrate the inner structure of the atom and in a sense violate that deepest mystery for trivial or destructive purposes, we may get power, but nature throws at us its most deadly consequences. We are still helpless with regard to what to do once we have broken into the mysterious recesses of nuclear power. Forces have been let loose far beyond anything we can manage.

Earlier I mentioned five conditions for the integral emergence of the Ecozoic

Era. Here I would continue with a sixth condition: that we understand fully and respond effectively to our own human role in this new era. For while the Cenozoic Era unfolded in its full splendor entirely apart from any role fulfilled by the human, almost nothing of major significance is likely to happen in the Ecozoic Era that humans will not be involved in. The entire pattern of Earth's functioning is being altered in this transition from the Cenozoic to the Ecozoic. We did not even exist until the major developments of the Cenozoic were complete. In the Ecozoic, however, the human will have a pervasive influence on almost everything that happens. We are approaching a critical watershed in the entire modality of Earth's functioning. Our positive power of creativity in the natural life systems is minimal; our power of negating is immense. Whereas we cannot make a blade of grass, there is liable not to be a blade of grass unless it is accepted, fostered, and protected by the human. Protected mainly from ourselves so that the Earth can function from within its own dynamism.

There is, finally, the question of language. A new language, an Ecozoic language, is needed. Our late Cenozoic language is radically inadequate. The human mode of being is captured and destroyed by our present univalent, scientific, literal, unimaginative language. We need a multivalent language, one much richer in the symbolic meanings that language carried in its earlier forms when the human lived deep within the natural world and the entire range of Earth phenomena. As we recover this early experience in the emerging Ecozoic Era, all the archetypes of the collective unconscious will attain a new validity as well as new patterns of functioning, especially in our understanding of the death-rebirth symbol and the symbols of the heroic journey, the Great Mother, the tree of life.

Every reality in the natural world is multivalent. Nothing is univalent. Everything has a multitude of aspects and meanings, the way sunlight carries within itself warmth and light and energy. Sunlight is not a single thing. It awakens the multitude of living forms in the springtime; it awakens poetry in the soul and evokes a sense of the divine. It is mercy and healing, affliction and death. Sunlight is irreducible to any scientific equation or any literal description.

But all these meanings are based on the physical experience of sunlight. If we were deprived of sunlight, the entire visible world would be lost to us and eventually immense realms of consciousness and all of life. We would be retarded in our inner development in proportion to our deprivation of the experience of natural phenomena, of mountains and rivers and forests and seacoasts and all their living inhabitants. The natural world itself is our primary language, our primary scripture, our primary awakening to the mysteries of existence. We might well put all our written scriptures on the shelf for twenty years until we learn what we are being told by unmediated experience of the world about us.

So too we might put Webster on the shelf until we revise the language of all our professions, especially law, medicine, and education. In ethics we need new words such as biocide and geocide, words that have not yet been adopted into the language. In law, we need to define society in terms that include the larger community of living beings of the bioregion, of the Earth, and even of the universe. Certainly human society separated from such contexts is an abstraction. Life, liberty, habitat, and the pursuit of happiness are rights that should be granted to every living creature, each in accord with its own mode of being.

I conclude with a reference to the Exodus symbol, which has exercised such great power over our Western civilization. Many peoples came to this country believing they were leaving a land of oppression and going to a land of liberation. We have always had a sense of transition. Progress supposedly is taking us from an undesirable situation to a kind of beatitude. So we might think of the transition from the terminal Cenozoic to the emerging Ecozoic as a kind of Exodus out of a period when humans are devastating the planet to a period when humans will begin to live on the Earth in a mutually enhancing manner.

There is a vast difference, however, in the case of this present transition, which is one not simply of the human but of the entire planet—its land, its air, its water, its biosystems, its human communities. This Exodus is a journey of the Earth entire. It is my hope that we will make the transition successfully. Whatever the future holds for us, however, it will be an experience shared by humans and every other earthly being. There is only one community, one destiny.

—1991

Hunter G. Hannum

Wagner and the Fate of the Earth:
A Contemporary Reading of *The Ring*

The essay that follows ventures into a realm of myth and high culture
largely overlooked by the others in this volume. Taking his cue from
visionary Thomas Berry, teacher, critic, and German translator Hunter
Hannum turns his attention to Richard Wagner's four-part opera *The
Ring of the Nibelung*. A lifelong scholar of German literature, he co-edited
the anthology *Modern German Drama*, widely used in American universities, and with Schumacher Society board member Hildegarde Hannum has translated many works, including *Unwritten Memories*, an informal autobiography of Katia Mann, the wife of Thomas Mann, and
several books by Switzerland-based psychotherapists Alice Miller and
Arno Gruen.

Interpreting *The Ring* as a modern allegory, Hannum reexamines it
as a possible source for forms of mythic healing and wholeness that
might balance the destructiveness of our time. If, as E. F. Schumacher
and others hold, European culture is the source of the present global
crisis, then studying one of the glories of that culture for messages
other than those of exploitation and domination is an innovative approach. It is possible that we may be, in Hannum's words, "made
aware of the connection between the violation of Nature and universal
destruction." Unlike previous interpreters of Wagner's classic, Hannum sees the emergence of the feminine and its link to natural order
as one of its major themes, one that holds the promise of redemption.
From the perspective of feminist and environmental scholarship, Dana
Jackson considered the conjoining of these issues as key to ending the

patriarchal domination of women and nature. Hunter Hannum traces a comparable path from within the high art of the culture. This largely ignored but integral part of our own cultural heritage—as opposed to mythic elements transplanted from other traditions— renders the potential for healing and wholeness all the more immediate and accessible.

Much of our trouble during these past two centuries has been caused by our limited, our microphase, modes of thought. We centered ourselves on the individual, on personal aggrandizement, on a competitive way of life. . . . A sense of the planet Earth never entered our minds. We paid little attention to the more comprehensive visions of reality. This was for the poets, the romanticists.—Thomas Berry, *The Dream of the Earth* (1988)

In a lecture sponsored by the E. F. Schumacher Society on October 19, 1991 (the ninety-fifth anniversary of the birth of Lewis Mumford), Thomas Berry—like Mumford a cultural historian with a catholic breadth of vision—called his audience's attention to the order of magnitude of the changes presently taking place on our planet.[1] These changes are not simply the sort of historical transitions we learned about in school—from the classical to the medieval world, for instance, or from the medieval to the Renaissance one; they do not even resemble the shifts that took place between the Neolithic and historical worlds. They are changes unique in the history of the planet, for heretofore all developments on Earth took place almost entirely without human intervention, whereas henceforth human beings will have a decisive influence on almost everything that happens on this planet.

We now find ourselves, Berry continued, in the terminal phase of the Cenozoic period, an era in geologic time beginning some seventy million years ago that saw the (relatively) rapid evolution of mammals as well as grasses, shrubs, and flowering plants. Our present phase is a terminal one because our scientific-technological orientation has come to dominate the entire planet during the past hundred years or so, wreaking almost irreparable damage on the biosphere. This is, of course, a well-known situation, familiar even to some politicians nowadays. What Berry emphasizes is the urgent need for a new vision to guide us into the ecological or what he prefers to call the "Ecozoic" age, a period in which humankind must learn once again to live in harmony with the Earth. Central to this vision must be a new "story." Berry anticipates new developments in the arts to express this story, especially in the art of the drama, because its essence is conflict—a distinguishing characteristic of our age of transition. In recent times, however, conflict in drama has been situated, Berry points out, exclusively within the human sphere (that is, individual pitted against individual or against society), whereas now the stage of conflict is much broader, encompassing extrahuman dimensions: the fate of the whole Earth is at stake.

Here a question arises: Does any past dramatic work portray our present terminal phase in its larger dimensions? Has any drama dealt with the destructive elements of our age, as represented by what Berry calls "the industrial-commercial plundering process," while at the same time offering a vision of healing and wholeness? I suggest that Richard Wagner's tetralogy Der Ring des Nibelungen is such a drama, the largest work—at least in terms of length—in the history of Western music. The Ring, of course, draws upon Teutonic mythology, and because myth traditionally treats a totality—subhuman as well as super-human worlds—The Ring fulfills one of Berry's criteria for "new" drama: its conflicts are not located merely in the sphere of human individuals.

Being a mythic work has other consequences: since its creation and first performance in the second half of the last century, The Ring has given rise to a multitude of interpretations, the scale ranging from Marxist to Jungian. This reaction was predictable, for myth—from early on in the Western tradition—has been subjected to explanation or "rationalization." Myth has, in other words, been read as allegory (the most famous example: the Old Testament was interpreted as a figurative anticipation of the New according to St. Augustine's formula, "In the Old Testament the New Testament is concealed; in the New Testament the Old Testament is revealed").

It is as a "modern" allegory that I want to read The Ring here, keeping in mind Thomas Berry's description of our age and his call for a new story.

This allegorical approach was taken as early as the end of the last century. In 1898, for example, George Bernard Shaw published his brilliant and influential study The Perfect Wagnerite: A Commentary on the Niblung's Ring, on the very first page of which he tells us: "The Ring, with all its gods and giants and dwarfs, its water-maidens and Valkyries, its wishing-cap, magic ring, enchanted sword, and miraculous treasure, is a drama of today, and not of a remote and fabulous antiquity. It could not have been written before the second half of the nineteenth century, because it deals with events which were only then consummating themselves."[2]

My reading will attempt to bring Shaw's up to date. In so doing, it also necessarily represents a correction of his interpretation in certain basic respects, for, after all, nearly one hundred years have passed since the "today" he refers to in his commentary, fateful years in the planet's history—in Berry's view "the terminal phase of the Cenozoic." Basically, I hope to show that Shaw's "today" is actually our day, which of course is in keeping with Wagner's assertion that myth "is true for all time, and its content . . . inexhaustible throughout the ages."

I shall try to get along with as little plot summary as possible (can anyone who has heard it ever forget Anna Russell's hilarious parody of The Ring's plot?) and restrict myself to salient points. In the first place—this phrase, as we shall see, is especially appropriate here—in Das Rheingold, the opera (or music drama, as Wagner preferred to call it) that functions as prelude to the three that follow, we find ourselves in the primeval element of water, in the depths of the Rhine River, where three water nymphs are keeping watch over a pristine hoard of gold. Alberich appears, a lustful dwarf who is at first attracted to the nymphs but, spurned by them, soon turns his attention to the glistening gold. The Rhinemaidens tell him that whoever forges it into a ring will attain limitless power, but at the cost of having to renounce love. Without hesitation the dwarf seizes the gold and rushes off with it.

A fundamental choice is made here that determines the subsequent action. In an entirely different context, contemporary psychoanalyst Arno Gruen sums up the situation succinctly: "Human development may follow one of two paths: that of love or that of power. The way of power, which is central in most cultures, leads to a self that mirrors the ideology of domination."[3] It is this "ideology of domination" that fuels the tragedy of Der Ring des Nibelungen. But domination over what? First of all, over Nature. Here Wagner's drama mirrors the drama of our age as Berry describes it, while at the same time departing radically from classical-traditional precedent and telling a "new story."

The Ring shares, it is true, a basic feature of classical-traditional dramatic narratives: in the beginning a crime or act of rebellion occurs that is followed by a curse; subsequently, the resolution of the narrative involves the expiation of this curse. But Wagner's is a new approach. We shall see the novelty of his Ring more clearly if we compare it with two works of comparable weight from the previous tradition, each of them a "world-poem" (Shaw's word for The Ring) because each of them, like it, deals with a larger stage than the merely human one. The first, Milton's Paradise Lost, also posits a primeval crime; yet "man's first disobedience" is not directed against Nature but, as the Judeo-Christian myth Milton follows instructs us, against the Deity, and this original sin can be atoned for only by the Deity Himself (cf. Paradise Regained). Whereas the second, Goethe's Faust—a great part of which was written in the same century as Wagner's Ring—is in many respects a transitional work, standing at the threshold between the classical-traditional and the scientific-technological ages (to use Berry's periodization again), it too takes its departure from an act of rebellion: in "The Prologue in Heaven" that precedes the drama proper, Mephistopheles, as a kind of surrogate for Faust himself, cynically calls the harmony and perfection of the Divine creation into question. Faust then ends with its eponymous hero, despite his human imperfections, being taken up into that Heaven we saw at the beginning of the play.

But back to Earth—that is, to the depths of the Rhine and to Wagner's new story. The curse of the Nibelung Alberich is actually a twofold one. As he wrests the gold from its natural setting beneath the waters, he curses love for the sake of attaining power. And another decisive curse is to follow: after his theft of the gold ("The Theft of the Rhinegold" was Wagner's original title for the work, pointing to the centrality of this crime in the composer's conception), Alberich returns to the land of the dwarfs and, using the power afforded by the ring he forges from the gold, forces his fellow creatures to amass immense wealth for him. When we experience *Das Rheingold*'s third scene, it is difficult not to believe we are witnessing an allegorical presentation of the first modern factories or mines: we hear the almost deafening din of forging, we see the ruddy glow of fires in murky subterranean darkness—in short, we witness the inhuman toil of creatures slaving under inhumane conditions. Shaw recognized this immediately, stating that Wagner's "picture of Niblunghome under the reign of Alberic is a poetic vision of unregulated industrial capitalism as it was made known in Germany in the middle of the nineteenth century by Engel's *The Condition of the Working Class in England in 1844*" (p. xvii). (Today, taking into account the fate of Eastern Europe and the lands of the former Soviet Union, we know it is not capitalism alone that is capable of creating the human and environmental inferno of "Niblunghome.")

In this netherworld appear Wotan, ruler of the gods, and his helper-companion Loge, god of fire and of cleverness and trickery. Wotan has a problem: he has overextended himself in the area of real estate and can't pay his bills. He has contracted with two giants to build him and his cohorts a sumptuous castle, Valhalla, and has promised as payment Freia, goddess of youth and love. When Wotan and the other gods discover they are unable to do without Freia, they search for ersatz payment. Cunning Loge comes up with the solution of seizing Alberich's wealth, whose source is the Rhinegold and the ring fashioned from it. With the help of Loge's trickery Wotan outwits Alberich, gains control of the hoard and ring, and pays off the giants with them.

It is now that Alberich pronounces his second curse. First he had cursed love in order to gain the gold; now he calls down a curse on anyone who in the future shall possess the ring made from it. This curse takes effect immediately: the giants Fasolt and Fafner accept the Nibelung's treasure (including the ring) as a substitute for the promised Freia and instantly start to quarrel over their payment, whereupon Fafner slays his brother and takes the entire treasure for himself. Just before this incident, which is anticipated to some degree in his Scandinavian sources,[4] Wagner inserts a significant invention of his own: when Wotan is reluctant to give up the ring (the means to wealth and power) even to save Freia (representative of Nature, of love and youth, and—as we shall see—of the Feminine), a female figure emerges from the depths of the Earth to warn him to relinquish it, for it augurs, she announces, a bleak day for the gods. Wagner names this character "Erda," the Old High German word for "Earth"; today she is better known by another ancient name, Gaia. Wotan heeds her warning and includes the ring with the Nibelung treasure he surrenders to the giants; nevertheless, he immediately starts making plans to recover it, thus helping to bring about the conflict—and tragedy—to come.

Instead of tracing the intricacies of The Ring's subsequent plot, let's examine some of that plot's underlying elements often overlooked by previous commentators. I have said the work deals with a primal crime against Nature, but this is too imprecise; put more concretely: a pure natural substance is brought up from its resting place in the depths and, misused, becomes the object of a tragic power struggle. (A linguistic point: at the conclusion of Das Rheingold the Rhinemaidens lament the loss of their gold to the unheeding ears of the gods, who are preparing to enter the splendid fortress of Valhalla. "Rheingold! Rheingold! Reines Gold!" they sing—an untranslatable play on words, for "rein"—pronounced like the name of the river—means pure, unadulterated, natural. It is this purity that of course is lost when the natural order is violated.)

As if to reinforce the importance of the theme of a primal injury to Nature, Wagner gives us a second instance of it: at the beginning of Die Götterdämmerung (The Twilight of the Gods), the final work in the tetralogy, three female figures, the Norns or Fates (daughters of Erda), review the history of the drama. They recount how Wotan once came to the World Ash tree to drink of wisdom from a spring at its roots; at the same time he broke from the tree a branch that he used to fashion a spear, the instrument (and symbol) of his power. This act caused the tree to wither and eventually die; now, the Norns inform us, the dry wood from that tree is heaped up in the form of mighty branches around Valhalla, awaiting the spark to ignite it and cause the conflagration that will spell the doom of the gods. Here, even more clearly, we are made aware of the connection between the violation of Nature and universal destruction.

It is hard to believe that another theme of prime importance in Wagner's fable has not received greater attention: the role of the Feminine in this drama of violation—of crime, curse, passion, and resolution. To speak solely of a primal crime against Nature in The Ring is, once again, imprecise. It is the Feminine as well that is the victim of violation throughout the work, and the villain is what we have come to recognize as the patriarchy, with all that this concept connotes for the contemporary mind. And once again Wagner's work contains an artistic expression of Thomas Berry's analysis of our terminal era and anticipates the "new story" Berry finds necessary for our time, for he describes a precondition for the emergence of the next era of Earth's history as follows: "The Ecozoic can come into existence only through an appreciation of the feminine dimension of the Earth, through a liberation of women from the oppressions and the constraints [of the past]" (from a list of conditions, handed out at Berry's lecture, that he proposes for entering the Ecozoic Era).

The Ring opens in the depths of the Rhine, in the female element of water, where the Rhinemaidens are guarding the gold. Its theft, which launches the plot, is at the same time a wrong done to its three guardians, one they lament to the gods. But Wotan wants to hear nothing of their complaint, and Loge advises them that, since their gold is gone, they should sun themselves henceforth "in the gods' new splendor"—in other words, in the reflected glory of the patriarchy. Yet the clever Loge realizes that this splendor is doomed: "They are hastening to their end, those [gods] who believe so firmly in their enduring power."

We have already observed a second offense against the Feminine, its use as a pawn in a commercial transaction. In Das Rheingold Wotan is almost willing to surrender Freia to the giants as payment for Valhalla, site and symbol of his power. Freia, who with her golden apples that insure the gods' youth and vitality, is clearly a fertility goddess and another incarnation of Gaia. Although Wotan heeds Erda's warning at the last moment and relinquishes the ring in order to save Freia, this consideration for the Feminine doesn't last. Indeed, it is the splitting off of this element in himself that contributes greatly to making him the tragic figure he is, for the exigencies of the plot later lead him to repudiate his beloved daughter Brünnhilde (offspring of his union with the Earth goddess Erda). From that moment—dramatized, with some of The Ring's most moving music, at the conclusion of Die Walküre—Wotan is a doomed and dying god. Analogously, the fact that Siegfried, grandson of Wotan, "forgets" his feminine half Brünnhilde—as the result of a magic potion—seals his fate.

And also from that time on, it is Brünnhilde, not her spouse Siegfried, who is the true hero(ine) of the work. For passion or suffering is not enough to make a figure into a tragic hero; there must be perception of the root of suffering as well (as Oedipus, for instance, must first realize and then atone for his crime). Brünnhilde exhibits this perception at the end of The Ring, in the last act of Die Götterdämmerung. One is tempted to say that the "Ring" is closed or "rounded off" here, that the "cycle" is completed. Two examples: at the end of Das Rheingold the Rhinemaidens "lamented" the theft of their gold to the gods, who ignored them. The word Wagner uses is "klagen," which means to complain or lodge a complaint as well as to lament. Then, at the end of Die Götterdämmerung Brünnhilde addresses the ruler of the gods (the patriarchy once again): "Meine Klage hör, du hehrster Gott!" (Hear my complaint, you august god!). It is the sins of the patriarchy—the rape of Nature, the denigration of the Feminine, and the ensuing struggle for profit and power—that have brought about her present grief, the murder of her spouse Siegfried. And all of this has happened, she adds, "dass wissend würde ein

Weib!" (that a *woman* might gain knowledge!—my emphasis). The immediate consequence of her (feminine) knowledge is the second instance of rounding off or closing the circle: Brünnhilde takes the gold ring from the finger of her dead husband and places it on her own, saying that she is taking back what is rightfully hers and bequeathing it to "the wise sisters who dwell in the depths of the waters"—the Rhinemaidens are to take it from her ashes after her self-immolation on Siegfried's funeral pyre.

It may seem like a relatively minor element in the total drama that Brünnhilde claims the ring as *her* "Erbe" (heritage, inheritance) and places it on *her* finger, that the Rhinemaidens retrieve it from *her* ashes and bear it back to its pristine place in the depths of the Rhine (after one last attempt on the part of Hagen, Alberich's son, to reclaim it for the patriarchy—in vain, for the Rhinemaidens pull him down into the water and drown him), yet I find it significant that Shaw, for all his perspicacity about the work, misses this point completely. Here is his description of the scene (he is very scornful of *Die Götterdämmerung* in general, seeing in it a surrender to the clichés of grand opera): "The hall of the Gibichungs catches fire, as most halls would were a cremation attempted in the middle of the floor (I permit myself this gibe purposely to emphasize the excessive artificiality of the scene) but the Rhine overflows its banks to allow the three Rhine maidens to take the ring from Siegfried's [sic] finger, incidentally extinguishing the conflagration as it does so" (p. 82).

Not accidentally, another scene in *Die Götterdämmerung* also elicits Shaw's disapproval: in Act One Waltraute, a sister Valkyrie, begs Brünnhilde to return to the Rhinemaidens the ring Siegfried has given her. Shaw's words again: "Clinging in anguish to Wotan's knees, she [Waltraute] has heard him mutter that were the ring returned to the daughters of the deep Rhine, both gods and world would be redeemed from that stage curse of Alberic's in The Rhine Gold" (p. 74). Shaw's comment: "[Waltraute] betrays her irrelevance by explaining that the gods can be saved by the restoration of the ring to the Rhine maidens. This, considered as part of the previous allegory, is nonsense." (pp. 80–81).

To understand Shaw's difficulties with the conclusion of *The Ring*—why he sees in it "the collapse of the allegory"—we must first be clear about what he takes to be the drama's meaning:

In the old-fashioned orders of creation [it is these "orders," incidentally, that are mirrored in *Paradise Lost* and *Faust*, those "world-poems" mentioned earlier], the supernatural personages are invariably conceived as greater than man, for good or evil. In the modern humanitarian order as adopted by Wagner, Man is the highest. In *The Rhine Gold*, it is pretended that there are as yet no men on the earth. There are dwarfs, giants, and gods. The danger is that you will jump to the conclusion that the gods, at least, are a higher order than the human order. On the contrary, the world is waiting for Man to redeem it from the

lame and cramped government of the gods. Once grasp that; and the allegory becomes
simple enough. Really, of course, the dwarfs, giants, and gods are dramatizations of
the three main orders of men: to wit, the instinctive, predatory, lustful, greedy people;
the patient, toiling, stupid, respectful, money-worshipping people; and the intellectual,
moral, talented people who devise and administer States and Churches [sic]. History
shows us only one order higher than the highest of these: namely, the order of Heroes.
(Shaw, pp. 28–29)

It is the defeat, the tragic fall, of this Hero, this Man (not entirely coinciden-
tally Shaw uses the capitalized masculine form) that Shaw cannot accept. In terms
of the drama, what is involved is the fall of Siegfried, whom he sees as a quasi-
Nietzschean superman, a proto-Protestant who, revolting against the powers
that be, extends the dominion of Man. Elements of Wagner's work that don't fit
into Shaw's scheme are dismissed as "irrelevant" or "nonsense," and prominent
among these, as we have seen, is the motif of the restoration of the ring to the
Rhinemaidens—he even misses the point that it is Brünnhilde who ultimately
performs this act! What Shaw fails to perceive here are the (linked) themes of
Nature and the Feminine that sound throughout the work and are "irrelevant"
neither to it nor to our time.

The question of time brings us back to a consideration of Thomas Berry's
periodization: when Shaw stated at the outset of his study that The Ring is "a drama
of today," we must ask ourselves what day that is in Berry's scheme and how it
differs from ours. According to Berry, Shaw (and of course Wagner) lived during
what might be called the heroic stage of the scientific-technological age, when
the West's "commercial-industrial plundering process" was beginning to spread
triumphantly over the entire globe. Now, by our day, that process has brought
about the terminal phase of a much longer period in Earth history, the Cenozoic;
as a result, we are in a much better position than Shaw was to understand the
prophetic fable of Die Götterdämmerung—indeed, the import of The Ring as a whole.

In the foregoing I have argued for a contemporary allegorical reading of The
Ring; now it is time for a brief recapitulation and amplification of that argument. A
pure natural substance is wrested from the depths and, misused, becomes the
object of a disastrous power struggle, one that destroys not only human beings
but Nature as well (cf. the fire and flood at the end of Die Götterdämmerung). To begin
with this ending: what struck Shaw as "irrelevant"—the female protagonist's
return of the substance in question to its natural setting—strikes us now as an
anticipation of today's movement toward "ecological restoration," the desire to
make amends to a violated natural order.[5] It is in this direction—and in this
direction only—that many believe humankind's "redemption" lies, and it is this
which gives meaning in contemporary terms to Brünnhilde's "redemptive" act.

(Out of the multitude of musical motifs that make up The Ring's score, two are dominant at its close: one is the motif of the Rhinemaidens; the other, generally designated "Redemption through Love," sounds—high in the first violins—over a final restatement of the Valhalla motif. Thus, in his conclusion the composer gives special emphasis once again to Nature and the Feminine vis-à-vis the power of the patriarchy.)

Now, one basic and meaningful way to define the century and a half since the birth of Shaw (1856) and the beginning of the composition of The Ring's music (1853) is to call it the Hydrocarbon Age. Once we accept this definition, it is difficult not to read The Ring as an allegorical account of that age and especially of the role of oil—sometimes referred to as "liquid gold" or "black gold"—in it. A fascinating and instructive book by Daniel Yergin, The Prize: The Epic Quest for Oil, Money, and Power,[6] demonstrates in detail how, inescapably and fatefully, practically every figure of outstanding wealth and political power for over a century has been closely connected with oil and its history; indeed, oil wealth and political power have sometimes even been united in one figure, such as that of the forty-first President of the United States, George Bush. Interestingly enough, the figurative language used to recount this history frequently recalls the actual figures on Wagner's stage: for example, one speaks of the "giants" Standard Oil and Royal Dutch/Shell, and Yergin's chapter describing Ida Tarbell's exposé of Standard Oil that led to the dismemberment of that monopoly is entitled "The Dragon Slain" (Siegfried's famous slaying of a dragon occurs in the third drama of the tetralogy)!

The struggle for control of "black gold" has of course involved not only wealthy and powerful individuals and corporations but entire nations as well, leading to conflicts as recent as the Gulf War. Even sober observers of the conflagration at the end of that war sound the same apocalyptic note we hear at the end of The Ring: "As we continued our journey from Saudi Arabia into Kuwait, we felt we were entering an environmental hell. In the oil fields, well fires and smoke clouds filled the horizon; oil spilling from the damaged wells formed extensive rivers and lakes; and soot covered the desert with a black crust. It looked like the end of the world" (emphasis mine).[7]

It is striking, then, how Wagner anticipates our present age and equally striking how Shaw failed to see, in his critique of Wagner, the implications of the scientific-technological phase of modern history. One might argue that, after all, most of Shaw's life was spent in the "heroic" period of that history, yet as recently as 1952 Theodor Adorno, a German Marxist critic of The Ring, shares essentially the view of the Irish dramatist, seeing in Wagner's work the regrettable retraction of a revolutionary impulse: "If we wanted to express the 'idea' of The Ring in simple words, we could say: man emancipates himself from the blind natural context from which he originates and achieves power over nature only to fall victim to it in the last analysis. The allegory of The Ring states the equivalence of domination over nature and subjection to nature."[8] Adorno saw this equivalence as a nihilistic one. To counter both his view and Shaw's and to reinforce Wagner's vision from a non-Western perspective, here are the words of Russell Means, a Native American: "All European tradition, Marxism included, has conspired to defy the natural order of all things. Mother Earth has been abused, the powers have been abused, and this cannot go on forever. No theory can alter that simple fact. Mother Earth will retaliate, the whole environment will retaliate, and the abusers will be eliminated. Things come full circle, back to where they started. That's revolution. And that's a prophecy of my people, of the Hopi people."[9]

In his assessment of the scientific-technological phase of recent history, Thomas Berry tells us that "[a] sense of the planet Earth never entered into our minds. We paid little attention to the more comprehensive visions of reality. This was for the poets."[10] If we had paid attention, I have tried to demonstrate, we might have appreciated the geocentric (Berry's word is "biocentric") vision of Wagner's Ring and been able to measure accurately its distance from Shaw's anthropocentric one (which, we have also seen, may be called a patriarchal one). In this same connection, we might also have paid some attention to another geocentric poet writing in German, Rainer Maria Rilke, whose poems abound with celebration of the plant and animal worlds, who passionately affirmed the Earth in his ninth Duino Elegy and lamented, at the time of the cataclysm of Western civilization represented by the First World War, that "the world has fallen into the hands of men."

If we pay close attention to The Ring today, I believe we will realize that it deals with what Berry calls those "larger contours of conflict [occurring] in this stupendous transition from the terminal Cenozoic to the emerging Ecozoic" and that it points to what he refers to in the same lecture as the healing "feminine dimension of the Earth."

Perhaps it is high time that we listen to the poets.

—1993

Notes

1 His lecture has been published in revised form as The Ecozoic Era (Great Barrington, Mass.: E. F. Schumacher Society, 1991).

2 Reprint of fourth edition (New York: Dover, 1967), p. 1. Subsequent page references to this edition will be given in the text. Shaw's view was shared by Patrice Chéreaux, who, in his famous and controversial 1976 production of The Ring at Bayreuth, set the scene in the modern industrial age.

3 The Betrayal of the Self: The Fear of Autonomy in Men and Women, trans. Hildegarde and Hunter Hannum (New York: Grove Press, 1988), p. 1.

4 For an exhaustive study of Wagner's sources and what he did with them, see Deryck Cooke, I Saw the World End: A Study of Wagner's Ring (London: Oxford University Press, 1979). His discussion of Erda is on pp. 226–32.

5 For concrete examples of this movement, see Stephanie Mills's Schumacher lecture Making Amends to the Myriad Creatures (Great Barrington, Mass.: E. F. Schumacher Society, 1991).

6 New York: Simon & Schuster, 1992.

7 Richard S. Golob, " 'It Looks Like the End of the World,' " Harvard Magazine, Sept.–Oct. 1991, pp. 30 and 32.

8 Versuch über Wagner (Frankfurt: Suhrkamp Taschenbuch Verlag, 1974), p. 126 (translation mine).

9 "Fighting Words on the Future of the Earth," Mother Jones, Dec. 1980, p. 30.

10 The Dream of the Earth (San Francisco: Sierra Club Books, 1990), p. 44.

Kirkpatrick Sale
Mother of All:
An Introduction to Bioregionalism

In the concept of bioregionalism lies the structural model for the application of the ideas advocated by E. F. Schumacher. Kirkpatrick Sale is
among its most eloquent proponents. In this lecture, which was expanded into his book *Dwellers in the Land,* he cites Schumacher's critique
of the market economy of the late twentieth century as having "erred
fundamentally because it erred against nature." Sale maintains that the
primary requirement for a bioregional economy is an informed and
profound sense of place. Its very foundation is the naturally defined
ecosystem—the interlocking network of climate, watershed, terrain,
soils, plants, lakes, streams, rivers, and coasts as well as all the fauna
from microscopic to human—of which it is comprised.

Fundamental to the concept of bioregionalism is the recognition
of what Sale considers the four basic determinants of any organized
civilization: scale, economy, politics, and society. Like Jane Jacobs,
he is an advocate of cities, but he sees a number of small cities as
more viable than the metropolis of today. The bioregional society he
foresees would foster a far greater symbiosis of urban and rural, industrial and agricultural, population and resources. In such a mix,
he contends, "the concern for place, the preservation of nature, the
return of such traditional American values as self-reliance, local control, town-meeting democracy" can flourish, blunting the edge of
many current political divisions. Given the upsurge in grass-roots
activism that is Frances Moore Lappé's current concern, the bio-

regional vision—with its emphasis on environmental issues—becomes increasingly feasible.

> To Gaea, mother of all of life and oldest
>   of gods, I sing,
> You who make and feed and guide all
>   creatures of the earth,
> Those who move on your firm and radiant
>   land, those who wing
> Your skies, those who swim your seas, to
>   all these you have given birth;
> Mistress, from you come all our harvests,
>   our children, our night and day,
> Yours the power to give us life, yours
>   to take away.
> To you, who contain everything,
> To Gaea, mother of all, I sing.
>   —Homeric *Hymn to Earth*

In the beginning, as the Greeks saw it, when chaos settled into form, there was a sphere, aloft, floating free beneath the moist, gleaming embrace of the sky and its swirling drifts of white cloud, a great vibrant being of green and blue and brown and gray, binding together in a holy, deep-breasted synchrony the temperatures of the sun, the gases of the air, the chemicals of the sea, the minerals of the soil, and bearing the organized, self-contained, and almost purposeful aspect of a single organism, even a living, breathing body, a heart, a spirit, a soul, a goddess—in the awed words of Plato, "a living creature, one and visible, containing within itself all living creatures."

To this the Greeks gave a name: Gaea, the earth mother. She was the mother of the heavens, Uranus, and of time, Cronus; she was the mother of the Titans and the Cyclops, of the Meliae, the ash-tree spirits who were the progenitors of all humankind; she was the mother of all, first of the cosmos, creator of the creators. She became the symbol of all that was sacred and the font of all wisdom, and at the fissures and rifts in her surface—at Delphi, especially, and Dodona and Piraeus—she would impart her knowledge to those oracles who knew how to hear it. And ultimately, inevitably, she became embodied in the language of the Greeks as the unit of life or birth or origination, combined into the word "genos" to give us, in English, "genesis," "genus," "genitals," "genetics," and "generation."

"Earth is a goddess," wrote Xenophon in the fourth century before Christ, "and teaches justice to those who can learn." Justice and compassion and prudence and appropriateness and harmony—all of what were later called the cardinal virtues: "The better she is served," Xenophon taught, "the more good things she gives in return."

All that seems obvious enough, at least to those who first inhabited the earth and created her cultures—which is why, in virtually every early preliterate society that we know of, the primary deity, worshiped before all others, was the earth. And even in those societies that eventually came to displace the earth goddess with other deities, most typically the sky god—a male figure, be it noted, and one adopted almost exclusively by those cultures (even the later Greeks) that simultaneously created empire, war, hierarchy, priesthood, and slavery—the earth was still considered a living being, sentient and organic, and still retained its character as a deity.

It was not until the development of European science, from about the sixteenth century on, that this animistic conception of the earth finally gave way, to be replaced by one supported by the new insights of physics, chemistry, mechanics, astronomy, and mathematics. The new perception held—in fact it proved—that the earth, the universe, and all within it operated by certain clear and calculable laws and not by the whims of any living, thinking being; that, far from being divine and omnipotent, these laws were capable of scientific prediction and manipulation; and that objects, from the smallest stone to the earth itself and the planets beyond, were not animate with souls and wills and purposes but were nothing more than the combination of certain chemical and mechanical properties. The cosmos was in no sense like a purposeful, pulsatory celestial thing alive but rather, in the Newtonian image, something more like a giant clock, its many parts moving in an ordered, kinetic, mechanical way. Europe's scientific revolution—in the triumphant words of the seventeenth-century physicist Robert Hooke—enabled humankind "to discover all the secret workings of nature, almost in the same manner as we do those that are the productions of [human] Art and are managed by Wheels, and Engines, and Springs."

Now, as I am sure you know, the history of ideas is just like the history of technologies: those that suit the powers-that-be are embraced, those that seem to have no utility are forgotten. The ideas of the new science were very quickly heeded and their creators rewarded and pantheonized by a European establishment that at the same time was in the process of creating other complementary attitudes and systems for which scientism provided both intellectual conditioning and practical guidance. For the scientific system was developed contemporaneously with—and, by no means accidentally, in aid of—the consolidation of the nation state, the growth of mercantile and then corporate capitalism, and the spread of global exploitation and colonialism. Its inherent message—the celebration of the quantifiable, the mechanistic, the physiochemical, and the tangible, as opposed to the organic, the spiritual, the creative, and the intangible—had immense importance, far beyond the laboratories, for the European society that developed out of the sixteenth century. And its ultimate governing principle—that humans should not merely understand but be capable of manipulating nature, and indeed, as Descartes put it for all of European science, be "masters and possessors of Nature"—became ingrained into not only the scientific but also all scholarly and most popular thinking in the Western world and now shapes the perceptions of our senses and the patterns of our psyches.

And if at the end of the twentieth century we see the earth as a static and neutral arena that is alterable by our chemicals and controllable by our technologies; if we see ourselves as a superior species, to whom is given the right to kill off as many hundreds of others as we wish and "have dominion over" the rest; if we believe we have the power to reorder earth's atoms and reassemble its genes, to contrive weapons and machines fueled by our own invented elements and capable of destroying forever most of its organic life; if we create technologies capable of plundering its resources, befouling its systems, poisoning its air perhaps irretrievably, and altering its eons-old processes to suit our wishes; and if those who are most especially devoted to the truest nature of the earth, those we call the ecologists, can choose to dress themselves in the cloaks of scientism to talk of nature's "entropy" or energy "production" or food "chains" or forest "management" or even displace the image of the biotic community with that of the mechanical eco-"system"—if this is our condition, it is so because, far from calling into question the scientific view of the universe in these past four centuries, we have accepted it virtually in its entirety. It has become the foundation and sustenance not only of our various social systems—education, agriculture, medicine, religion, energy, communication, transportation—but of our most basic economic and political institutions as well.

To be sure, the scientific worldview is not without its values, its uses, its triumphs even, and I think we may want to call the world a better place for our knowledge of hygiene, say, or radiotelegraphy or immunology or electricity. But its shortcomings, its failures, its calamitous dangers have by now become obvious, and it is surely safe to say that the path of sanity, perhaps survival, is to regain the spirit of the ancient Greeks, to once again comprehend the earth as a living creature. We need to recover the sense, as Schumacher puts it in *Good Work*, "that man is the servant of this world, or at least a trustee," a concept that has been "organized out of our thinking," as he put it, "by the modern world," and we must listen again to the two great teachers, one "the marvelous system of living nature" and the other "the traditional wisdom of mankind," teachers we have "rejected and replaced by some extraordinary structure we call objective science." And we must re-envision humans as participants and not masters in the biotic community, as only one among many species, special perhaps in having certain skills of information-gathering and communication but not for that reason superior to those with other skills—for the human being, as Mark Twain might have said, is different from other animals only in that it is able to blush. Or needs to.

In *The Interpreters*, a book by the Irish author known as AE, written at the height of the Irish Revolution, there is a passage in which a group of disparate men, all prisoners, sit around discussing what the ideal new world should look like. One of them, the poet Lavelle, argues fervently against the vision put forth by one prisoner, a philosopher, of a global, scientific, cosmopolitan culture. "If all wisdom was acquired without," he says, "it might be politic to make our culture cosmopolitan. But I believe our best wisdom does not come from without, but arises in the soul and is an emanation of the earth-spirit, a voice speaking directly to us as dwellers in the land."

To become "dwellers in the land," to regain the spirit of the Greeks, to fully and honestly come to know the earth, the crucial and perhaps only and all-encompassing task is to understand the place, the immediate, specific place, where we live: as Schumacher says, "In the question of how we treat the land, our entire way of life is involved." We must somehow live as close to it as possible, be in touch with its particular soils, its waters, its winds. We must learn its ways, its capacities, its limits. We must make its rhythms our patterns, its laws our guide, its fruits our bounty.

That, in essence, is bioregionalism.

Now, I must acknowledge that "bioregionalism" is not yet quite a household word—you're writing a book on what? my friends say—and when the Schumacher Society board of trustees decided to use it as the theme of this forum, we knew we ran the risk both of alienating the uninvolved and perplexing the sympathetic. But I believe bioregionalism to be a concept so accessible, so serviceable, so productive—and, after about five years now, so impelling as to have created a momentum of its own—that I feel quite confident in its use. For there is really nothing so mysterious about the components of the word—"bio," from the Greek for life; "regional," from the Latin for territory to be ruled; "ism," from the Greek for doctrine—and nothing, after a moment's thought, so terribly strange in what they convey. If it initially falls oddly on our ears, that may perhaps only be a measure of how far we have distanced ourselves from its wisdom—and how badly we need it now.

Let me take a little time to excavate this concept of bioregionalism a bit, baring and examining its several layers as one might in looking at the strata of the earth.

All aspects of the bioregional society—and, one might imagine, a bioregional world—take their forms from that of Gaea herself. One of Gaea's many offspring, the first of all her daughters, was Themis, the goddess of the laws of nature and the mother of the seasons, and it is by a diligent study of her—her laws, her messages, her patterns as they have been established over these many uncounted millennia—that we can guide ourselves in constructing human settlements and systems. This is not, of course, an easy undertaking, for the lessons of nature can sometimes seem confusing, even contradictory, and perhaps I have read them wrong; perhaps only more time and more opportunity to be closer to nature, as close as the preliterate peoples who have twenty words for snow and distinguish thirty kinds of annual seasons, will allow us to learn these lessons properly. But I think I have at least the outlines right, and I am bolstered by the knowledge that they seem to accord well with the findings of many others who have looked in this direction, not the least of whom was Fritz Schumacher himself.

I would offer, then, what it seems to me are the bioregional guidelines bearing upon what I regard as the four basic determinants of any organized civilization: scale, economy, politics, and society.

Scale

I will, if I may—I always do—start with scale: the size, the dimensions of the bioregion as set by the characteristics of the earth, by the "givens" of nature. A bioregion is a part of the earth's surface whose rough boundaries are determined by natural rather than human dictates and is distinguishable from other areas by attributes of flora, fauna, water, climate, soils, landforms, and the human settlements and cultures those attributes have given rise to. The borders between such areas are usually not rigid—nature works with more flexibility and fluidity than that—but the general contours of the regions themselves are not hard to identify and indeed will probably be felt, understood, sensed, or in some way known to many of the inhabitants, particularly those rooted in the land—farmers, ranchers, hunters and fishers, foresters and botanists, and most especially, across the face of America, tribal Indians, those still in touch with a culture that for centuries knew the earth as sacred and its well-being as imperative.

Now, one rather interesting thing about all this is that when you start to look closely at how nature is patterned—and I have spent a considerable amount of time doing this for North America in the past few months—you discover that you are dealing with something almost, appropriately enough, organic. For just as bioregions normally merge with one another without hard-edged boundaries, so they overlap and even subsume one another in a complex arrangement of sizes depending upon the detail and specificity of natural characteristics. The whole matter is complex, and I do not wish to go into all its intricacies today, but let me suggest the labels with which I propose to describe (and, I hope, to popularize) the various kinds of bioregional gradations.

The widest region, taking its character from the broadest measures of native vegetation and soil contours, may be called the "ecoregion" and will generally cover several hundred thousands of square miles over several states; it is possible to determine somewhere between forty and fifty such areas across North America. But within these ecoregions it is easy to distinguish other coherent territories that define themselves primarily by their surface features—a watershed or river basin, a valley, a desert, a plateau, a mountain range—and that we may call the "georegion." And within these georegions, in turn, one can often locate still smaller areas of perhaps several thousand square miles, discrete and identifiable with their own topographies and inhabitants, their own variations and human culture and agriculture, to which we may give the name "vitaregion."

Using that terminology for our location today, we would say we are in an ecoregion that could be thought of as the Northeastern Hardwood, stretching (in conventional terms) from mid-New Hampshire and mid-Vermont to mid-New Jersey, an area characterized by birch and beech in addition to conifers, largely podzolic and blue podzolic soils, and a July-maximum, January-minimum rainfall. Within this territory are a number of obvious georegions—the Hudson watershed, the Berkshires, the Massachusetts Bay systems—and South Hadley, Massachusetts, site of today's Lectures, is solidly within the Connecticut River georegion, a long fertile valley running between the Green and Taconic Mountains on the west and the White Mountains on the east all the way down to Long Island Sound. But there are obvious distinctions to be made within this georegion, too, for the valley here as it broadens out from the Deerfield River on down to the Meshomasic foothills south of Hartford is quite different from the stretch up north to the Ammonoosuc or south in the pinched and hilly course to the Sound; and within this vitaregion clear differences from surrounding areas in both agriculture—tobacco, for example, and potatoes—and homoculture can be seen.

But I do not wish to dwell on such distinctions, to elaborate this cosmography, for I think at this stage of bioregional consciousness it is more important to stand a bit aside and appreciate the broad contours of the concepts than to plunge headlong into the briarbush of elaborate differences and definitions. Whether we speak of ecoregion or georegion or vitaregion, after all, we speak of bioregions, and it is this essential archetype that is most important to comprehend. For once that is done on any significant scale, then the matter of making distinctions among bioregions and creating human institutions to match them can safely be left to the inhabitants, the dwellers in the land, who will always know them best.

In the discussion to follow, therefore, we may imagine that bioregionalism will apply in its initial and formative phases to the largest territory, the ecoregion, and thereafter, in an evolving organic process narrowing in scale as the perceptions become sharper and the tools more finely honed, to smaller and smaller territories, to the vitaregion and perhaps beyond, moving closer and closer to the specifics of the soil and those who live upon it.

Economy

The economy that comes into being within a bioregion also derives its character from the conditions, the laws, of nature. Our ignorance is immense, but what we can be said to know with some surety after these many centuries of living on the soil has been cogently summarized by Edward Goldsmith, the editor of The Ecologist, as the laws of Ecodynamics—to be distinguished, of course, from the laws of Thermodynamics.

The first law is that conservation—preservation, sustenance—is the central goal of the natural world, hence its ingenerate, fundamental resistance to large-scale structural change; the second law is that, far from being entropic (that's an image rightly belonging to physics, errantly borrowed by scientific ecologists), nature is inherently stable, working in all times and places toward what ecology calls a "climax," that is, a balanced, harmonious, integrative state of maturity, which, once reached, is maintained for prolonged periods. From this it follows that a bioregional economy would seek to maintain rather than exploit the natural world, accommodate to the environment rather than resist it; it would attempt to create conditions for a climax, a balance, for what some economists have recently taken to calling a "steady state," rather than for perpetual change and continual growth in service to "progress," a false and delusory goddess if ever there was one. A bioregional economy would, in practical terms, minimize resource use, emphasize conservation and recycling, avoid pollution and waste. It would adopt its systems to the given bioregional resources: energy based on wind, for example, where nature called for that, or on wood, where that was appropriate, and food based on what the region itself—particularly in its native, pre-agricultural state—could grow.

And thus this kind of economy would be based, above all, on the most elemental and most elegant principle of the natural world, that of self-sufficiency. Just as nature does not depend on trade, does not create elaborate networks of continental dependency, so the bioregion would find all its needed resources—for energy, food, shelter, clothing, craft, manufacture, luxury—within its own environment. And far from being deprived, far from being thereby impoverished, it would gain in every measure of economic health. It would be more stable, free from boom-and-bust cycles and distant political crises; it would be able to plan, to allocate its resources, to develop what it wanted to develop at the safest pace, in the most ecological manner. It would not be at the mercy of distant and uncontrollable national bureaucracies and transnational governments and thus would be more self-regarding, more cohesive, developing a sense of place, of community, of comradeship, and the pride that comes from stability, control, competence, and independence.

In what was perhaps one of his most prescient perceptions Fritz Schumacher realized that the market economy of twentieth-century capitalism erred fundamentally, because it erred repeatedly, against nature. "It is inherent in the methodology of economics to ignore man's dependence on the natural world," he wrote. "The market represents only the surface of society and its significance relates to the momentary situation as it exists there and then. There is no probing into the depths of things, into the natural or social facts that lie behind them." And this is why, as he points out, conventional economics makes no distinctions at all between primary goods, "which man has to win from nature," and secondary goods manufactured from them or between renewable and nonrenewable resources or the environmental and social costs of developing one against the other.

A bioregional economy, in sharpest contrast, makes—in fact is grounded in—these vital distinctions.

Politics

Political principles on a bioregional scale are also grounded in the dictates presented by nature, in which what is forever valued are not the imperatives of giantism, centralization, hierarchy, and monolithicity but rather, in starkest contraposition, those of scale, decentralization, division, and diversity.

Nothing is more striking in the examination of a natural setting than the absence of the forms of authoritarianism, domination, and sovereignty that are taken as inevitable in human governance; even the queen bee is queen only because we designate her so. In a healthy econiche the various sets of animals—whether themselves organized as individuals, families, bands, or communal hives—get along with one another without the need of any system of authority or dominance—indeed, without structure or organization of any kind whatsoever. No one species rules, not one even makes the attempt, and the only assertion of power has to do with territory, with a particular area to be left alone in. Each set, each species, in the system has its own methods of organization, but none attempts to impose them on any other or to set itself up as the central source or power or sovereign. Far from there being contention and discord, the pseudo-

Darwinian war of all-against-all, there is for the most part balance and adjustment, cooperation among communities, integration into the environment, variety, complexity, and flexibility.

The lessons are of course obvious and suggest immediately the design for a bioregion as well as for a continent of bioregions. Each unit, of the size that the natural settings promote, may be unified and cohesive—let us imagine, for a start, a neighborhood, a community, a small town—and yet live side by side with others in a settled and mutual pattern, together comprising a vitaregion; and that vitaregion may have its own unification and cohesiveness, its own method of governance, and yet live side by side with other regions, organized as they may see fit; and so on, outward, in self-sufficient collaboration, unit upon unit, for so long as the natural boundaries may permit and the natural affinities be kept intact.

Similar lessons may be derived from the patterns of human nature, and in the matter of political relations it is only fitting to factor those in as well. Throughout all human history, even in the past several hundred years, people have tended to live in separate and independent groups, a "fragmentation of human society" that Harold Isaacs, the veteran MIT professor of international affairs, has described as something that is akin to "a pervasive force in human affairs and always has been." Even when nations and empires have arisen, he notes, they have no staying power against the innate human drive to fragmentation: "The record shows that there could be all kinds of lags, that declines could take a long time and falls run long overdue, but that these conditions could never be indefinitely maintained. Under external or internal pressures—usually both—authority was eroded, legitimacy challenged, and in wars, collapse, and revolution, the system of power redrawn."

I feel I must add here a note that may be painful for those whose allegiance to the precepts of fragmentation and diversification tends to crumble halfway through. Bioregional diversity means exactly that. It does not mean that every region of the Northeast or of North America or of the globe will build upon the values of democracy, equality, liberty, freedom, justice, and other suchlike "desiderata." It means rather that truly autonomous bioregions will likely go their own separate ways and end up with quite disparate political systems—some democracies, no doubt, some direct, some representative, some federative, but undoubtedly all kinds of aristocracies, oligarchies, theocracies, principalities, margravates, duchies, and palatinates as well. And some with values, beliefs, standards, and customs quite antithetical to those that the people in this room, for example, hold dearest.

Schumacher somewhere quotes with favor Gandhi's remark that it is worthless to go on "dreaming of systems so perfect that no one will need to be good." But that is exactly what I think *is* necessary: systems so perfect, or at least good enough, that they accommodate people who are not good. There's no point, it seems to me, in dreaming that people will be good, not merely because that would produce a fairly vapid society, I should think, but because there's every reason to suppose that it is simply not likely to take place on this planet in this galaxy. We must dream of systems, rather, which allow people to be people in all their variety, to be wrong upon occasion and errant and bad and even evil, to commit the crimes which as near as we know have always been committed— brutality, subjugation, even war—and yet systems in which all social and civil structures will work to minimize such errancies and, what is even more important, hold them within strict bounds should they occur.

Bioregionalism, properly conceived, is such a construct, for it provides a scale at which misconduct is likely to be mitigated because bonds of community are strong, and material and social needs for the most part fulfilled; a scale at which the consequences of individual and regional actions are visible and unconcealable, and violence can be seen to be a transgression against the environment and its people in defiance of basic ecological common sense; a scale at which even error and iniquity, should they happen, will not do irreparable damage beyond the narrow regional limits and will not send their poisons coursing through the veins of entire continents and the world itself. Bioregionalism, properly conceived, not merely tolerates but thrives upon the diversities of human behavior and the varieties of political and social arrangements those give rise to, even if at times they may stem from the baser rather than the more noble motives. In any case, there is no other way to have it.

Society

When asked recently to name the seven wonders of the world, the renowned biologist Lewis Thomas led off with the extraordinary phenomenon of the oncideres beetle and the mimosa tree. It seems that when she wishes to lay her eggs, the female oncideres beetle unfailingly picks out the mimosa tree from all others in the forest, crawls out on one of its limbs and cuts a long lengthwise slit into which she drops her sacs. Then, because in the larva stage the offspring cannot survive in live wood, she backs down the branch a foot or so and cuts a neat circular slit through the bark all around the limb, which has the effect of killing the branch within a very short time, whereupon it falls to the ground in the next strong wind and becomes the home for the next generation of oncideres beetles. But, interestingly, this process also has the effect of pruning the mimosa tree, a rather valuable ancillary result because, left alone, a mimosa has a lifespan of twenty-five to thirty years, but pruned in just this simple way it can flourish for a century or more.

Dr. Thomas seems to regard this relationship as sufficiently extraordinary to be regarded as a wonder of the world, particularly worthy, he writes, because such things "keep reminding us of how little we know about nature." Well, perhaps confession on the side of ignorance is wise in these matters, and yet I do think it is permissible to point out that, far from being unusual, this sort of biological interaction is in fact commonplace throughout every phase of nature, and, more-over, it is found with such regularity that it should indicate at least one lesson we know very well. The relationship is called, of course, symbiosis, and its per-sistence and pervasiveness in the natural world should be allowed to suggest to us, if we will but let it, a fundamental principle. From the very mitochondria that float about in our cells, infinitesimal creatures with their own DNA and RNA who live on us as we live on them, right on to the giant clam, which lives off the photosynthesis created by the plant cells it engulfs and actually incorporates into its body, where the cells live happily in a protected environment that even in-cludes small lenses in the clam's tissues particularly adapted to increase their needed sunlight—from the smallest to the largest, the recurrence of the phenom-enon of symbiosis provides a model for, if it does not strongly suggest the need for, a reordering of human society along similar lines, with families and neigh-borhoods and communities and cities operating within a bioregion on the basis of collaboration, exchange, cooperation, and mutuality rather than contention, competition, and selfishness.

The prime example of such an interaction on the bioregional scale would be the social symbiosis between the city and the countryside, the urban and the rural—a correlation that has been celebrated by philosophers from Aristotle on and brilliantly analyzed by the woman with whom I share the platform this morning, Jane Jacobs, and the demise of which has been tellingly bemoaned by most of the giants of our century from Mumford to Borsodi to Bookchin. Listen to Fritz Schumacher:

Human life, to be fully human, needs the city; but it also needs food and other raw materials gained from the country. Everybody needs ready access to both countryside and city. It follows that the aim must be a pattern of urbanization so that every rural area has a nearby city, near enough so that people can visit it and be back the same day. No other pattern makes human sense.

Actual developments during the last hundred years or so, however, have been in the exactly opposite direction: the rural areas have been increasingly deprived of access to worthwhile cities. There has been a monstrous and highly pathological polarization of the pattern of settlements.

"Pathological polarization": the mixedness of the metaphor aside, that is obviously the exact opposite of symbiosis and is equally obviously, as Schumacher saw, the condition of our time. Could we imagine a sadder comment?

In a bioregional society the division between urban and rural, industrial and agricultural, population and resources, would be replaced by an equilibrium, a symbiosis. On the one hand, the city would be necessary as a producer of certain kinds of goods, as a center of artistic culture, as a source of the assembled civic virtues, though the city need not be of immense size—indeed, no larger than fifty thousand or one hundred thousand people—and in fact would ideally replicate rather than grow so that instead of a single metropolis there would be a multiplicity of cities of modest sizes scattered throughout the bioregion. And on the other hand, the country would of course be necessary as the prime source of food and water and the materials of shelter, clothing, artisanship, and trade, and especially as the embodiment of the bioregional spirit of Gaea, whose presence should be felt daily by the inhabitants of every settlement, of whatever size. This equilibrium should not suggest some sort of polarization of its own, for in a bioregional society living with, on, for, and around the earth as a necessitarian matter of course, the countryside would become part of the city, not merely in the sense of parks and woodlands and greenswards and open waterways—as fundamental as they are—and not merely in the sense of backyard and rooftop gardens and floral displays and tree-lined streets and plaza fountains—as desirable as they are—but through the integration into every urban process of a total understanding of ecological principles until the smallest child knows that water does not come from a pipe in the basement and that you can't throw anything away because there is no "away."

Now, it would be possible to continue this description of the bioregional civilization as derived from the laws of nature—as it relates to energy, for example, and agriculture and health and defense and much else besides—and of course such a task is ultimately necessary for the bioregional citizenry to undertake, with study, in depth, over time. But in discussing those four primary determinants— scale, economy, politics, and society—I hope I have suggested to you the outlines of such a project and something of what it might evolve into—some of the bones, a little of the meat—so that you can appreciate its validity at least as a philosophical approach.

To my perception, honed through these many years of mid-century turmoil, bioregionalism satisfies the principal conditions of an effective political project, most particularly in these respects: it is rooted in the historical realities of the past, it accords with the visible patterns of the present, and it provides desirable and workable visions of the future. I would like, all too briefly, to touch upon each of these.

Historical Realities

There is nothing more fundamentally supportive of the validity of bioregional-ism than its being the modern version of a very old perception of the world held not merely by the Greeks but, as I indicated before, by virtually every preliterate society of which we have knowledge. It must mean something that the early human societies which occupied the earth held this perception as a truth so profound that it could be accurately described as almost innate; it must have significance that in most subsequent societies until quite recently the earth and its behavior formed the basis of all folk knowledge, not merely in matters of agricul-ture and nutrition but in medicine, religion, art, and even government. And as Schumacher says—it is indeed the ultimate sentence of *Small Is Beautiful*—"The guidance we need for our work cannot be found in science or technology, the value of which utterly depends on the ends they serve; but it can still be found in the traditional wisdom of mankind."

But the historical validity of this concept—the provenance I might say, as the art dealers do in describing the history of an artwork to establish its authenticity—can be certified in an even more concrete way, closer to this time and place. Re-gionalism, whether conceived of as sectionalism, localism, separatism, agrarian-ism, states rights, or nullificationism, has a fine and venerable tradition in this country and is by any reckoning as American as—depending on your region—apple, peach, Boston cream, Jefferson Davis, sweet cactus, German cherry, or Key lime pie.

Frederick Jackson Turner, the great Wisconsin historian, knew it, and it formed the basis of a lifetime of studies culminating in *The Significance of Sections in American History*, where he showed that only by a consideration of American sectional, or regional, differences could one understand the patterns of settle-ment, migration, architecture, literature, and economic and political life: "We in America are in reality," he concluded, "a federation of sections rather than states."

Lewis Mumford knew it when he put together the Regional Plan Association in 1923, an ambitious—and for a decade successful—attempt to create regional plans along geographical lines, which would, in his words, mean the "reinvigo-ration and rehabilitation of whole regions so that the products of culture and civi-lization, instead of being confined to a prosperous minority in the congested cen-ters, shall be available to everyone at every point" and so that we may "eliminate our enormous economic wastes, give a new life to stable agriculture, and [though I blush to say it] set down fresh communities planned on a human scale."

Howard Odum knew it when he started a highly honored and remarkably multidisciplinary school of regionalism at the University of North Carolina in the 1930s and over two decades produced a series of scholarly books highlighted by the massive 1938 study *American Regionalism*, all to the point of showing, as he put it, that "regionalism . . . represents the philosophy and technique of self-help, self-development, and initiative in which each area unit is not only aided in, but is committed to the full development of its own resources and capacities."

And even the United States government, *mirabile dictu*, knew it when in 1934 it authorized a National Resource Committee to study the regions of America and discovered that "regional differentiation may turn out to be the true expression of American life and culture [reflecting] American ideals, needs, and viewpoints far more adequately than does State consciousness and loyalty." It was out of this exhaustive study, more than fifteen reports in all, that the Tennessee Valley Authority was born, America's greatest—though in some respects most distorted—experiment with regionalism.

Much there is today that goes against the grain of regionalism, of course, much forcing the nation away from its natural contours toward the artificial unanimity of a monolithic plasticized government. And yet . . . and yet . . . even in an age such as this the historical realities of regionalism, as perceived by those several generations of scholars and planners, cannot be erased. And that is why—just to touch on a small part of this complex subject—there are today more than twenty-five specialized regional governments on the TVA model operating in the United States, more than a thousand metropolitan regional districts, almost five hundred substate planning districts, and more than a hundred multi-county regional associations. That is why regional planning, particularly since the 1970s, has become an established academic and governmental profession and all but ten states have active regional planning departments, some of which are now beginning to be responsive to bioregional imperatives. And that is why there are real and persistent rivalries among national regions for such things as defense contracts, army bases, public works projects, businesses, conventions, sports franchises, and the like, a competition so strong these days that the *Wall Street Journal* this spring ran a front-page story declaring that "another war between the states is raging."

Contemporary Trends

Another salient measure of the validity of the bioregional enterprise is that it accords well with the most basic—and complementary—political processes in the world today: first, the pressure from a series of mounting national and global crises that threaten to end with nothing less than the collapse of the established order and, second, the concurrent trend toward the disintegration of imperial, continental, and national arrangements—what is called separatism, decentralism, or, to use a supposedly derogatory term, Balkanization.

I do not need to belabor the evidence of the crises threatening the contemporary industrial system. It is sufficient, I think, merely to say that those who are predicting some sort of near-term calamity and collapse range through all the academic disciplines, from physics to philosophy, and through all the political positions, from anarchist to authoritarian. To be sure, that is not enough to guarantee that such a disintegration will in fact take place, but it is accompanied by plentiful signs of the failure of the established orders to satisfy the most basic human needs of large portions of the population, signs of the apparently unstoppable disintegration of America's cities, signs of the rising tides of poverty, disease, ignorance, anomie, suicide, violence, and crime, even in this most affluent of nations. And it is interesting that, whatever form the collapse in fact will take, we already possess a wide variety of labels for it: Schumacher's "the degeneration of the industrial system," Robert Nisbet's "twilight of authority," the Club of Rome's "oncoming age of scarcity," Arnold Toynbee's "the end of the frontier," and, variously, "the coming dark age," "the twilight of capitalism," "the biological time bomb," "overshoot," "the end of the American Era."

I remember Fritz Schumacher, during the height of the oil crisis, looking up at the skyscrapers of New York and remarking, "I wonder how many people will want to climb to the fortieth floor when there is not enough electricity to run the elevators," and going on to suggest that the human limit for climbing is about four or five stories. In his typically gentle way he was encapsulating a truth: that the disintegration of the present system is coming about virtually by itself as the era of industrial capitalism, based on the exploitation of unending frontiers and non-renewable resources or, as William Catton puts it, on its ability to steal from elsewhere and elsewhen, reaches its inevitable end; and when it does, the whole face of industrial society—the height of its buildings, the size of its cities, the extent of its markets, the reach and power of its governments, the nature of its institutions—will be forced to change, and change drastically. There is no escaping this eventual transformation, for its inevitability is programmed into the very genes of this society, part of its capitalistic DNA if you will, and as Schumacher wrote in his final work, "It is no longer possible to believe that any political or economic reform, or scientific advance, or technological progress could solve the life and death problems of industrial society."

The alternative society that may rise from its ashes—or, if we are terribly lucky, that will evolve before the fires of destruction actually begin and create those ashes—the one that could logically be thought of as befitting the coming age, attuned to the conditions that will prevail after the industrial society runs its course, is the bioregional one. But in a sense we do not necessarily need to wait until then, however near that "then" is, because at least one form of the bioregional society is already taking shape in the nascent separatist movements that have come into being in almost every corner of the globe within the last generation. They too represent an organic, I would argue an inevitable, response to the disintegration of the contemporary order, a growing centrifugal force as industrialism spins more wildly about. As a global phenomenon the current rise of these movements is something quite without precedent in history; it is, according to Eric Hobsbawm, "the characteristic nationalist movement of our time" and "an unquestionably active, growing and powerful socio-political force." An exhaustive elaboration would be exhausting; it should be enough to note only the most active movements just within Europe, the continent where it might have been presumed that nations were the oldest, strongest, and most cohesive: there are the Bretons, Corsicans, Occitanians, and Alsatians in France; the Catalonians, Andalusians, and Basques in Spain; the Welsh, Scots, and Cornish in Britain; the Sicilians and Tyrolians in Italy; the Waloons and Flemish in Belgium; the Latts, Lithuanians, Estonians, Ukrainians, Georgians, and a variety of Asians in the Soviet Union; the Turks and Greeks in Cyprus; the Croatians, Serbs, Bosnians, Macedonians, and Montenegrins in Yugoslavia—and that, I remind you, is the bare surface.

It is truly remarkable. The undeniable trend of these past forty years has not been toward larger and more consolidated arrangements but, everywhere in the world, toward smaller and more decentralized ones. In the words of Harold Isaacs: "What we are experiencing is not the shaping of new coherences but the world breaking into its bits and pieces. . . . We are refragmenting and retribalizing ourselves."

What is so interesting in this amazing process is the clear expression of the bioregional idea. For though it has long been acknowledged that the cultural aspects of these separatist movements are grounded in their special regional histories, from which they take their obvious and cherished differences of language and dress and music, the fact is that their political and social characters are every bit as rooted in the long, intimate, and knowledgeable association with their particular bioregion and its history. And the truths these movements embody, the apparently unquenchable truths, are in every case the product of the land they hold sacred.

Desirable Visions

In treading upon the insubstantial ground of the future we take certain risks, and we must face the fact that the word "utopian" has become an epithet, a chastisement, for those who would dream of things that never were and imagine that they still might be. Yet it is a necessary part of any political construct that it offer an image of the future that can be regarded as positive and liberatory and realistic and energizing. This, I submit, bioregionalism succeeds in doing.

For what the bioregional vision suggests is a way of living that not merely can take us away from the calamities of the present and the diseases of our quotidian lives but can provide its own indwelling enrichments and satisfactions, a widening of human possibilities. Imagine, if you will, the joy of knowing, as we can imagine from the scholarly record, what the American Indians knew: the meaning of the changes of the wind on a summer afternoon; the ameliorative properties of everyday plants; the comfort of tribal, clannic, and community ties throughout life; the satisfaction of being rooted in history, in lore, in place; the excitement of a culture understandable because immanent in the simple realities of the surroundings. Imagine a life primarily of contemplation and leisure, where work takes up only a few hours a day—an average of fewer than four, according to the studies of nonliterate societies—where conversation and play and lovemaking become the common rituals of the afternoon, and there is no scramble for the necessities of life because they are provided regularly, equally, joyfully, and without charge. Imagine a life—and here I am paraphrasing an anthropologist's description of a California Indian tribe—where people do not feel themselves to be independent, autonomous individuals but rather deeply bound together with other people and with the surrounding nonhuman forms of life in a complex interconnected web of being, a true community in which all creatures and all things can be felt almost as brothers and sisters and where the principle of nonexploitation, of respect and reverence for all creatures, all living things, is as much a part of life as breathing.

Yet I think, however enchanting that image might be, the bioregional vision is even more important in that it actually has an air of the practical, the doable, the achievable: it has the smell of reality about it.

For one thing, the idea of the bioregion is accessible to people, all kinds of people, for as Kevin Lynch notes in his Managing the Sense of a Region, "Our senses are local, while our experience is regional." Lee Swenson, an early bioregionalist, has reported that when he took his bioregional slide show across the country, it didn't take long for his audiences to come up with some rough consensus about the territories they lived in that pretty well matched any ecological definition of their bioregions. If true, this suggests that the process of organizing around this issue, especially among those outside of the usual constituencies for social change, is made much easier.

Then, too, bioregionalism joins—or at least has the potential to join—right and left (or, perhaps more precisely, it ignores right and left), thus uniting the communard and the NRA hunter, the homesteader and the conservationist, the antinuclear activist and the antipowerline farmer. The concern for place, for the preservation of nature, the return to such traditional American values as self-reliance, local control, town-meeting democracy—these things can ally many different kinds of political people; in fact, they have a way of blunting and diminishing other and less important political differences.

Bioregionalism also has the virtue of gradualism, for it suggests that the process of change—or organizing, educating, energizing a following and reshaping, refashioning, recreating a continent—is, like the overarching processes of Gaea herself, not revolutionary and cataclysmic but, like the drift of the continents on their tectonic plates, steady, slow, continuous, regular, and inevitable. One does not imagine a bioregional civilization coming about by revolutionary decree—no matter whose revolution—or even, in truth, by legislative or administrative fiat. If one had to dictate or legislate the bioregional future, it would never happen, because it would be resisted and sabotaged as crazy and utopian and impractical and un-American; it is only by the long and steady tenor of evolution that people will ease themselves into such a society as the alternative futures gradually come to seem senseless and the bioregional prospect becomes the only sane choice.

And finally, the bioregional vision does not demand elaborate wrenching of either physical or human realities. It does not posit, on the one hand, the violent interference with nature that so many of the scientific technofix visions of the future do—those, for example, that ask for icebergs to be floated into deserts or the Great Plains to be given over to concentrated nuclear power plants (it does not, for that matter, have anything to do with nuclear fission, the single most unnatural project humankind has ever devised) or rockets full of people to be fired millions of miles away into space colonies around the sun. And it does not imagine, on the other hand, the creation of some kind of unlikely and never-before-encountered superbeings as do so many of the reformist and radical visions of the future—those, for example, that promise "a new socialist man" without motives of greed or self-interest or that plan by education or religion or therapy to create a populace living in aquarian harmony without human vices. On the contrary, bioregionalism insists on taking the world "as it is" and people, as I have indicated before, as they are.

I hope I do not suggest with all of this that the bioregional project is blind to the chances of failure—or what is worse, half-failure—or is unmindful of the pains that might attend the accomplishment of its ends. Just because I am suggesting hope and desirability, I am not suggesting sanguinity or quiescence or detachment or passivity. I mean merely to underscore that element of the project which speaks to the Biblical admonition, "Where there is no vision the people perish."

Lewis Thomas concludes his fascinating *Lives of a Cell* with this observation:

"Viewed from the distance of the moon, the astonishing thing about the earth, catching the breath, is that it is alive. The photographs show the dry, pounded surface of the moon in the foreground, dead as an old bone. Aloft, floating free beneath the moist, gleaming membrane of bright blue sky, is the rising earth, the only exuberant thing in this part of the cosmos. If you could look long enough, you would see the swirling of the great drifts of white cloud, covering and uncovering the half-hidden masses. . . . It has the organized, self-contained look of a live creature, full of information, marvelously skilled at handling the sun."

And just one year later, in 1975, the British atmospheric chemist James Love-lock described in the magazine *New Scientist* a perception of the world that had come to him and his colleagues one day: "It appeared to us that the Earth's biosphere is able to control at least the temperature of the Earth's surface and the composition of the atmosphere. Prima facie, the atmosphere looked like a contrivance put together cooperatively by the totality of living systems to carry out certain necessary control functions. This led to the formulation of the proposition that living matter, the air, the oceans, the land surface, were parts of a giant system which was able to control temperature, the composition of the air and sea, the Ph of the soil and so on as to be optimal for survival of the biosphere. The system seemed to exhibit the behaviour of a single organism, even a living creature."

Out of this new perception Lovelock and his colleagues created a whole new scientific hypothesis on the nature of the biosphere. Or should I say a very old hypothesis? For when they went in search of a name for this hypothesis, they sought out William Golding, the novelist who just recently was honored with the Nobel Prize. And what did he suggest immediately? As Lovelock writes, "He suggested Gaia—the name given by the ancient Greeks to their Earth goddess."

So after all, there seems to be no doubt about it. The earth, the biosphere, is alive, a living creature, one and visible, containing within itself all living creatures. Like any living entity it can be stressed or injured or diseased, as it surely is now. But it will live—of that we can be sure—one way or another, and it will resettle itself, restore itself, with humankind or without. It behooves us, as nothing in the long history of humankind, I believe, has so far behooved us, to come to this literally most vital understanding and, before it is too late, give up those demonic practices that threaten our fundamental forms of existence and ultimately our existence itself. We must make the goddess Gaea part of—no, I want to say the whole of—our lives, even though that may be, as John Todd has suggested, a change of consciousness as profound and as wrenching as that which accompanied the origination of agriculture some 10,000 years ago. But then, what other choice, really, do we have?

—1983

David W. Orr

Environmental Literacy:
Education as if the Earth Mattered

David Orr ranks among today's leading environmental educators. He is the founder of the Meadowcreek Education Project and professor of environmental studies at Oberlin College, where he teaches the information, the thinking, and the practical skills that will enable students to undertake the work of protecting and restoring the environment. His is one of the most lucid voices speaking on behalf of ecology and educational reform at the end of the twentieth century.

He is the author of *The Global Predicament: Ecological Perspectives on World Order*, *Ecological Literacy and the Transition to a Postmodern World*, *The Campus and Environmental Responsibility*, and *Collected Essays on Conservation Biology*. Lincolnesque in his expression of the complexity of our time and the means by which we may tackle the problems that confront us, his writing is characterized by timeliness, depth, and wonderfully welcome horse sense. Although his humor is quintessentially American, his message, like Schumacher's, is planetary, visionary, and practical.

Orr defines the encroaching planetary crisis as one of "mind, perception, and values"—and therefore traceable to education. Like David Ehrenfeld, he sees the partnership between the university on the one hand and power and commerce on the other as destroying the integrity of the academy. Yet Orr is convinced that "the ways in which people are educated make all the difference." What is called for, he urges, is a form of education that will change our vision of reality and cultivate environmental intelligence and ecological imagination. Through his programs at Oberlin and Meadowcreek, his books, and his lectures, he is bringing this change about.

After due reflection on the state of education in his time, H. L. Mencken concluded that significant improvement required only that the schools be burned to the ground and the professoriate hanged. For better or worse, the suggestion was largely ignored. Made today, however, it might find a more receptive public ready to purchase the gasoline and rope. Americans, united on little else, seem of one mind in believing that the educational system K through Ph.D. is too expensive, too cumbersome, and not, on the whole, very effective. They believe that it needs radical reform. They are divided, however, on how to go about it. On one side of the debate are those who argue that the failure is due mostly to the lack of funding for laboratories, libraries, equipment, salaries, and new buildings—a view held most avidly, not surprisingly, by professional educators. On the other side are those such as Benno Schmidt, the former president of Yale University, who propose to abandon much of the present system of schools and create a national system of for-profit schools.

Both sides of the debate agree, nonetheless, on the basic aims and purposes of education, which are, first, to equip our nation with a "world class" labor force in order to compete more favorably in the global economy and, second, to provide each individual with the means for maximum upward mobility. About these, the purposes of education both higher and lower, there is great assurance.

Yet there are better reasons to rethink education that have to do with issues of human survival, which will dominate the world of the twenty-first century. The generation now being educated will have to do what we, the present generation, have been unable or unwilling to do: stabilize a world population that is growing at the rate of a quarter of a million each day; stabilize and then reduce the emission of greenhouse gases, which threaten to change the climate—perhaps disastrously; protect biological diversity, now declining at an estimated rate of one hundred to two hundred species per day; reverse the destruction of rainforests (both tropical and temperate), now being lost at the rate of one hundred and sixteen square miles or more each day; and conserve soils, now being eroded at the rate of sixty-five million tons per day. Those who follow us must learn how to use energy and materials with great efficiency. They must learn how to utilize solar energy in all its forms. They must rebuild the economy in order to eliminate waste and pollution. They must learn how to manage renewable resources for the long term. They must begin the great work of repairing, as much as possible, the damage done to the earth in the past two hundred years of industrialization. And they must do all of this while addressing worsening social and racial inequities. No generation has ever faced a more daunting agenda.

For the most part, however, we are still educating the young as if there were no planetary emergency. Remove computers, a scattering of courses throughout the catalog, and a few programs, and the curriculum of the 1990s looks a lot like that of the 1950s. But the crisis we face is first and foremost a crisis of mind, perception, and values—hence, a challenge to those institutions presuming to shape minds, perceptions, and values. It is an educational challenge.

More of the same kind of education that enabled us to industrialize the earth can only make things worse. This needs to be stated strongly to underscore the fact that the environmental crisis is not primarily the work of the ignorant and uneducated; rather, it is that of so-called well-educated people who, in Gary Snyder's words in The Practice of the Wild, "make unimaginably large sums of money, people impeccably groomed, excellently educated at the best universities—male and female alike—eating fine foods and reading classy literature, while orchestrating the investment and legislation that ruin the world." These are people who have been educated to think that human domination of nature is our rightful destiny. I am not making an argument against education but rather an argument for the kind of education that prepares people for lives and livelihoods suited to a planet with a biosphere that operates by the laws of ecology and thermodynamics.

The skills, aptitudes, and attitudes necessary to industrialize the earth are not necessarily the same as those that will be needed to heal the earth or to build durable economies and good communities. Resolution of the great ecological challenges of the next century will require us to reconsider the substance, process, and purposes of education at all levels and to do so, as Yale historian Jaroslav Pelikan writes in The Idea of a University: A Reexamination, "with an intensity and ingenuity matching that shown by previous generations in obeying the command to have dominion over the planet." But Pelikan himself doubts whether the university "has the capacity to meet a crisis that is not only ecological and technological, but ultimately educational and moral." Why should this be so? Why should those institutions charged with the task of preparing the young for the challenges of life be so slow to recognize and act on the major challenges of the coming century?

A clue can be found in a small book called Universities and the Future of America, written by Derek Bok, the former President of Harvard University. Bok writes:

Our universities excel in pursuing the easier opportunities where established academic and social priorities coincide. On the other hand, when social needs are not clearly recognized and backed by adequate financial support, higher education has often failed to respond as effectively as it might, even to some of the most important challenges facing America. Armed with the security of tenure and the time to study the world with care, professors would appear to have a unique opportunity to act as society's scouts to signal impending problems. . . . Yet rarely have members of the academy succeeded in discovering emerging issues and bringing them vividly to the attention of the public. What Rachel Carson did for risks to the environment, Ralph Nader for consumer protection, Michael Harrington for problems of poverty, Betty Friedan for women's rights, they did as independent critics, not as members of a faculty.

This observation appears on page 105 of the book and is not mentioned again. It should have been on page one and would have provided the subject for a better book. Had Bok gone further, he might have been led to ask whether the same charge of lethargy might be made against those presuming to lead American education. He might also have been led to rethink old and unquestioned assumptions about liberal education. For example, John Henry Newman, in his classic *The Idea of a University*, drew a distinction between practical and liberal learning that has influenced education from his time to our own. Liberal knowledge, according to Newman, "refuses to be informed by any end, or absorbed into any art." Knowledge is liberal if "nothing accrues of consequence beyond the using." "Liberal education and liberal pursuits," he wrote, "are exercises of mind, of reason, of reflection." All else he regarded as practical learning, which had no place in the liberal arts.

To this day Newman's distinction between practical and liberal knowledge is seldom transgressed in liberal arts institutions. Is it any wonder that faculty, mindful of the penalties for transgressions of one sort or another, do not often deal boldly with the kinds of issues that Bok describes? I do not wish to excuse faculty, but I would like to note that educational institutions more often than not reward indoor thinking, careerism, and safe conformity to prevailing standards, all of which maintain the split between liberal and practical knowledge.

Harvard philosopher and mathematician Alfred North Whitehead had a different view of the liberal arts. "The mediocrity of the learned world," he wrote in 1929 in *The Aims of Education*, could be traced to its "exclusive association of learning with book-learning." Real education required "first-hand knowledge," by which he meant an intimate connection between the mind and "material creative activity." Others, like John Dewey and J. Glenn Gray, reached similar conclusions. In *Re-Thinking American Education*, Gray wrote, "Liberal education is least dependent on formal instruction. It can be pursued in the kitchen, the workshop, on the ranch or farm . . . where we learn wholeness in response to others." A genuinely liberal education, in other words, ought to be liberally conducted, aiming to develop the full range of human capacities. And institutions dedicated to the liberal arts ought to be more than mere agglomerations of specializations.

Had Bok cared to proceed even further, he would have had to address the loss of moral vision throughout higher education. Stan Rowe writes in *In Home Place: Essays on Ecology* that the university has "shaped itself to an industrial ideal—the knowledge factory. Now it is overloaded and top-heavy with expertness and information. It has become a know-how institution when it ought to be a know-why institution. Its goal should be deliverance from the crushing weight of unevaluated facts, from bare-bones cognition or ignorant knowledge: knowing in fragment, knowing without direction, knowing without commitment." Many years ago William James saw this coming and feared that the university might one day develop into a "tyrannical Machine with unforeseen powers of exclusion and corruption." We are moving along that road and should ask why this has come about and what can be done to change it.

One source of the corruption is the marriage between the academy and the worlds of power and commerce. This marriage was first proposed by Francis Bacon, though it was not fully consummated until the Manhattan Project during World War II. But marriage, implying affection and mutual consent, is perhaps not an accurate metaphor. The union is, rather, a cash relationship, beginning with a defense contract here and a research project there. At present, not a few university departments still work as adjuncts of the Pentagon and others as adjuncts to industry, in the hope of reaping billions of dollars in fields such as genetic engineering, nanotechnologies, agribusiness, and computers. Even where this is not true, it is difficult to escape the conclusion that much of what passes for research, as historian Page Smith writes in *Killing the Spirit*, is "essentially worthless . . . busywork on a vast almost incomprehensible scale."

Behind the glossy facade of the modern academy there is often a vacuum of purpose waiting to be filled by whomever and whatever. For example, as reported by Gene Logsdon (in "Death of a Sacred Cow," *Ohio*, May 1992) the College of Agriculture at a land-grant university of note claims to be helping "position farmers for the future." But when asked what farming would be like in the twenty-first century, the Dean of the College replied, "I don't know." Asked, "How can you [then] position yourself for it?" the Dean replied, "We have to try as best we can to plan ahead." This reminds me of the old joke in which the airline pilot reports to the passengers that he has good news and bad news. The good is that the flight is ahead of schedule; the bad is that they are lost. Ironically, in a time of eroding soils and declining rural communities, "turf grass management" is the hot new item at this College of Agriculture.

Finally, had Bok so chosen, he would have been led to question how we define intelligence and what that might imply for our definition of an "educated" person. From an ecological perspective it is clear that we have often confused cleverness and intelligence. Cleverness, as I understand it, tends to fragment things and to focus on the short term. The epitome of cleverness is the specialist whose intellect and person have been shaped by the demands of a single function. Ecological intelligence, on the other hand, requires a broader view of the world and a long-term perspective. Cleverness can be adequately measured by SAT and GRE tests, but intelligence is not so easily computed. In time, I think we will come to see that true intelligence tends to be integrative and often works slowly while mulling things over. Further, intelligence can be inferred, according to Wendell Berry in *Standing By Words*, from the "good order or harmoniousness of [one's] surroundings." In other words, the consequences of our actions are a measure of our intelligence, and the plea of ignorance is no good defense. Because some consequences cannot be predicted, the exercise of intelligence requires forbearance and a sense of limits. Ecological intelligence, in contrast to mere cleverness, does not presume to act beyond a certain scale at which effects can be known and unpredictable consequences would not be catastrophic.

The modern fetish with smartness is no accident. The highly specialized, narrowly focused intellect fits the demands of instrumental rationality built into the industrial economy, and for reasons described by Brooks Adams eighty years ago (and quoted by Page Smith in *Dissenting Opinions*), "Capital has preferred the specialized mind and that not of the highest quality, since it has found it profitable to set quantity before quality to the limit the market will endure. Capitalists have never insisted upon raising an educational standard save in science and mechanics, and the relative overstimulation of the scientific mind has now become an actual menace to order." The demands of building good communities within a sustainable society within a just world order will require more than the specialized, one-dimensional mind and more than instrumental cleverness.

For perspective, let me add that the only people who have lived sustainably on the earth without damaging it could not read. This does not mean they were ignorant. To the contrary, they had enormous amounts of knowledge. Indigenous peoples' knowledge of their ecosystems is extensive. We will never be able to match it. Some ancient agricultural systems were exquisite ecological creations. The ways in which people are educated make all the difference. All previous peoples who had sustainable cultures wove education and research together within the vessel of community. Our culture has taken education and research out of community and broken that vessel.

Looking ahead to the twenty-first century, the task of building a sustainable world order will require dismantling the jerry-built scaffolding of ideas, philosophies, and ideologies that constitutes the modern curriculum. Five measures are necessary to do this.

First, we must develop more comprehensive and ecologically solvent standards for truth. The architects of the modern worldview, notably Galileo and Descartes, assumed that those things that could be weighed, measured, and counted were more true than those that could not be quantified. If it couldn't be counted, in other words, it didn't count. Cartesian philosophy was full of potential ecological mischief, a potential that Descartes's heirs developed to its fullest. His philosophy separated humans from the natural world, stripped nature of its intrinsic value, and segregated mind from body. Descartes was at heart an engineer, and his legacy to the environment of our time is the cold passion to remake the world as if we were merely remodeling a machine. Feelings and intuition have been tossed out along with those fuzzy, qualitative parts of reality such as aesthetic appreciation, loyalty, friendship, sentiment, charity, and love.

These assumptions are not as simple or as inconsequential as they might have appeared in Descartes's lifetime (1596–1650). A growing number of scientists now believe, with Stephen Jay Gould, that "we cannot win this battle to save [objectively measurable] species and environments without forging an [entirely subjective] emotional bond between ourselves and nature as well—for we will not fight to save what we do not love" ("Enchanted Evening," *Natural History*, Sept. 1991).

If saving species and environments is our aim, we will need a broader conception of science and a more inclusive rationality that joins empirical knowledge with the emotions that make us love and sometimes fight. Alfred North Whitehead noted the difference in *Science and the Modern World*: "When you understand all about the sun and all about the atmosphere and all about the rotation of the earth, you may still miss the radiance of the sunset. There is no substitute for the direct perception of the concrete achievement of a thing in its actuality. We want concrete fact with a high light thrown on what is relevant to its preciousness."

Karl Polanyi described this as "personal knowledge" in his book of that name, by which he meant knowledge that calls forth a wider range of human perceptions, feelings, and intellectual powers than those presumed to be narrowly "objective." Personal knowledge, in Polanyi's words, "is not made but discovered. . . . It commits us, passionately and far beyond our comprehension, to a vision of reality. Of this responsibility we cannot divest ourselves by setting up objective criteria of verifiability—or falsifiability, or testability. . . . For we live in it as in the garment of our own skin. Like love, to which it is akin, this commitment is a 'shirt of flame,' blazing with passion and, also like love, consumed by devotion to a universal demand. Such is the true sense of objectivity in science."

Cartesian science rejects passion and personality but, ironically, can escape neither. Passion and personality are embedded in all knowledge, including the most ascetic scientific knowledge informed by the passion for objectivity. Descartes and his heirs simply had it wrong: there is no way to separate feeling from knowledge or object from subject; there is no good way to separate mind or body from its ecological and emotional context. It may even be the case—as Donald Griffin, among others, is coming to suspect and as he states in *Animal Minds*—that intelligence is not a human monopoly. Science without passion and love can give us no good reason to appreciate the sunset, nor can it give us any purely objective reason to value life. These must come from deeper sources.

Second, we must challenge the hubris buried in the *hidden* curriculum which assumes that human domination of nature is good, that the growth economy is natural, that all knowledge, regardless of its consequences, is equally valuable, and that material progress is our right. Because we hold these beliefs, we suffer a kind of cultural immune-deficiency anemia that renders us unable to resist the seductions of technology, convenience, and short-term gain. In this perspective, the ecological crisis is a matter of discerning between "life and death, blessing and cursing," as the writer of Deuteronomy put it, and of learning to choose life. It is a test of our loyalties and of our deeper affinities for the living world, what E. O. Wilson calls "biophilia."

Third, we must address the fact that the modern curriculum teaches a great deal about individualism and rights but teaches little about citizenship and responsibilities. The ecological emergency can be resolved only if enough people come to hold a bigger idea of what it means to be a citizen, and this knowledge will have to be taught carefully at all levels of education. Unfortunately, a pervasive cynicism about our higher potentials and collective abilities now works against us. Even my most idealistic students often confuse self-interest with selfishness, a mistake that allows both Mother Theresa and Donald Trump to be described as self-maximizers, both merely "doing their thing."

This is not just a social and political problem. The ecological emergency is about the failure to comprehend our citizenship in the biotic community, as Aldo Leopold noted in *A Sand County Almanac*. From the modern perspective we should see clearly how utterly dependent we are on the wider community of life. Our political language gives little hint of this dependence. The word "patriotism," for example, is devoid of ecological content. It must come to mean how we use our land, forests, air, water, and wildlife. To abuse natural resources, to erode soils, to destroy natural diversity, to waste, to take more than one's fair share, or to fail to replenish what has been used must someday come to be regarded as unpatriotic. And "politics" must once again come to mean, as Vaclav Havel puts it in *Summer Meditations*, "serving the community and serving those who will come after us." Our notions of citizenship and politics are anemic in large measure because our language has been corrupted by those who have stood to gain a great deal if words could be compromised. A primary task of educators and teachers is to restore integrity to language in order that we might reclaim the commonwealth that rightfully belongs to all of us.

Fourth, we must question the widespread assumption that our future is one of constantly evolving technology and that this is a good thing. Those who call this faith into question are dismissed as Luddites by people who, as far as I can tell, know little or nothing about the real history of Luddism. Faith in technology is built into nearly every part of the curriculum as a kind of blind acceptance of the notion of progress. When pressed, however, true believers describe progress to mean not a consciously chosen path but a mindless, uncontrollable technological juggernaut moving through history. Increasingly, such assumptions are being incorporated into our methods of pedagogy without much serious question. Computer literacy, for instance, has become a national goal, pushed more often than not by people who have something to sell. This technological fundamental-ism deserves to be questioned. Is technological change taking us where we want to go? What effect does technology have on our imagination and particularly on our social, ethical, and political imagination? And what net effect does it have on our ecological prospects?

We need an ecological imagination with which we can envision restored landscapes, renewed ecosystems, and whole people living in a whole biosphere. Yet in a technological age it should come as no surprise that our imagination is increasingly confined to technological possibilities: faster and more powerful computers, television, virtual reality generators, genetic engineering, nanotech-nologies. In *The Road to Wigan Pier* George Orwell warned that the "logical end" of technological progress "is to reduce the human being to something resembling a brain in a bottle." Behold, fifty years later there are now those who propose to develop the necessary technology to "download" the contents of the mind into a robot-like machine / body (Moravic). Orwell's nightmare is coming true, thanks in no small part to research conducted in our proudest universities. Such research stands in sharp contrast to our real needs. We need decent communities, good work to do, loving relationships, stable families, and a way to transcend our inherent self-centeredness. Our needs, in short, are those of the spirit, yet our imagination and creativity are overwhelmingly aimed at *things*.

There is a fifth challenge looming on the horizon, one that strikes at the oldest and most comfortable assumption of all: that education can take place only in "educational" institutions. During a recent social gathering I was bluntly in-formed by a Fortune 500 executive that corporations, now engaged in what they take to be education, will put many schools and colleges out of business in the next two decades. This is a warning to which teachers and administrators should listen, and for the same reasons that General Motors should have listened had a

Toyota executive said something similar around, say, 1970. Colleges and universities are expensive, slow moving, often unimaginative, and weighted down by the burdens of self-congratulation and tradition. They offer a discipline-centric curriculum that corresponds modestly with reality. The grip colleges and universities now have on "education" will be broken when young people discover alternatives that are far cheaper, faster, and better adapted to economic realities. The rub is that corporations will not educate liberally. Instead, they will offer something more akin to hi-tech job training. But that will not matter much to the growing number unable to afford the expense of a liberal arts education; it will matter, however, in terms of our larger prospects, whether people are trained narrowly or educated liberally.

"No important change in ethics," Aldo Leopold once wrote, "was ever accomplished without an internal change in our intellectual emphasis, loyalties, affections, and convictions." Ecological education aims to bring about that change in emphasis, loyalties, affections, and convictions necessary to heal the breach between humanity and its habitat. It is less a reform tinkering at the margins of the status quo than a jailbreak from old assumptions, from the straitjacket of discipline-centric curricula, and even from confinement in classrooms and school buildings.

Ecological education will, first, require the reintegration of experience into education, because experience is an indispensable ingredient of good thinking. One way to do this is to use the campus as a laboratory for the study of food, energy, materials, water, and waste flows. Research on the ecological impacts of a specific institution reduces the abstractness of complex issues to manageable dimensions, and it does so on a scale that lends itself to finding solutions, which is an antidote to the despair felt by students when they understand problems but are powerless to effect change. Campuses need to take a closer look at the economic potential of their regions to find out how their money could be spent locally and invested locally to help move the world in a more sustainable direction. For example, students researching food purchases at several liberal arts colleges helped their food services to replace distant suppliers with ones closer to the campus while reducing costs, improving food quality, and helping the local economy. At Oberlin, where I teach, students acquired data on campus resource flows and presented recommendations to the administration, some of which are being implemented at a significant savings to the college.

We need to go further. The old curriculum is shaped around the goal of extending human dominion over the earth to its fullest extent. The new curriculum must be organized around what can be called the "ecological design arts," around developing the analytic abilities, ecological wisdom, and practical wherewithal essential to making things fit in a world of microbes, plants, animals, and entropy. Ecological problems are in many ways design problems: our cities, cars, houses, and technologies often do not fit in the biosphere. Ecological design requires the ability to comprehend patterns that connect, which means looking beyond the boxes we call disciplines to see things in their larger context. Ecological design is the careful meshing of human purposes with the larger patterns and flows of the natural world; it is the careful study of those patterns and flows to inform human purposes. Competence in ecological design requires spreading ecological intelligence—knowledge about how nature works—throughout the curriculum. It means teaching students the basics of what they will need to know in order to stretch their horizons, to create a civilization that runs on sunlight; uses energy and materials with great efficiency; preserves biotic diversity, soils, and forests; develops sustainable local and regional economies; and restores the damage inflicted on the earth throughout the industrial era.

But we must go further still. I do not know who first proposed dividing the world up into disciplines, but the time has come to think about how we might reconnect things. To do so, I propose that we dedicate part of the curriculum at all levels to the study of a thing or a place in our environment such as a river, a mountain, a valley, a lake, soils, a marsh, a particular animal, birds, the sky, the seashore, or even an entire small town. A course on a local river, for instance, could begin with a float trip down the river to acquaint students with the thing itself. Students might then select different aspects of the river to study, including its evolution, human settlements, ecology, fish and aquatic life, the effects of pollution on it, the laws governing its use, and so forth. The course ought to conclude with a second float trip, during which students describe to one another what they've learned.

Things like rivers are real, disciplines are abstract. Real things engage all of the senses, not just the intellect. To understand a river one must master most of what is in the curriculum and some things that are not. To know a river well, moreover, one must feel it, taste it, smell it, swim in it, see it in its different moods, and converse with other people who know it well. Disciplinary knowledge tends to be isolated from tangible realities and is often difficult to connect with concrete ecological realities. I am proposing that students learn to appreciate, respect, and perhaps even love a specific part of the created world before we give them the power implicit in purely abstract, decontextualized knowledge. If students learn to understand how the world works as a physical system and why this understanding is important for their life prospects and their means of livelihood, they will also know how to make an economy that works.

The point here has to do with where to attach the cart to the horse. Defenders of the conventional curriculum believe that mastery of a discipline leading toward specialized knowledge is an end in itself. I recommend reversing this priority in order to place knowledge in a specific ecological context, to engage all of the senses of the student, not just the intellect, to initiate a romance with the natural world, and perhaps also to teach the limits of knowledge relative to a specific feature of the natural world, which is the beginning of ecological wisdom.

Ecological education will also require changes in the operations and priorities of schools and colleges. For example, in the survey of institutional resource flows mentioned above, students discovered ways in which their institution could reduce costs, improve services, lower environmental impacts, and help the local economy. The principle here is simple: those institutions that purport to induct the young into responsible adulthood ought themselves to be responsible stewards of the world the young will inherit. Colleges and universities often measure themselves by such indicators as endowment per student or percentage of faculty with Ph.D.s; from an ecological perspective, another set of indicators of institutional quality would include:

1. emission of CO_2 per student;
2. percentage of materials recycled;
3. percentage of recycled materials purchased;
4. use of toxic materials;
5. percentage of renewable energy consumed;
6. percentage of organic wastes composted;
7. water use per student;
8. percentage of food served that was organically grown;
9. beef consumed per student.

Beyond reducing environmental impacts, educational institutions could make use of their large budgets to help leverage the emergence of sustainable local and regional economies. Thus, a decision to buy food grown organically on local farms could provide an incentive to local farmers to shift production toward environmentally sustainable methods of farming. The same principle is true throughout most of the range of goods and services purchased by educational institutions. These institutions are visible and well respected. What they do and how they do it matters to a public looking for leadership. Were a college or university to announce that it intended to take the future seriously enough to reduce its environmental impacts to the lowest possible level and to do so while improving the local and regional economy, people would take notice.

These principles apply similarly to the management of endowment funds. Arguably, the best investment an endowment manager can make is not in one stock or another but in energy efficiency on the campus. For example, reducing the amount of energy consumed by campus lighting to what is now technically possible can return 40 to 60 percent on the investment (equivalent to a payback time of roughly two years). Other improvements in energy efficiency also offer attractive, if lower, rates of return. After installation, the savings from energy efficiency operate much like an endowment for only the cost of maintenance. The same opportunities may be possible for investments in local energy efficiency. In either case, the point is to use the financial power of educational institutions to leverage the emergence of environmentally sound local economies.

We think that education occurs mostly in buildings, yet apparently we believe that the design and operation of those same buildings have nothing to do with education. This assumption is a mistake, partly because it overlooks the hidden curriculum in academic architecture. The design of academic buildings is a kind of crystallized pedagogy full of hidden assumptions about power, about how people learn, how they relate to the natural world, and how they relate to one another. But the design and operation of buildings provide an educational opportunity as well: the art of ecological design encompasses such fields as landscape architecture, solar engineering, the ethics of material selection, the economics of life-cycle costing, and the design of closed-loop waste systems.

Architecture need not convey messages of human dominance over nature. A solar aquatic wastewater system designed by John Todd for the Bourne School in Toronto is a beautiful example of ecological architecture. The system, one of Todd's "living machines," is an ensemble of plants and animals in a series of translucent fiberglass tanks located within the school. Wastewater enters the tanks and exits cleaner than conventionally treated wastewater. Instead of being hidden, the system is prominently visible. It looks like an indoor greenhouse, and, more importantly, it teaches students in a visible and powerful way about closing waste loops and about how to use natural means to solve real problems that all too often appear beyond remedy.

Finally, I would like to add a word about the goals of ecological education. The value of an education cited most often by its vendors is that it increases the graduate's upward mobility and lifetime earnings.

Accordingly, we aim to prepare the young for what guidance counselors call "careers." We rarely mention what used to be described as a "calling." In a larger perspective, this is foolish. Students ought to be encouraged first to find their calling: that particular thing for which they have deep passion and which they would like to do above all else. A calling is about the person one wants to make of oneself. A career is a coldly calculated plan to achieve security and have a bit of "fun" that turns out, more often than not, to be deeply unsatisfying, whatever the pay. A calling is not the product of calculation but of an inner conversation about what really matters in life and what difference one wants to make in the world. A calling starts as a hunch. It is risky. It operates more by inspiration than by premeditation. A career is a test of one's IQ; a calling not only tests for intelligence but for one's wisdom, character, loyalty, and moral stamina as well. A person can always find a career in a calling, but it is far more difficult later in life to find a calling in a career. Once a person opts for safety, the die is cast. A career is, finally, a failure of imagination and a sign that one believes the world to be poor in possibilities.

We ought to encourage our students to find their calling in good and necessary work. The best and most necessary work for our age involves in a thousand ways the recalibration of humanity's values, institutions, behaviors, and expectations with those of the Earth. This is the task of education in our time.

I would like to close with the words of E. F. Schumacher: "Education which fails to clarify our central convictions is mere training or indulgence. For it is our central convictions that are in disorder, and, as long as the present anti-metaphysical temper persists, the disorder will grow worse. Education, far from ranking as [our] greatest resource, will then be an agent of destruction." I think these words are prophetic. It is time to address the ecological emergency as, in fact, a crisis of mind and of education.

—1992

Wes Jackson
Call for a Revolution in Agriculture

It is not an exaggeration to consider the thinking of Wes Jackson in the area of sustainable prairie agriculture as characterized by genius. He is among the more colorful of Schumacherians; his deliberately culti-vated rural persona and Kansas drawl do little, however, to conceal his brilliant mind. First and foremost a farmer and a man of the prairie, Jackson is also a widely respected plant geneticist and teacher whose ecological approach to agriculture is gradually penetrating govern-ment policy.

When Wes Jackson cofounded the innovative Land Institute in Sa-lina, Kansas, in 1976, E. F. Schumacher was invited to be the first hon-orary member of the board of trustees. Schumacher managed, in the brief interval between its founding and his death, to find the time to visit the fledgling institute. Like Schumacher, Jackson is a persuasive writer and communicator. He is the author of *New Roots for Agriculture: Meeting the Expectations of the Land, Altars of Unhewn Stone,* and *Becoming Native to This Place.*

At the crux of Jackson's work is his recognition of the extraordinary resource represented by what he calls nature's information systems. He begins here by tracing the devastation wrought over millennia by harmful agricultural practices. Like John McKnight, whose lecture ap-pears in the second section, he makes the connection between an origi-nal ecosystem—the unspoiled integrity of wilderness—and the manip-ulated systems of agriculture. His particular focus is the prairie; his phrase "the wisdom of the prairie" refers to the inherent intelligence of the ecological information system evolved by and in the prairie landscape and its biota over eons of time. It is not only our responsibility to adapt or readapt to that system, he argues; we fail to do so at our peril.

It is Wes Jackson's goal to create an agriculture that recognizes and co-evolves with natural systems. At The Land Institute his research is directed toward how best to accomplish this. Among his most radical and promising experiments is the development of perennial grains that could be harvested with minimal disruption of the soil. In such work lies the hope, not of the outmoded human effort to transcend nature but rather, through mutual healing, of the grace of redemption.

"Call for a Revolution in Agriculture," which appears here in abridged form, contains Jackson's further thinking on the ideas expressed in *New Roots for Agriculture*.

I think the number one environmental problem, aside from nuclear war, is agriculture. Industrial pollution and material and energy resource depletion are serious, but even though industrial society seems likely to collapse one day, to a point almost beyond recognition, agriculture would not have to sink to such depths if we can keep our soil and water resources intact. The fact that till agriculture sends soil seaward and destroys the water-holding capacity of the soil is, in my view, the problem of agriculture.

In the long run, contamination and loss of the soil resource will lead to the loss of ourselves as people. Soil loss is a problem for the short run too. From a summation of numerous studies done in the corn-belt states one can conclude that with a two-inch soil loss, yield is reduced 15 percent; with a four-inch loss, 22 percent; six inches, 30 percent; eight inches, 41 percent; ten inches, 57 percent; and twelve inches, 75 percent. The consequence varies with the area, of course, and depends on the type of soil and how deep it is. During an extreme downpour, such as occurred in southwest Iowa in May 1950, up to 250 tons may be lost per acre. We are now losing from two to four billion tons of soil each year, depending on which estimate we accept. If it is four billion tons, it is equal to the loss of seven inches of soil per year; more than fifty million tons of nitrogen, phosphorous, and potassium are lost. Without replacement this eventually leads to reduced quality in the crop produced. We should remember that more than nitrogen, phosphorus, and potassium is lost in erosion, and if we supply only these elements, some nutrients will be severely lacking. The problem is relatively simple when we are dealing with measurable characteristics. Less measurable properties are equally important, as for example when infiltration rate and water-holding capacity are reduced and the soil's water conservation plan slips away.

The story is as old as civilization. Two thousand years before Christ, the Tigris and Euphrates rivers of Mesopotamia watered an area so rich that early Biblical scholars believed that somewhere in this area was the Garden of Eden. An extensive irrigation system brought life to remarkable cities and helped sponsor a very complex civilization. The area today is a desert of shifting sand around the great buildings and monuments of Babylon, including the palace of the arrogant King Nebuchadnezzar. In one sense it wasn't all that long ago, for bristlecone pines alive today in the White Mountains of the American West were contemporary with forests on the hills surrounding the great valley of this river system. The hillsides are bare now, as they have been for centuries. The ditches that carried the lifeblood of the land are quiet and the harbors of their commerce filled in.

The hillsides of Syria grew magnificent trees, which became lumber for ships and cities, more than one hundred cities, in fact—cities that now have their foundations and doorsteps several feet above where one stands. The record shows that this was an area that once exported huge quantities of olive oil and wine to Rome.

King Solomon, three thousand years ago, purchased so many cedars from the King of Tyre that he had to marshal eighty thousand lumberjacks to cut the trees and seventy thousand men to skid the logs from Lebanon to build the temple. The cut area was placed under cultivation following log removal, and in an historical instant the soil began its search for a new home in the sea. One wonders if the King of Tyre ever recognized that he had made such a disastrous deal.

Both to the east and west of this region the story is similar in all but the minute details. China's Yellow River is called yellow for good reason. Though its basin once supported countless rich and prosperous people, at flood stage soil now accounts for 50 percent, by weight, of the river's flow. In the delta area, which is four hundred miles wide, the channel runs forty or fifty feet above the fields. The Chinese now value their soil so much that they carry it uphill in buckets.

We can take a trip around the Mediterranean and see the source of civilization as well as its ultimate failure. Perhaps most notable are the Spanish, who spent much of what was left of their forests and soils to plunder foreign lands. Then it was England's turn, for after the defeat of the Spanish armada she spent her forests to rule the waves for three hundred years.

Why have some nations managed to prosper with limited resources by saving their soils while others, with much more ecological capital originally, have vanished? The answer is complex, I suspect, but it might be that many peoples who had few land and water resources to begin with set the cultural pattern for those who followed. People of the Netherlands, who claimed swamps and mud flats from the sea, and Indians of the high Andes have preserved their precious natural heritage and indeed improved on the local environment.

In our own country we suffer in part from a history of early abundance and in part from the legacy we brought from western and northern Europe, where rains are so light that soil runoff is scarcely a problem. When our ancestors found the virgin woodlands and prairies of our continent, they encountered for the first time the thunderstorms and associated quick drenchings and then went on to introduce such soil-exposing row crops as corn, cotton, and tobacco. The age of fossil fuel has allowed erosion to accelerate at an even faster pace. At the same time this ancient fuel has allowed us to mask almost completely the old telltale symptoms of agricultural decay. Huge tractors and their equipment can now completely wipe out the clear evidence of decay which resides in a large gully. Natural gas, which provides the feedstock for most of the nitrogen fertilizer applied to the field, can restore partial fertility.

Civilization *has* brought its prophets, most of them eloquent, impassioned, knowledgeable; but almost all peoples have somehow been unable to respond adequately to their warnings. In a real sense, prophecy has failed. So have the organizations, including the U.S. Soil Conservation Service. Saddest of all, perhaps, is the failure of stewardship. There is little financial incentive to practice stewardship now. Pimentel has noted that soil-conserving crop rotation in northeastern Illinois costs $39 per acre. I know that Mennonite farms in Kansas fare little if any better than those of their non-Mennonite neighbors. I saw Amish farms in Pennsylvania this summer that also seemed to be experiencing erosion, though not nearly so severely as their non-Amish neighbors. It is bad enough to realize that the economic system is insensitive to ecological necessity. Even so, it is somewhat easy for all of us to deal with, for we can vent our frustration by being angry at the economic system. But the fact that many of the most ecologically correct stewards have fields losing their soil presents another problem.

Our generation is certainly not the first to contemplate the numerous dimensions of agricultural failure. From my point of view this failure at agriculture is the primordial germ of the "human condition," which philosophers and thinking people have long discussed. I believe it is the very essence of the human condition as outlined in Genesis.

How Did We Get Into This Fix?

We will never precisely know how we got ourselves into such a fix. There is the Genesis version, but most who read that story see it as something separate from our fall into agriculture. I don't, and I want to talk about it a little later. The orthodox view is that we were gatherers and hunters who started to till, and as we increased the amount of food energy, fewer died, which freed some people up and also led to more population and eventually to villages and cities.

Jane Jacobs, in *The Economy of Cities*, has developed the interesting theory that it is the other way around. She cites evidence that agriculture evolved out of an urban environment and contends that cities predate the agricultural communities. She thinks cities formerly were amply supplied with wild game and wild plants. The problem of enough space to store the animal and plant materials in an urban setting then led to plant and animal domestication.

It may be unimportant here to know which arose first, cities or fields, but what is important is that within a very short time we moved away from gathering and hunting in tribal groups to agriculture. If our non-agricultural ancestors were like what modern "primitive" tribes are today, the work week was less than twenty hours and for less than half the adult population. I think it is important to remember that agriculture did bend our straight backs for so long during the day that it was not pleasant. *Homo sapiens* is now a species out of context, and the most out-of-context activity, it seems to me, is the very production of food. So unpleasant is extensive till agriculture on a large scale that when the human did it all, it is easy to imagine that the draft animal was a welcome substitute for the traditional toil necessary to grow and harvest food. As gatherers and hunters we could take our food without thought for the morrow. Early gardening changed all that but perhaps not very much at first. It is easy to imagine that in our hemisphere one group or another probably supplemented its diet with such annual crops as potatoes, beans, amaranth, and corn, while in the old world wheat, rice, barley, oats, etc., were featured. As the patches in the garden increased in size to become a field, two big new things were happening: a new global language was being born, and there was a realization that we could no longer take without thought for the morrow. Perhaps one to seven days is all the gatherer and hunter had to plan for actively, but the serious planter had to think of an entire year at the minimum. The Biblical Joseph became famous as the first secretary of agriculture by planning on seven years—or perhaps, more accurately, fourteen years. Recall his prophecy of seven years of plenty to be followed by seven years of famine. It is quite a shock for a paleolithic mind to have to plan ahead, not for one day or seven days or even 365 days, but for 2,500 to 5,000 days.

With all our planning ahead, even for fourteen years, none of our cultural information was adequate to arrest the wasting away of the soil resource. Both Plato and the Biblical Job knew the score, but it made little if any difference. By then one of nature's beings, meant to be a gatherer-hunter, was so deep into agriculture that there was no turning back, not without massive trimming of the numbers.

If We Are to Get Ourselves Out of Such a Fix, What Can We Trust?

Most people I know who have thought seriously about the overall problem conclude finally that there is no escape, that we are locked into a Whiteheadian dramatic tragedy. Agriculture is necessary to feed people, they realize, even though it has disastrously depleted the very soil that makes it possible. If we leave it at that, we have our answer, for what makes something tragic, of course, is that the human can not do anything about it! In my view agriculture will remain a tragedy so long as it is kept separate from the problem of the human condition. And the human condition will remain a tragic problem as long as it is kept separate from the problem of agriculture. Look over the recommendations of the major philosophers throughout history or even the economic theoreticians and social planners of the late twentieth century and anyone else worried about the human condition, and I think you will see that their recommendations appear almost like random ideas of what to do next. I know I'm not comforted when I look over the list. The philosophical "fix it" people are the most consistent, for they almost always anchor their utopian idea in some former time, some past that sounded better and maybe was better. They would have us recapture that time, add a wrinkle or two, and be home free.

I don't believe that any solution which is more the product of civilization than the product of nature is trustable. Of course we must have both. But I don't believe we can understand the human as a product of civilization or as a product of agriculture nearly as well as we can understand the human as a product of nature. To learn about the nature of the human by relying exclusively on the history of civilization is a bit like studying patients in an asylum and trying to develop some synthetic theory of human behavior. You can gain information, but the context is so out of joint with the environment which shaped us that the value of the information is limited. I jokingly tell my historian friends that if one is interested in abnormal social psychology, one studies history or becomes a cultural historian.

But I have been holding back from you the most important consideration in this talk. To get at both what the human is and what agriculture is, I think we must study and understand what Wendell Berry calls the "natural integrities" that preceded agriculture. For my part of the country that would be the abundant prairies, which had supported the Indians and greeted the settlers. Here in the east it would be the deciduous forests. I am by no means the first to suggest this connection between agriculture and wildness. In England Sir Albert Howard, author of The Soil and Health, talked about the structure of the forest and promoted an agriculture in which the forest served as an analogy.

In our own country a man born and bred in this region, Henry David Thoreau, asked, "Would it not be well to consult with Nature in the outset, for she is the most extensive and experienced planter of us all." Thoreau wanted to see the wild forest return, but as a practical matter he knew we could not turn our backs on the food-growing process. In his journal of 1859 and again in his "Huckleberries" lecture, Thoreau struck a compromise that he felt would satisfy the demands of civilization *and* natural order. He suggested that each town preserve within its borders "a primitive forest" of five hundred or one thousand acres, "where a stick should never be cut for fuel, a common possession forever, for instruction and recreation." This was for the purpose of people learning how *nature's economy* functions. This is a rich insight that I hope we can elaborate on here.

In 1864 George Perkins Marsh, a countryman from Vermont, brought out *Man and Nature*, the most extensive work on land management to that date. He had observed New England farming first hand and had read the works of numerous authors in Europe—naturalists, geographers, foresters, hydrologists. Marsh, a contemporary of Thoreau, came to the conclusion that "the equation of animal and vegetable life is too complicated a problem for the human intelligence to solve, and we can never know how wide a circle of disturbance we produce in the harmonies of nature when we throw the smallest pebble in the ocean of organic life." Even with respect to wildlife he advised the farmer to err on the side of caution.

Marsh believed, though not in these words, that for life forms to grow and flourish, information is required, lots of information, more information than we can possibly imagine or comprehend. And that is the first lesson: it *is* more than we can comprehend! With agriculture we destroy some of the information of nature and substitute cultural information, much of which, Wendell Berry points out, lends itself more to habit than to thought. Thank God for that, for if it were the other way around, as Wendell has said, we would have perished long ago. Even so, cultural information too is incomplete, partly because the time frame of the human is never as long as the time consideration of patient nature. We have not had to worry about the morrow. Furthermore, because we live three score and ten, our concern weakens as the distance in time, both forward and backward, lengthens. We pay more attention to our grandparents, for example, than to our great, great grandparents. The best of people pay more attention to the future our grandchildren will inherit than to the future of our great, great grandchildren. It seems to be a symmetrical thing.

No matter how incomplete it is, cultural information will always be necessary. The acquisition of cultural information doesn't happen overnight. It requires time, as Wendell has emphasized, lots of time with lots of people staying put. We have a clear example in the history of the Exodus—Hebrews who took over the land of Canaan—of the problems associated with quickly acquiring cultural information to perform agriculture efficiently. What the Hebrews brought with them from the wilderness experience, and before that from a life of slavery in Egypt, was the religion of Yahweh. Here was a people about to pass from a nomadic to an agricultural and urban life. It is instructive to study how they handled this passage. Fresh in from the desert, they were forced to deal immediately with the Canaanites, a people whose Gods were farm-gods, known as Baals, which means "owners or possessors of the soil." Every little square foot of fertile ground in Canaan owed its fertility to the presence of some Baal, who either imparted or withheld the fertility power of the soil. There were places to worship the numerous Baals all over Canaan—in valleys, on elevated ground, at springs and wells. Each city had its patron Baal, whose name would be hyphenated with that of the city. Bull images and bronze snakes were popular representations of fertility.

The Hebrew herdsmen to the south had little trouble accepting Yahweh, who had guided them in the wilderness, but those who took up agriculture had a problem. Because they were ignorant of agriculture and the art of husbandry, they had to learn nearly everything from scratch. This was not a time when one could learn soil chemistry, and to learn the spirit-lore was an efficient and effective and more complete way, besides, to learn an encoded language for agricultural behavior. It is easy to imagine that the ten tribes of the fertile and agricultural north had more temptation to practice Baal worship than the shepherd Hebrews of the rocky south. Though northern Israelites did not lose their faith in Yahweh, they did engage in the practice of giving their first fruits to the local Baals, and they observed the festivals of their Canaanite neighbors. Gradually they became convinced that Yahweh controlled agriculture also. The put-down of the Baals was not as total or as emphatic as one might gather from a superficial reading of the scriptures. Maybe Yahwehism in the north was successful in direct proportion to its eventual ability to accommodate itself to that scattering of local Baals.

I am relating this history in which religious language becomes elaborate because, as I said earlier, if one is to practice till agriculture, a great deal of information present in the original wilderness must have been destroyed ahead of time. That is, the DNA of thousands of species, both plant and animal, that had evolved in the natural ecosystem of the area was eliminated. This monumental loss of information stored in the genetic code is not due to species extinction only but includes the huge loss of information within many of the species left. We can be happy that some of these species are yet represented around cemeteries or on some rocky slopes that have escaped grazing or tilling, but most of them have had their genetic information severely reduced. Though we are rightly concerned about a narrowing of the genetic base in our major crops now, it is an extension of the human's venture into agriculture early on.

I believe the Genesis version of the fall is substantially correct but would expand it to state explicitly, as I have already said, that the root of our fall is inherent in the root of agriculture and the root of agriculture can be found in a garden, more specifically in any of the innocent patches of the garden. Small must have been beautiful in the earlier garden, which featured people working patches in an appropriate ratio. But as the size of the patch increased to the size of a field, even a field much smaller than a modern American field, the number of people engaged in the food-getting process diminished. They took up other jobs as civilization flourished, and their descendants eventually gave us a Renaissance, seen generally as good. Perhaps it wasn't pleasant to work those patches at times, but it must have been less pleasant for those working in the fields. Even so, we might imagine that those left to the field work were most likely the ones with the psychological makeup not to mind it so much and during certain times even to enjoy it. The cultural language necessary to keep the food coming, even from patch-type agriculture, was necessarily fine-grained. But as patches were aggregated into fields, the cultural language was inadequate to rejuvenate the area, and the next thing we know, the irrigation canals were silted in. Invaders did contribute heavily to the destruction of Babylonian agriculture, but this is further testimony to the fragility of cultural information acting alone. One Plains Indian tribe of hunters and gatherers might wipe out the village of a rival, but they did not wipe out the ability of the land to supply food in the future.

But I am getting ahead of my story. What seems important to remember is that essentially all the canals that had contributed to the greatness of Babylon and Canaan did silt in, and we can see for all practical purposes a repeat of this history time and time again, down to our own time as we observe the soils that are becoming salted in southern California or in the lower stretches of the Colorado. It would take a super-duper Baal to say, "Dam my rivers and I'll salt your valleys" or "Give my mountain waters advice with systems of ditches and I'll fill them in." The natural system never had to know such truths because the total completeness of its information system prevented the problem from developing in the first place.

My contention, then, is that we can get away with destroying a certain amount of nature's information and maintain high-yield patches without resorting to a slave economy. Patch-type agriculture is within the limits of ecosystem redemption. Perhaps what is important is that we can work the patch and not get too bored by such work and have it enhance rather than reduce our aesthetic dimension. At the patch level both humans and nature can accommodate the products and makers of civilization in high-yielding annual crops such as rice, corn, wheat, and soybeans or carrots, spinach, broccoli, potatoes, sweet corn, green beans, and the like. But when we move to the field level with any of these crops, we will want, if not need, slaves as we always have. They can be human slaves, fossil or uranium energy slaves, draft animal slaves, or fields themselves as slaves, for they would be the source of alcohol energy for tractor slaves or for the draft animal. Human slavery is out. Fossil or uranium energy will soon be going out, leaving the field as the energy source and the draft animal to help us. But at the field level, on sloping ground, in most cases there is still no information system complete or compelling enough to prevent the rush of useful atoms toward the sea with each rain or insistent wind.

If We Must Have Fields, What Will Nature Require of Us?

At The Land Institute we are working on the development of mixed perennial grain crops. We are interested in simulating the old prairie or in building domestic prairies for the future. Our current agricultural system, which features annuals in information-poor monoculture, is nearly opposite to the original prairie or forest, which features mixtures of perennials. If we could build domestic prairies, we might one day be able to have high-yielding fields that are planted, say, only once every twenty years or so. There would be mostly harvest after establishment, and from then on we would be counting on the species diversity that breeds dependable chemistry. This above-ground diversity has a multiplier effect on the kinds of seldom-seen teeming diversity below. Bacteria, fungi, and invertebrates live out their lives reproducing by the power of sun-sponsored photons captured in the green molecular traps set above. If we could adjust our eyes to a power beyond that of the electron microscope, we would reel at the sight of a seemingly surrealistic universe of exchanging ions, where water molecules dominate and where colloidal clay plates are held in position by organic thread molecules important in a larger purpose but regarded as just another meal by innumerable microscopic invertebrates. The action begins when roots decay and above-ground residues break down and the released nutrients begin their downward tumble through soil catacombs to start all over again. And we who stand above in thoughtful examination, all the while smelling fresh dirt and rolling it between our fingers and thumbs, distill these myriads of action into one concept—soil health or balance—and leave it at that.

Conventional agriculture still coasts on accumulated principle and interest, hard earned by nature's life forms over those millions of years of adjustment to dryness, fire, and grinding ice. Lately, agriculture has been coasting on the sunlight trapped by floras long extinct. We pump it, process it, transport it over the countryside as chemicals, and inject it into our wasting fields as chemotherapy. Then we watch the fields respond with an unsurpassed vigor, and we feel informed on the subject of agronomics. That we can feed billions is less a sign of nature's renewable bounty and our knowledge than of her forgiveness and our discounting of the future. For how opposite could the annual condition in monoculture be from what nature prefers? Roots and above-ground parts alike die every year, so through much of the calendar year the mechanical grip on the soil must rely on death rather than life. Mechanical disturbance, powered by an ancient flora, imposed by a mined metal, may make weed control effective but the farm far from weatherproof. In the course of it all, soil compacts, crumb structure declines, soil porosity decreases, and the loss of a wick effect for pulling moisture down diminishes. Monoculture means a decline in the range of invertebrate and microbial forms. Microbial specialists with narrow enzyme systems make such specific demands that just any old crop won't do. We do manage some diversity through crop rotation, but from the point of view of various microbes, it is probably a poor substitute for the greater diversity which was always there on the prairie or in the forest. Monoculture means that botanical and hence chemical diversity above ground is also absent, which invites epidemics of pathogens or epidemic grazing by insect populations that can spend most of their respiratory energy reproducing, eating, and growing. Insects are better controlled if they are forced to spend a good portion of their energy budget buzzing around, hunting among a mixture of many species for the plants they evolved to eat.

Some of the activity found in the pre-turned sod can be found in the human-managed fields, but the plowing sharply reduced many of these soil qualities. Had too much been destroyed, of course we would not have food today. But then, who can say our great grandchildren will have it in 2081? It is hard to quantify exactly what happened when the heart of America was ripped open, but when the shear made its zipper-sound, the wisdom the prairie had accumulated over the millions of years was forgotten in favor of the simpler, more human-directed system.

I think it is possible to return to a system that is at once self-renewing like the prairie or forest and capable of supporting the current human population.

Much scientific knowledge and narrow technical application have contributed to the modern agricultural problem. Nevertheless, because of advances in biology over the past half-century, I think we have the opportunity to develop a truly sustainable agriculture based on mixtures of perennials. This would be an agriculture in which soil erosion is so slight that it is detectable only by the most sophisticated equipment, an agriculture that is chemical-free, or nearly so, and certainly an agriculture that is scarcely demanding on fossil fuel.

Succession in Agriculture

Probably the most radical idea in our work at The Land is that of introducing succession to agriculture. Annual monoculture has been compelling in the short run because we have successfully denied our fields the opportunity to go beyond the first stage or two of succession. Agriculture has depended upon successfully fighting what nature wants to do. But if we are serious in our inquiry into what nature will require of us, then we must be prepared to reject an agriculture based on the near complete denial of succession, except at the patch level.

Succession is an ecosystem's way of obeying a fundamental of biology. We know that for all levels of biological organization there is a juvenile stage, eventually a mature stage, and the inevitable senescence and death. It happens to cells. It happens to tissues, organs, organ systems and to the individual. It also happens to species, by the way, which evolve—often with a flourish—settle down, and then become extinct before changing into something else or dying out. It seems to be a property of life, part of the great round. Information systems have evolved to accept numerous realities: the reality of failure of a species to adapt, failure of a particular genetic ensemble to live forever even if we are talking of the predominantly non-sexual blue-green algae or bacteria, for mutation pressure is always there to change the information program.

When it happens to plant and animal communities, we call it succession. Weaver and Clements of Nebraska studied this phenomenon thoroughly early in this century, and Henry David Thoreau was preoccupied with the idea in the nineteenth century. I think that revolutionary agriculturists will need to have it central in all considerations.

I began to think seriously about ecological succession in agriculture as we considered putting together mixtures of species which would stimulate the tall, mid, and short grass prairies. We had thought that perhaps there would be two harvests: one to collect the seeds of the cool season plants in early summer and a fall harvest. We would need to breed these plants to set seed in synchrony for one harvest or another. Maybe there will be cool season prairies and warm season prairies. All plants in the mix need not be seed producers but may be there for other purposes such as nitrogen-fixation or to serve as a host for a critical mass of certain insects which stand ready to nip a potential epidemic of other insects. Somewhere in all this I was forced to think about ecological succession for agriculture. We looked over our inventory of plants in the herbary and picked some to be sown in four feet by twenty feet plots for another experiment. We noted, however, that the aggressive types, such early-stage invaders as Illinois bundle flower or Maximillian sunflower, had few weeds in the plot material. On the other hand, the plots of the climax species, purple prairie clover, which were not weeded, were mostly taken over by cheat and other weeds. The unweeded plants were not worth harvesting.

These two conditions represent extreme ends of succession on the prairie. But what of the intermediate species, which represent other stages between the invaders and the climax? The question now became, can we be participants in succession? Thinking ahead a century, perhaps some of our descendants will plant their bare fields with high-yielding luxuriant perennial mixes—aggressive species, which would then be followed closely with other species, plants with less total potential energy but with more elaborate information systems and perhaps a more efficient use of energy overall. After twenty years perhaps they would have accelerated their domestic prairie to the climax stage with a relatively high yield along the way. The presumption here is that the information system of the prairie has evolved to accept succession as a reality just as much as gravity.

This is one of many possibilities for an ecological agriculture, but I want to conclude with a word of caution.

A Problem with Ecological Agriculture

I have promoted an ecological agriculture that emphasizes our maximizing the information of nature's storehouse beyond what the human is capable of comprehending. I hope I have given due emphasis to the fact that cultural information must evolve to accommodate this system, though it need not be as complex as what is necessary for a sustainable agriculture using the crops and the livestock of the late twentieth century. We may feel that we are on the verge of a new "tomorrow" if only agriculture embraces ecology. But we've got to watch out.

Toward the end of Nature's Economy, a fine paperback on the roots of ecology, the author, Donald Worster, builds a strong case for mistrusting ecology as an operating paradigm for future human action. He takes pains to show that the science of ecology has been studied and understood in the language of economics and industry by people who, whether they know it or not, not only betray their belief in the economic system in which we now operate but also betray their belief in the industrial society. As early as 1910 one of the pioneers of modern ecology said: "Bio-economically speaking, it is the duty of the plant world to manufacture the food-stuffs for its complement, the animal world. . . . Every day, from sunrise until sunset, myriads of [plant] laboratories, factories, workshops and industries all the world over, on land and in the sea, in the earth and on the surface soils are incessantly occupied, adding each its little contribution to the general fund of organic wealth." We may think, "Well, that was a long time ago when such language was used in describing nature," but less than fifteen years ago a noted ecologist at the University of California, Berkeley, said, "Like any factory the river's productivity is limited by its supply of raw materials and its efficiency in converting these materials into finished products." The metaphors used in understanding ecology, Worster says, are more than casual or incidental, for they express the dominant tendency in the scientific ecology of our time. Nature has been transformed into a reflection of the modern corporate industrial system.

Unfortunately, ecology has had little or no influence on economics; rather, economics has tainted ecology. It's been a one-way street.

What do we do now that ecology is saddled with such words or phrases as food chain, producers, consumers, and *ecosystem*, now that the whole is broken down, now that energy is the medium of exchange, now that efficiency is such a virtue it is seen as the key to biological order? That working ecologists see little wrong with the use of such language is a sign that the completely economic world view is coming closer. Perhaps science is such an inherently alienating force that we have to begin anew.

The problem is, where do we begin? What do we build on? I think that a long time ago nature gave us two important ecological concepts that became religious philosophy and that both will need emphasis in a new ecology. Both are central to the Judeo-Christian tradition, though in recent times they have been understood in rather shallow ways. These concepts center around the idea of redemption and the idea of transcendence. Regarding the first, nature has shown us that we can damage an area, yet it will redeem itself—perhaps not completely, perhaps not for a long time, but eventually and to some degree. This idea of redemption is a source of hope: abuse a hillside and the sins of the father will visit the sons even unto the third and fourth generations but not necessarily forever, for redemption of the wasted hillside is possible if loving care is given it.

The idea of transcendence is one that even the most ardent zealot of reductionist science can't ignore. For example, there is nothing about the properties of hydrogen and oxygen that gives a clue about the properties of water. The properties of both are completely transcended by what water can do and how it figures in our lives. We can move up the hierarchy of the sciences and see that at every step of the way more is different. As we approach the cultural level, more specifically the agricultural level, we have a clear example of the power of transcendence in the Amish as compared to the conventional farmer of today. Wendell has written of the fifty-six-acre Amish farm in Indiana that grossed $43,000 and netted $22,000 in one year. As Wendell sees it, the reason a conventional farmer could not do that is because, as the circles widen and move away from the farm, much of the harmony of the organic world is left behind, incorporating more of the non-harmonious industrial world. Useful information, much of which we are unaware of or cannot comprehend, is lost. The conventional farmer may have a degree in agronomy or agricultural economics, but he turns out to be a terribly unsophisticated farmer in the sense that his way is not sustainable. The Amish farmer probably never had a single vocational agriculture course in high school. The Amish simply believe that the highest calling ordained by God is to be stewards of the land, and this duty is tightly tied to an aesthetic ideal. Because economics is not foremost in their thinking, they are *able* to make sound economic decisions. By being obedient to a higher calling, "All these other things are added unto them." This is a practical kind of transcendence that *all* can experience. It requires no guru or priest or minister. That the consistently sound economic decisions are made by people who do *not* make economics primary should be no more surprising than the fact that water is more than the combined properties of hydrogen and oxygen. The idea of transcendence cuts through all and is essential to an ecological agriculture. It can go a long way toward helping us temper the unfortunate language we are saddled with, the reductionist language of economics and industry, which has been applied to ecology. It should help us soften the utilitarian point of view.

If we do one thing that is ecologically right, we have reason to expect *more* than a multiplicative effect, indeed a transcending effect, just as when we do something that is ecologically wrong, it works in the other direction. If what we are talking about is not real, as the rigorous reductionists insist, then neither is water.

The implications of an ecological agriculture in which some of nature's information is allowed to operate are unforeseeable at the moment, but it is nevertheless something we *can* trust. This approach to agriculture is clearly in the spirit and teachings of our brother E. F. Schumacher, who really was talking about transcendence in his descriptions of meta-economics.

It is both interesting and important that Schumacher, economist that he was, was very much interested in ecology. He was president of the Soil Society of England. He was a strong advocate of planting and caring for trees, which he saw as more than bearers of fruit, for he thought of them as symbols of what he called "permanence," which he used as a synonym for sustainability. He was a man who grew a garden, which by definition consists of patches. A man whose primary message was *transcendence* of the economic world saw the perennial trees as *redeemers* of the landscape.

—1981

John Todd
An Ecological Economic Order

At a conference in the mid-1970s a skillful organizer arranged that a lecture by John Todd directly follow one by E. F. Schumacher. After Schumacher had spoken convincingly of the need for stewardship of the land, sustainable agriculture, renewable energy, and intermediate and appropriate-scale technologies, Todd rose to report on the work then underway at the New Alchemy Institute. He described the research being done to address basic human needs for food, energy, and shelter without impinging further on deteriorating ecosystems. He showed slides of productive, flower-rimmed gardens, young orchards, aquaculture ponds, windmills, and solar collectors, all being tended by people obviously thriving on the work they were doing. It was the embodiment of the ideas just advocated by Schumacher. The two talks fitted together, as one observer noted, like a golf ball on a tee. Todd's book *From Eco-Cities to Living Machines: Principles of Ecological Design*, written with Nancy Jack Todd, discusses these concepts at greater length.

Like Wes Jackson, biologist John Todd turns to the information systems of the natural world for the knowledge he seeks to restore damaged ecosystems and evolve sustainable economies. Subsequent to his work at New Alchemy, through his research at Ocean Arks International, he has continued to apply the embedded evolutionary intelligence of natural systems to human problems. Among his most recent innovations is the living machine, the generic name for what is now a family of licensed, ecologically based technologies for bioremediation. A living machine consists of a series of contained ecosystems engineered to perform specific tasks. These can include waste treatment, food production, temperature regulation, and the integration of the built environment with the natural world. Such design represents a marriage between human ingenuity and evolutionary processes of nature.

The ecological design represented by living machines is rapidly becoming a cornerstone of the ecological economic order Todd refers to in the following lecture. Through the firm Living Technologies, Inc., living machines for environmental repair have been installed across the United States and in Canada, Europe, Australia, and South America. Not only are they proving cost effective in purifying polluted waters but many of the installations are producing byproducts in the form of plants and fish, which serve to increase the economic viability of the system. Perhaps the most encouraging result of the research with living machines, however, is the news that the worst of polluted soils and waters can be successfully restored with this type of ecologically based technology. It constitutes further proof that the human and natural worlds can co-evolve to the enhancement of both.

Not long before E. F Schumacher's death, I was with him at an appropriate technology congress on the Indonesian island of Bali. Although I had known Fritz for years, my most cherished memory of him is from the Balinese countryside. We were visiting an international development project that included a modern fish culture facility. Unlike the rest of the food culture on the island the demonstration fish farm seemed alien—with fences, rectangular ponds, and its separation from the agriculture and the villages. Like a prison in our society, it was removed from the interwoven fabric of Balinese culture.

Later that day we visited a temple. The water, trees, architecture, and gardens expressed a deep harmony and what seemed to me a merging of mind, nature, and the sacred. As the sun fell, Fritz spoke of trees as the most powerful of transformative tools, of their planting and tending as fundamental acts. For him, trees were the starting point for creating social and biological equity between peoples and regions of the earth.

Our conversation inspired some of the ideas I am going to present to you now. I owe E. F. Schumacher a debt of gratitude for helping me see economics as if people and nature mattered. Subsequently I have come to believe that a new sustainable economic order can be established with ecologically based enterprises. Further, the conceptual bases of these enterprises are similar whether applied in rich industrial nations or in poorer tropical countries. If this thesis is correct, applied ecology has the intrinsic potential to dissolve old divisions between north and south, industrial and agrarian, rich and poor. This is so because ecological knowledge can be applied universally and, equally important, can often lead directly to substitutes for capital and for nonrenewable resources. In the sense that Fritz Schumacher meant, it has the ability to increase equity on a global scale.

Ecology as the basis for design is the framework of this new economic order. This approach needs to be combined with a view in which the earth is seen as a living entity—a Gaian worldview—and our obligations as humans are not just to ourselves but to all of life. Earth stewardship then becomes the larger framework within which ecological design and technologies exist. One day it may be possible for political and social systems to mirror the broad workings of nature, and current divisions of left versus right, centralist versus decentralist, expansionist versus steady state, bioregional versus nation-state will be transformed into a systemic Gaian world organization and order.

But change, even on a Gaian scale, has to begin with small, tangible, and concrete steps. When I first began working at the New Alchemy Institute with ecological concepts that might serve humanity, my associates and I started with two questions: "Can nature be the basis of design? If so, are there ecological models to prove this?"

We started with food and agreed that the contemporary mechanistic agricultural model would in the long run fail to feed the planet. We looked for other models to guide us. The larger workings of nature provided us with clues. We sought out several places where nature is extremely bountiful and made a shopping list of the attributes unique to those places. As patterns gradually emerged, this effort proved directly fruitful. We also sought out places that, under the guiding hand of humans, have been bountiful for millennia. This was significant because humans normally destroy their biological capital. We wanted to learn what stable cultures know about caring for their lands.

A farm in central Java, near Bandung, was rich with clues. It had maintained and possibly increased its fertility over centuries. The farm was located on a hillside that was particularly vulnerable to erosion, which was prevented by mimicking nature's most efficient erosion-control strategy, namely, tree-covered slopes. It was not a wild forest but a domestic one in which the biota were fruit, nut, fuel, and fodder trees useful to humans; nevertheless, it had some of the structural integrity found in the wild. Without the trees on the hills it would have been very hard to sustain the land's fertility. The farm received its water from an aqueduct flowing across the slope half way up. The water came from a farm higher up and arrived in a clean, relatively pure state. Upon reaching the lower farm it was, within a short distance, intentionally polluted by passing it directly under slatted livestock barns first and then under the household latrine.

Although it might appear shocking at first glance, the livestock and household sewage was utilized in a clever way. The solids were "digested" by a few fish, whose sole function was to provide primary waste treatment. The nutrient-laden sewage was then aerated and exposed to light by passing over a low waterfall. Secondary and tertiary treatments were agricultural. The sewage was used to irrigate and fertilize vegetable crops planted in raised beds. The nutrient-rich water flowed down channels and dispersed laterally into the soil to feed the crop roots. It is important to note that the secondary sewage was not applied directly to the crops but to the soil. Having fertilized the soil, the water emerged from the raised-bed crop garden in an equivalent condition to standard tertiary treatment. It then flowed into a system that required pure water, namely, a small hatchery for baby fish. Here in the hatchery pond, the young fish began the enrichment cycle again by slightly fertilizing the water with their wastes. This triggered the growth of algae and microscopic animals that helped feed the young fish. These biota were also carried along with the current to add nutrients and feeds to the larger fish cultured in grow-out ponds below, highly enriched ponds that fertilized the rice paddies just downstream. The rapidly growing rice used up the nutrients and purified the water before releasing it again to a community pond in the basin below.

The intriguing thing about the farm was that it represented a complete agricultural microcosm. There was a balance not seen in Western farming. The trees, soils, vegetable crops, livestock, water, and fish were all linked to create a whole symbiotic system in which no one element was allowed to dominate. Such a system, while beautifully efficient and productive, can also be vulnerable to abuse. One single toxin, like a pesticide, will kill the fish and unravel the system. The lesson here is that we can create ecological agrisystems and let nature do the recycling, or we can manage a complex system chemically and ultimately destroy its underlying structure. At New Alchemy, when we began to design food-growing ecosystems, we tried to keep intact the biological relationships we had observed on the Java farm.

Comparable lessons can be drawn from all over the world, even from endangered places; as Shakespeare said, there are "sermons in stones, books in the running brooks, and good in everything."

Perhaps one of the single greatest challenges facing humanity is the restoration and re-creation of soils. Without healthy soils, human economies cannot be sustained for long, and yet they are dying around the world. Deforestation, overgrazing, and erosion are the primary villains. Soils need to be given back their organic matter, humus, and moisture-retaining qualities.

To understand the importance of soils and how we are joined with them, we need to realize that soils are alive; they are meta-organisms comprised of myriads of different kinds of living creatures. When stripped of plant cover and exposed to blowing winds and mineralization, they become increasingly lifeless and porous, losing their ability to retain rainwater near the surface. Most of the world's spreading deserts follow in the wake of soils becoming more porous and devoid of rich microscopic life.

Several years ago we visited an atoll in the Seychelles in the middle of the Indian Ocean. Because the soils of coral islands are notoriously porous, they don't hold water. Rainwater percolates quickly through the soil and collects in underground lenses. On the atoll we visited, the one hundred villagers had almost pumped their fresh water lens dry, and salt water was beginning to intrude and contaminate their water. Within a few years the inhabitants would have to abandon their islands.

Their seemingly intractable problem could be solved if somehow impermeable basins could be created to capture rainwater during the monsoons. The soils, however, were too porous for the idea of a surface pond to be considered as a viable option. I remembered the research of two biologists who had discovered a strange anomaly in Russia: they noticed that on top of hills comprised of rubble mounds, ponds or small lakes would occasionally be found. Because the underlying soils were incapable of holding rainwater, there had to be some mechanism that sealed these ponds so that they could capture and hold rain. They then discovered a comparatively rare process in which microorganisms, in concert with organic matter, combine to produce a biological sealant. This sealant formed a liner in the natural basins, which then held water. They called the process gley formation.

We decided to mimic the process discovered by the Russians but in the very different environment of a tropical coral island. We hoped gley formation might take place quickly if the conditions were just right. The challenge was to make them right. We started by digging a small lake with a backhoe; then we found the necessary carbon and fiber component in coconut husks, which we shredded and placed in a six-inch layer over the bottom and sides. For our source of nitrogen we collected the wild papaya trees, ubiquitous on the atoll, and chopped up their stems, branches, and fruits, placing them in a six-inch layer above the husks. Finally, six inches of sand was added on top of the husks and papaya. The Russians had found that gley forms in the absence of oxygen, so the basin was packed down to drive out oxygen. A small amount of water from the lens was pumped in to flood the bottom. To our great pleasure, when the monsoon rains came shortly thereafter, the basin filled with water, and it stayed.

The pond became a source of irrigation water, a home to cultured fish, and a haven for wild fowl, including migratory birds. This experiment opens up a whole range of ecological and economic possibilities. Not only can coral islands become ecologically and socially diversified, but the very same process can be used wherever there is a need to store seasonal rains. I can foresee, throughout the world, previously barren landscapes now nurtured by small gley-created impoundments that are the epicenters for restoration of damaged environments.

The new fresh-water source inspired an experiment to make the atoll's alkaline and nutrient-poor soils capable of growing economic crops other than coconuts. The soils had been degraded by fires and oceanic storms, and their composition was up to 90 percent calcium carbonate. Stuart Hill, the creative Canadian soil ecologist who was with us, believed that the island's soils could be made productive through the application of compost. Compost can be used to refurbish soils or can even function as a soil substitute, because it mimics the cation exchange of good soils and is a very stable form of organic matter. It can also bring other gifts to coral islands, such as the release of plant hormones, especially cytokinins, which in turn stimulate plants to produce larger and more branched roots. Compost is also a prime substrate for nitrogen-fixing bacteria and blue-green algae, thereby providing atmospheric nitrogen for plants. The blue-green algae are an excellent source of nutrients, too.

Stuart Hill was able to show us that compost can play one other crucial role in alkaline soils: it releases organic acids which, if applied at the right time in the decomposition process, neutralize the soil. As a result of Stuart's work, vegetables and fruits are now being grown to diversify the diet of islanders. He found that the island lacked several essential minerals—specifically manganese, boron, and iron—which had to be imported. Our long-term strategy would be to locate local oceanic creatures that concentrate these substances and add them to the compost. Nutrient independence is an important objective, particularly for regions of the world where foreign exchange is scarce or non-existent.

The teachings from Java and the experiments on the Seychelles are but two examples of what has informed the work at New Alchemy and, since 1980, our newer organization, Ocean Arks International; solar-based technologies are proving themselves even in the cold Canadian climates. These technologies all borrow their design features from a blend of ecosystem knowledge, materials science, and the wisdom of the Javanese farmer or the skills of the ancient Mayans of Central America, who, with their "chinampa" or "floating" agriculture, fed densely settled cities. We now refer to our ecologically designed technologies as living technologies. A single unit is called a living machine.

One such technology is the aquatic farming module. Its development began at the New Alchemy Institute in 1974 under my direction. An aquatic farming module, or living machine, is a translucent solar-energy-absorbing cylinder of up to one-thousand-gallon (3,785 liter) capacity that is filled with water and seeded with over a dozen species of algae and a complement of microscopic organisms. Within these cylinders phytoplankton-feeding and omnivorous fishes are cultured at very high densities. The species selected depend upon climate, region, and market opportunities. The range of species we have studied is broad, including African tilapia, Chinese carp, and North American catfish and trout.

Dense populations (up to one fish per two gallons or one fish per 7.6 liters) of actively growing fish produce high levels of waste nutrients, beyond the capability of the ecosystem to take them up. The module, however, eliminates these nutrients in four ways: (1) fish growth; (2) plankton proliferation; (3) partially digested algae—which flocculate out, settle to the bottom, and can then be periodically discharged through a valve to fertilize and irrigate the surrounding horticulture; and (4) a modern chinampa system, the uptake of nutrients by vegetable crops rafted on the cylinder surface; the root systems of the plants take up the nutrients before they reach toxic levels, secondarily capturing detritus and functioning as living filters that purify the water.

These modules can be productive—with fish yields, depending on species and supplemental feeding rates, of over 250 pounds (113.5 K) annually in a twenty-five-square foot (2.3 square meter) area. At the same time each unit can produce eighteen heads of lettuce weekly for an annual production of over nine hundred heads. Tomato and cucumber crops can also be cultured on the surface for even higher economic yields. The modules have the additional beneficial attribute of conserving water. Evaporation is almost eliminated from the surface so that makeup water rates are based on plant evapotranspiration and the amount of module water released to irrigate and fertilize the adjacent area.

Aquatic farming modules are an agro-ecology that requires initial seed capital to construct and establish, but to a large extent they are a substitute for heavy tillage and for fertilizing, irrigating, and harvesting equipment that would otherwise have to be used to establish and operate a farm. Not only are the modules space-conserving and less costly, they can be employed in urban centers (in greenhouses in northern climates) and as a key component in the process of restoring damaged environments.

Within a given land-restoration project the living-machine modules are established in rows in the most highly degraded areas. Trees are planted on the shaded side of the cylinders and subsequently nurtured by the periodic release of water and nutrients. On the sunny side of the modules, a variety of short-term economic crops could be established to add to the produce from the module. This module-based agriculture could provide the skilled labor pool to tend the emerging ecosystems.

The aquatic living-machine approach, when used for ecosystem reclamation, need not be static—that is, the modules, having fed and watered the newly emergent vegetation, including trees, through their most vulnerable stages, can be shifted to new locations to repeat the process. In this way the short-cycle biotechnology can spread its benefits to surrounding ecosystems over a larger geographic area.

A related ecological technology has been developed for arid regions. We have, for instance, designed a bioshelter system to assist with ecological diversification on the Atlantic coast of Morocco. The bioshelter is a transparent climatic envelope or greenhouse housing the fish and vegetable modules. Our prototype is a circular geodesic structure that functions as a solar still and as an "embryo" for the early stages of the ecological diversification process. These bioshelters can operate even where there is no fresh water: the aquaculture modules are placed inside the climatic envelope, and water from the sea is pumped through them. During the day the structure heats up and the temperature differential between the sea water in the tanks and the surrounding air is great enough to cause the tanks to sweat fresh water, irrigating the ground around them. Tree seedlings are then planted in this moist zone. At night the moisture-laden air is cooled by the desert sky, and water droplets form over the interior skin of the climatic envelope. We found that drumming on the prototype structure's membrane in the early morning caused it to "rain" inside, thus allowing the whole interior to be planted. In addition, this process permits drought-tolerant trees to be established around the outer periphery of the structure. Inside, marine fish and crustacea such as mullet and shrimp can be cultured to form the basis of an economy. After a few years the original cluster of climatic envelopes can be moved to a new location to repeat the cycle, leaving behind an established semi-arid agro-ecosystem.

These are two biotechnological examples drawn from a range of options that can help reverse environmental degradation and restore diversity and bounty to a region. These advanced living technologies may well prove to be essential tools in creating sustainable environments.

Most modern societies are faced with the crisis of waste accumulation. The natural world is threatened by our inability to integrate our agriculture and industry within the great planetary cycles. Industrial cultures are cancerous, yet they need not be. I see the cleansing of water as one point of intervention. Sewage treatment plants, as an example, are expensive but do not purify water. When they work, which is not often enough, they kill the "bugs" and remove the solids; they do not remove nutrients or toxic materials. This need not be so if wastes are seen as resources out of place and if concepts like total resource recovery underlie the design of waste purification systems.

Ecologically based resource recovery can alter the economics of recycling. Sewage treatment plants are a financial drain on communities whereas ecosystem water purification could provide a basis for economic activity. Sewage can be made into drinking water, and the by-products of the process can have economic value. To demonstrate this, I was initially involved in the design and development of a living-machine waste treatment facility, which is part of a joint venture by a governmental body, the Narragansett Bay Commission; our research organization, Ocean Arks International; and a new company established to take advanced ecological concepts into the market place.

The Narragansett Bay Commission of Rhode Island operates the city of Providence's huge sewage treatment facility. It is also concerned with protecting Narragansett Bay and its abundant but highly threatened marine resources. The Providence sewage plant does not remove nutrients or toxins other than those in the sediments; our living-machine-based treatment removed these on a demonstration scale of fifty thousand gallons a day. It also produced marketable by-products ranging from flowers to fish and served as a hatchery for over a million striped bass. Striped bass are a marine fish whose populations have collapsed because spawning grounds and nurseries have been destroyed by pollution.

The Providence facility is comprised of a solar-heated greenhouse within which flow two parallel streams separated by 180 of the living machines previously described. The modules trap and store solar energy, creating a year-round semi-tropical environment inside the building. They also serve to house and nurture the young striped bass. Sewage enters at one end of the structure and over a five-day period flows slowly through the facility. The two streams contain four sequentially arranged aquatic ecosystems, each with an essential task in the purification process and each housing biologically active food chains fed initially by the sewage.

The sewage is pretreated by means of ultra-violet sterilization, then charged with oxygen through aeration. Introduced air is essential at each stage. The first ecosystem has an algae base, algae being the penultimate utilizers of nitrogen, phosphorous, and other nutrients. The second ecosystem is dominated by floating aquatic plants, including water hyacinths, which trap the upstream algae in their filamentous roots. They also continue to remove nutrients and take up toxic materials. (The city of San Diego has found that water hyacinths remove most organic solvents as well as heavy metals. San Diego, together with NASA, has pioneered water hyacinth purification of sewage.) The third ecosystem is made up of clear water with artificial habitats on the bottom, where microscopic shrimp-like animals graze on the algae and bacteria resident on the attached substrate. They are fed upon by mosquito-fish and fundulus, which in their turn are fed to the bass in the adjacent aquatic farming modules. The fourth and final ecosystem is a marsh comprised of reeds and bulrushes planted in a gravel filter. These taller plants, twenty to thirty feet high in the greenhouse, remove any remaining organisms and toxins. They also polish the water. (Reed and bulrush waste treatment was the brainchild of Dr. Kaethe Seidel of the Max Planck Institute in Germany. She discovered that marsh plants could transform sewage into potable water. Her research findings have given new meaning to the protection of wild marshes.) After the water passes through the marsh filter in the living-machine waste treatment facility, it is ready for re-use. In the case of the Providence prototype it will be used for local industrial needs.

Ecologically designed treatment demonstrates the value of ecological integration and illustrates how nature's bounty can be applied to human needs. Most Third World countries, plagued with sewage-born diseases, cannot afford industrial waste treatment. Even if they could, they would be robbed of precious resources in the process. Living technologies can be designed to control disease and at the same time serve as epicenters for the production of fertilizers and the cultivation of plant materials, including trees for reforestation.

Our hope is that solar-based living technologies will be the catalyst for a new commitment to caring for water as our most precious and elemental resource. Stewardship needs to be extended to our ground waters, lakes and streams, and oceans. It is our sacred trust to the planet, to Gaia, to redefine our values so that our first order of business is to cleanse the waters, protect the soils, and tend to the trees.

I am aware that ours is a world of violence, hunger, environmental degradation, and inequities. For most of us, points of action and intervention to relieve these ills on behalf of the planet and ourselves may be hard to find. But I believe this will change if our economies become ecological. Then work and stewardship will be one.

I have sketched here only a few of the ideas and technologies derived from ecology, but I hope I have demonstrated that an ecological economic order has the intrinsic potential to allow each culture to explore the new frontier in its own way so that some of the old divisions between peoples and places can be removed. Fritz Schumacher worked for greater equity and justice in this regard, and so must we.

—1985

Stephanie Mills
Making Amends to the Myriad Creatures

It could well be said of Stephanie Mills, as it once was of Bob Dylan, that she burst on the scene already a legend. This occurred in 1969 with her commencement address, "The Future Is a Cruel Hoax," at Mills College in California. As valedictorian she drew attention to the global population crisis by pledging that she would never bear children. So radical a declaration by a young and beautiful woman instantly caught the attention of the media and subsequently the nation. She has maintained that commitment, devoting herself to environmental causes ever since.

A consummate wordsmith, the focus of much of Mills's work has been in writing and editing. She has served as the editor of a number of environmentally oriented publications, including *Co-Evolution Quarterly* and *Not Man Apart*. Her 1989 book *Whatever Happened to Ecology?* has been called a singular voice of sanity and a moving personal statement by a woman on the ecological frontiers. She has since edited and contributed to the anthology *In Praise of Nature*. "Making Amends to the Various Creatures" became part of her most recent book, *In Service of the Wild: Restoring and Reinhabiting Damaged Land*.

Like Kirkpatrick Sale, Mills is an eloquent proponent of bioregionalism. Such is her conviction that a number of years ago she abandoned a life of urban celebrity to make her home in the Leelanau peninsula of Michigan, where she is learning to reinhabit her chosen ecosystem. In a single sentence she characteristically sums up the need for and the means to practice this Schumacherian way of life: "Now, because our upstart civilization's insensitivity to the personhood of each member of an ecosystem threatens to bring the whole domicile down, we're being forced to remember how to belong and how to cherish."

"To me," she concludes, "it seems as though the fundamental ground of mind, god, and being is now shifting."

A few weeks ago a friend of mine, Lowell Cate, invited me over to his place, a couple of miles northeast of my home, to have a look at his woods. He's an older man, living near land his great-grandfather homesteaded in 1868. A lot of Lowell's acreage had been farmed, much to its detriment, and was recuperating under straight, evenly spaced rows of white pine. The hardwood groves where we walked were on slopes steep enough that the stout maples, beeches, bass-wood, birches, and hemlock had escaped being cleared. Over the years the slopes had been cut selectively, for firewood. Here and there stood some quite substan-tial trees, beeches mostly. My friend seemed familiar with them all, and he recalled the years and storms in which others had blown down and awaited the arrival of the chainsaw and the service of the flame. In Lowell there's no dearth of feeling for the woodland and no want of lore of local land use. He has knowledge of his home place.

As we were emerging from the leafy shadows to a spot where he'd had a commercial pulpwood harvester clear a stand of popples, I noted a burdock on the sunny margin and asked if he minded if I pulled it up, because it's a weed ("just a plant out of place" by one definition). I explained that I'd recently learned a new rule of thumb to justify this act. Every alien species that establishes itself in a new surrounding displaces about ten natives. This was news to Lowell, something he immediately began to ponder.

My point is that most of us, even people like Lowell who know and love and daily walk bits of land they have presumed to be well and carefully used, usually don't see the non sequiturs or gaps in those ecosystems. It's quite understandable. The tendency of our civilization from its beginnings has been to simplify the natural world in the process of taking it over for human occupancy. In the modern era this human tendency to transform landscapes for short-term advan-tage has, with the accelerations provided by rapid population growth, global transportation, and the juggernaut of mechanization, annihilated hundreds of thousands of species and made lonely relics of countless others in their vanishing habitats. Absent some special interest in botany or ecology, we haven't much sense of all that is lost in the domestication of wild places.

Our topic today, "Human Habitat and the Natural World," could be taken to suggest a dichotomy between the two realms it mentions—the world of nature and the dwelling places of Homo sapiens. The difference between the whole world and one's own home or settlement is a fairly recent concept in our life as a species, along with the human self-concept of existing in a transcendent category, apart from and superior to rude nature. For traditional peoples, we are told, nature and human habitat were one big, if not always happy, family wherein the two-leggeds were just folks along with coyote, raven, old doctor loon, fox, deer, and waterbug, and it was only polite to beg the pardon of the grass people upon whom you slept in the course of your journeys. Now, because our upstart civilization's insensitivity to the personhood of each member of an ecosystem threatens to bring the whole domicile down, we're being forced to remember how to belong and how to cherish. It may not be too late to begin to cultivate what Thomas Berry has phrased, with apt simplicity, a *courtesy* towards the other beings of the living world.

There is a developing practice of such courtesy, a discipline of learning how to beg the pardon of damaged and decimated landscapes. It is work called *ecological restoration.* Barry Lopez, who writes with a solemn sense of the relations of nature and the human, extols it as "so rich in the desire to make amends!"

Ecological restoration is an experimental science, seeking the ways and means of healing damaged landscapes, attempting to reinstate their original plant and animal communities and revive their chartless web of interactions. Restoration ecology recognizes that habitat is complex, distinct, and essential to the preservation of diverse species. Conceding that nature is unsurpassed in its genius at evolving niches, creatures, and adaptations beyond counting, this science cares to be faithful to nature's ways. By stoop labor, painstaking observation, and much applied intelligence, ecological restoration pays more than lip service to the sacredness of Earth's phenomenal—and threatened—biodiversity.

Beyond that, I would say it is work which has the potential to engender true love for the land: not romantic love, not blind love or possessive attachment but a love which declares, "I want thee to be." Not platonic but erotic, it is a love that increases one's knowledge of the beloved—the myriad creatures—and of the self, through deeds.

Part of the work lies in gaining eyes to see the detail of the natural world. It involves discerning an ecosystem's past, noticing significant events in the course of its recovery, and foreseeing a time of future thriving. Restoration ecology demands ingenious historic research and a little bit of schooled inference.

The motives for ecological restoration are various. Some restoration is required by law, as in mine-spoils reclamation or the trade-offs implied in the "no net loss of wetlands" doctrine. Mandating and bureaucratizing restoration projects can lead to low-bid jobs and very crude approximation of original

conditions, vastly inferior by any measure. Satisfaction of minimum requirements is the wrong philosophical context for the mission.

It remains to be seen whether even the most exacting restoration ecology will
succeed in its aspiration to restore successional processes and cradle evolution's
continuity. From 1936 to 1941 Theodore Sperry oversaw the pioneering restoration of the Curtis Prairie at the University of Wisconsin's arboretum at Madison.
He is perhaps restoration ecology's most venerable practitioner. Asked how long
it would take to restore the prairie, Sperry replied, "Roughly . . . a thousand
years."

However appropriate a millennium is as a time frame for work of evolutionary
dimensions, it is a rather longer while than engineers generally abide waiting for
permission to dredge or agencies are willing to wait to approve the results.
Despite this fundamental mismatch of frames of reference, a legal requirement
for some kind of restoration or mitigation of damage to ecosystems marks a
positive turn.

Some ecological restoration is seemingly altruistic, although given our absolute dependence on the functioning of the biosphere, nothing we do to maintain
its diversity and health fails to benefit our species. If we employ a narrower
definition of self-interest, though, spending one's weekends in pursuits like lop-
ping European buckthorn saplings, scything weeds, torching grassfires, or gathering, threshing, labeling, storing, then sowing and raking seed from hundreds
of different varieties of rare plants is altruistic. In Chicago and many other venues,
hundreds of volunteers do this work for the sake of the plants' survival as well as
to gain a new sense of relatedness with the wild plant, insect, and vertebrate
communities they serve. In the Chicago region, in a scattered group of sites called
the North Branch Prairies, this seasonal round of restoration activity is conducted
under the watchful eyes of volunteer preserve managers. All are engaged in
stewardship of lands that are either publicly owned or held in trust by nonprofit
organizations. Chief among these is the Illinois Nature Conservancy, whose basic
mission is ecosystem preservation. In northern Illinois the key endangered ecosystems are prairies and savannas or oak woodlands.

One of the greatest threats to ecosystems—and a subtler menace than the
omnipresent bulldozers, backhoes, and paving machines—is biological pollution: invasion by exotic plants or animals. When botanical immigrants find
themselves in a happy land where no predators or pests have co-evolved with
them, they proceed to outcompete the natives and to simplify and generalize
the landscape. Damage to the natural community offers invaders advantages, as
the scruffy plant gangs crowding vacant lots and edges of woods and fields
demonstrate.

Hence, much of the work of ecological restoration is plain old weeding—

removing exotics, restoring enough of the original structure and function so that earlier, more diverse communities can regain their ground and stand resilient against future incursions. Restoration is inescapably labor-intensive, requiring a horticultural "eyes-to-quadrats" ratio that makes Wes Jackson's eyes-to-acres proposals sound casual by comparison. Sometimes this tending involves the application of herbicides or mowing. Often it entails just going into the thickets and weed-whipping the garlic mustard or girdling the soft maples in an oak woodland so that they will die and admit the light that the understory natives have been dying without.

Another and far more spectacular part of ecological restoration is "ritual pyromania." The prairies and oak woodlands of Illinois are particularly fire-dependent ecosystems, communities that require periodic burning of grasses and leaf litter for a reduction of thatch, a release of nutrients, and the killing of invading herbs and brush. If left to their own vegetal devices, these few species of weedy herbs and woody perennials rise up and shade out the natives—scores of grasses, sedges, rushes, and forbs that make a dappled tapestry of life in which no single color or thread predominates. Only lately have we learned how fire was the evolutionary force that maintained that original richness.

When the plow broke the soil in the upper Midwest, prairies and savannas began to vanish in favor of the relentless order of agriculture, which depends on planting lots of the same damn thing and then defending it from all the plagues that assail even-aged monocultures. Years later, the advance of suburban residences guaranteed that wild prairie fires would occur no more. Thus, if the rare plant communities lurking in the shadows on preserves and parks are to be revived, there have to be fires, and the fires must be managed with exceeding care. Benign neglect will not suffice to ensure a future for these species.

One day this past summer, through meeting John and Jane Balaban, who are stewards of the Harm's Woods and Bunker Hill Prairie, I was able to learn a little about the spirit of restoration as practiced in Chicagoland. That same day, touring a savanna restoration site with Steve Packard, science director of the Illinois Nature Conservancy, I caught a sense of the entrancement in the work. What all three conveyed is that there's a mutuality in stewardship. It begins with a desire to know the plants, to be able to name them and to get a sense of their ethology. Plants live as any organism must—within ecosystems. Because there has been so much degradation of the ecosystems, we've reached a turning point in evolutionary history; now the plants need us.

"The idea that these sites can't exist without our help anymore," convinced the Balabans of the importance of their stewardship. "You can't just preserve something by building a fence around it," as John Balaban put it, noting "how dependent that [ecosystem's] structure is on our interference."

As volunteers like the Balabans deepen their involvement in prairie and sa-vanna restoration, they may wind up propagating rare plants in their garden, like the threatened small sundrops or pale Indian plantain a few beds away from the tomatoes. Stewardship is a further invitation to dedicated amateurs to share in a crucial botanical endeavor and the beginning of a postmodern interpenetration of human habitat and the natural world. For restoration ecology is scientific work, involving thousands of people in what is certainly the most critical array of biological experiments now underway.

With Packard I visited the Somme Woods Preserve near Chicago. The Preserve, not so long before, had been a suburban wasteland. Back in 1913 a generous and farsighted act established these Cook County Forest Preserves for the people of the Chicago region and their posterity. But ecological ignorance and neglect led to the degeneration of these lands. In their weedy, brushy, overgrown state they were used and abused the way vacant lots generally are—for kids' rendezvous, party spots, random trash heaps. Therefore, thirteen years ago a preliminary step in restoring the site was to haul away soggy car seats and quantities of broken bottles. The earlier, commonplace desolation was not hard to imagine and made the present, resurrected beauty all the more miraculous.

Accustomed as we are to seeing rare plants and animals only in the confines of botanic gardens and zoos, it's quite a wonder to be able to tread careful trails past healthy populations of these plants in natural communities. Such was my experi-ence that day in the Somme Woods Preserve. It was marvelous to gaze across a pelage of dozens of rare, and some endangered, plant species lifting their faces to the same sky that arched over their ancestors' arrival on the empty loess plains left ten thousand years ago after the glacier's retreat. In these open settings, restora-tion has become liberation.

The knowledge and ethic that had to be arrived at to effect these prairie and savanna restorations are of a peculiarly contemporary sort. They could only have been called forth in, and by, this unprecedented moment of life's peril. The volunteers and organizers of the North Branch Prairie Project live urban and suburban lives in built environments, depending on store-bought food and the stability of the existing economic order. Their restoration work is avocation or occupation but not a matter of immediate personal or community survival.

If we were still gatherers of wild food and if our medicines were still provided directly by the herbs and flowers of the field and forests, we would grow up with eyes to recognize the distinct character and value of the hundreds and hundreds of plants in a woodland's understory or on an open prairie. Our knowledge of our vital stake in an interdependence with the life and health of each individual plant and animal would be second nature (or so the theory goes).

The will to return to something like that unselfconscious human belonging to, and direct dependence on, the local ecosystem is another motive for restoration ecology. Bioregionalists call it reinhabitation. One inspiring saga of ecological restoration with reinhabitory purpose is unfolding in the Mattole River watershed in northern California. Its aim is to reinstate a native race of king salmon and with it to regain what some Mattolian activists refer to as "natural provision."

For timeless time in the Mattole River watershed, far north on the Lost Coast, the water made its way down to thousands of nameless rills, then scores of creeks, to become the Mattole River and flow past the bar to the Pacific. Its descent was just enough governed by the forest community of Douglas fir and tan oak and their compatriots that the water's action maintained long-term equilibrium in the watercourses. There were shaded creeks, clean gravel beds, deep pools, and a cool estuary whose banks were overhung with vegetation.

So the Mattole, like all the rivers pouring into the Pacific from Hokkaido to Monterey Bay, supported its own races of wild salmonids—unique stocks of king and silver salmon and steelhead trout. Historic descriptions of the abundance of these Pacific Northwest salmon runs beggar the imagination, just like the accounts of flocks of passenger pigeons millions-strong blackening the skies. But since those early days in the Mattole, clearcutting has removed nine-tenths of the forest. The roadbuilding that goes with logging and homesteading, as well as excessive grazing, have degraded lands upslope and hence salmon habitat downstream. These human activities have pushed the Mattole king salmon to the brink of extinction.

Tales of the Mattole Restoration Council's response to this, which I had been following during my years in San Francisco, were impressive in their own way. The stories, pride of bioregionalism in northern California, suggested a respectful, if strenuous, remarriage of humans and the natural world—"becoming native again to place." Some of the Council members see this as reinhabitation. Salmon support and watershed restoration in the Mattole declare a long-haul intention of working not merely in historic but geologic terms.

Whenever we try to pick something out of the universe, John Muir said, we find it hitched to everything else. So when, in the course of bioregional events, some 1970s back-to-the-landers settling in the Mattole Valley took it upon themselves, for mythopolitical reasons, to support their native race of king salmon, they learned that they had to follow the salmon's existential logic upstream into the fabric of the land and down into the root hairs.

If you want to restore the fish, it turns out that you have to heal a three hundred square mile watershed—deal with vegetation loss, erosion, social fragmentation, the works. To begin with, the Mattole Watershed Salmon Support Group, which is a member of the Mattole Restoration Council, set about to enhance the reproductive success of the king salmon. The erosion problems upstream had radically changed the seaward reaches of the Mattole, widening the bed, stripping away shade-providing riparian brush, and filling in the pools where the salmon, which love cool depths, could come to oceangoing maturity. Torrents carried away nesting gravels from creeks, and silts washed down and smothered them. So to save the world—of the salmon, at least—the support group devised hatch boxes, simple nurseries where eggs taken from, and fertilized by, wild salmon trapped in the river could be incubated. Of this part of the work Freeman House, a founder of the Council, writes: "To enter the river and attempt to bring this strong creature out of its own medium alive and uninjured is an opportunity to experience a momentary parity between human and salmon, mediated by slippery rocks and swift currents. Vivid experiences between species can put a crack in the resilient veneer of the perception of human dominance over other creatures. Information then begins to flow in both directions and we gain the ability to learn: from the salmon, from the landscape itself."

In the clean filtered water flowing through the hatch boxes and rearing pond, the salmon fry could grow, enjoying regular meals and an absence of predators, and then be released into their parent creeks. Over the decade that the hatch-box program has been going on, the valley's elementary school children have been invited to participate in the release, trekking down to the streamside and dumping big buckets of little fish into the free waters of their destiny. It's part of restoring salmon's presence to local culture.

On the terrestrial front the Mattole Restoration Council members have over the years planted thousands of Douglas fir seedlings on the cut-over lands upstream; they practiced "restoration geology," sometimes employing heavy equipment to remove boulders and debris dams from creek beds. They sweated to armor creek banks with head-sized cobbles, to transplant handfuls of native vegetation that could stabilize sections of eroded stream banks, to plant willow cuttings and alder seedlings along the river to shadow the water and foster the insect life on which the young fish feed. Through visits to the public school and creation of an alternative high school, the reinhabitors began to educate valley children and adolescents about the life of the salmon. Perhaps most importantly, during this time the Mattole Restoration Council has been learning how better to find consensus with neighbors whose sense of domain may be threatened by the very idea of restoration activity. They also are learning how to work with the various state, federal, and county agencies that govern land use and fisheries. This is endangered species-cum-habitat preservation in a place where much of the land is privately owned and can't be managed as a refuge.

In the winter of 1990 I finally made my way to the Mattole Valley. During my visit, Freeman House took me to a spot on Mill Creek where the Salmon Support Group had worked to restore a silver salmon habitat. In the Mattole, as on the

North Branch Prairies and in scores of other restoration projects around the planet, part of what is gained is a physical sense of the laboriousness of an entropy battle. How very difficult it is to attempt to put a casually torn-apart ecosystem back together! As more people feel that difficulty ache in their bones and sinew, perhaps there'll be less tearing apart and fiercer defense of life-places.

The restoration site is California-steep—not quite a ravine, overhung by gnarled leggy tan oaks, these festooned with lime-green mosses. Fallen Douglas firs caught and bridged the glen, which was musical with the sound of rushing water. As on many tributaries of the Mattole, logging-caused erosion had scoured Mill Creek and steepened its fall. Its waters flowed too swiftly to drop out the right-sized gravel for salmon redds. To slow it down required a structural solution. First the Salmon Support group had to winch a windfall Douglas fir log down from the cut-over north bank. Then a portable saw mill was hauled down to quarter the log lengthwise. The four pieces were placed across the stream to capture gravel and sediment and to moderate the gradient. The water falling over the straight edges of the dams scours out congenial pools beneath. This little project, all for the sake of thirty mating pairs of silver salmon, occupied eight or ten people for perhaps a thousand hours.

That wildly beautiful spot in Mill Creek was irrevocably human-altered. First, for the worse, by the frenzied logging upstream. But the healing, brought about by changing the stream gradient with those Douglas fir dams, left evidence of human activity, too. Another trace of human presence is the rock armoring, a dry masonry that artfully shores up the creek banks. Now there are patterns, straight lines, right angles, and planed faces of logs over which cascade linear waterfalls. The very music and echo of the creek have been tuned by Homo sapiens.

The Mattole River watershed is no set-apart wilderness preserve. One object of the human artifice invested there is to ensure that the salmon will again feed the people and that the natural community will supply immediate human needs once more. That humans could participate conscientiously in an ecosystem, not as atavistic but as postmodern hunter-gatherers, bespeaks visionary thinking.

Entering into that place and the vision it spawned, I began, most reluctantly, to question the conservationist dogma of a hands-off policy towards "wild" nature. To relinquish the fictive absolute distinction between wild and tame implies moving on past Homo sapiens's guilt and recrimination over the destruction that human action in the landscape has produced. The conscious gamble of those working in the Mattole is that cultural interaction with the watershed will inculcate a moral restraint on the impulse to control and determine, to expand and exploit. They can imagine a maturation beyond remorse into a sustainable way of life.

Peter Berg says that bioregionalism is about "more than just saving what's left." Clearly, reinhabitation in the Mattole means more than just preserving the spark of biodiversity embodied in the native races of salmon. As salmon recovery and all that it entails begins to make sense to the entire community, Mattole Restoration Council activists would see restoration becoming just as much an indigenous occupation as forestry, farming, fishing, or ranching—yet another livelihood connected to the locale.

In a vision statement in the inaugural issue of the Mattole Restoration Newsletter in 1983, House and David Simpson wrote:

> If you can, imagine starting from the ridgetops and headwaters . . . planting trees and grasses for slope stability and future timber . . . as roads get built and maintained so that erosion slows rather than increases. The river gradually flushes itself and stabilizes. Vegetation begins to seal it in a cooling shade again. Work in salmon enhancement begins to pay in visible increases in spawning runs. Silt washed off the upland slopes begins to deposit itself permanently in rich alluvial flats. Grains and vegetables grow in soil that was formerly swept to sea.
>
> A generation from now, our children reap a harvest not only of fine timber, abundant fish, productive grasslands and rich and varied plant and animal communities—but also a tradition which will assure the same harvest for their children.

One noble creature animates that vision and ennobles another at a time when for humans to entwine their destiny with wild species is a supremely risky form of romance. After a decade of dealing with the minute particulars of watershed recovery, the Mattolians are wondering where the king salmon are.

"Our river has had the same near non-existent runs as other rivers draining the Northern California Coast Range, a phenomenon made spookier still by the fact that it is occurring only two years after some of the biggest salmon catches in forty years," House wrote in spring 1991. This year, in a game departure from their usual fund-raiser, the Mattole Restoration Council had a "No-Salmon Barbecue," and in a hilarious refusal to accede to despair, Simpson and his wife Jane Lapiner, with their Feet First dance theater group, went on the road with "Queen Salmon: A Biologically Explicit Musical Comedy for People of Several Species."

David Simpson writes that the musical's "message of community responsibility seems to be working. We've made great headway recently in forming a working coalition with ranchers, timber interests, large and small landowners and restoration people . . . Now if Bush and his entire entourage would resign, the white Aryan brotherhood turn to rebuilding the inner cities, Earth First! open AIDs clinics, and NASA give up space shots to save the ozone layer . . . Ah well . . ."

Saving the world—one watershed, one species, one community at a time—is a good and honorable dream, and most of the people who engage in this hard work both love it and have some very bad days. The stories of human ingenuity and dedication which characterize restoration work are heartening. They suggest the correct behavior vis-à-vis evolution's precious mystery of species. People who make a firm commitment to their life-places are among my greatest heroines and heroes. I admire their relationship to the land, or at least what I am able to observe and infer of it.

The only relationship to land I can claim intimate knowledge of, however, is my own. And so as not to leave you with an unqualified idea of what such tenure can feel like, I mean to let some of my own experience twine around the trunk here, like a bittersweet vine whose leaves also seek the light.

One of the most poignant truths Aldo Leopold ever expressed is that "one of the penalties of an ecological education is that one lives alone in a world of wounds." Although ecological concern is growing, and living it is less lonely now, there are days when having even an inkling of what the average pre-Columbian ecosystem might have been like and how depauperate the North American landscape now is casts a hue of melancholy over the sight of my pleasant expanse of Michigan countryside. Sometimes sobering, sometimes bracing, is the awareness that one of the responsibilities attaching to an ecological education is envisioning how those wounds might be healed and the suppleness of the land restored.

Thanks to very little forethought and no hard work on my part, I own thirty-five acres of Section 26, Kasson Township, Leelanau County. The acreage is like a foster child, and at least one of us has posttraumatic stress syndrome. By the standards of my township it's a middling-sized parcel. According to the 1880 survey the woods in Section 26 consisted of sugar maple, beech, elm, ash, basswood, and hemlock. Of their descendants there are only a couple of little patches of second or third growth trees remaining, huddled along my property's south boundary. Most of the county was clear-cut for timber that fueled Great Lakes steamer traffic. When the trees were gone, the land was farmed—with varying degrees of success, varying with what the glaciers had deposited where. Corrugations just faintly perceptible in the central clearing and as yet undecomposed cornstalks suggest what the last crop growing on this land was before it came into the possession of certain feckless hippies. Four-fifths of my part of the holding is in teen-aged Scotch pines planted straight and regular as corn to hold down the naked, sandy soil.

I must confess, to my shame, that at times my relationship to this land has been quite as judgmental as my relationship with some significant human others in my life. I have "known" it to be damaged, impoverished, manipulated, and not the best. I have coveted my neighbor's woods. I have felt mortally threatened by all these scruffy, marginal pines, many of them sick and dying, rightfully fearing that in a drought year it wouldn't take much to kick off a conflagration. From time to time I have resented the self-imposed obligation to do something about all these Earth-afflictions, being in my essential character a sedentary type facing a grim thirty-five acres of yard work. I envy those whose love of their land is unalloyed. Maybe it's the curse of having as yet mainly restoration eyes but no restoration skill, back, or upper body.

Despite watching the weeds encroach, sand blow, pines pine, and fire hazard mount, I love visiting the patchwork of places out back of the house. Spectral dead pines grizzled with lichen. Stemmy, deer-browsed young maples, nuzzling right up to the Scotch pines for shelter. Small, secret clearings. Blackberry thickets and popple copses. A stand of some new shrub with a red stem and pinnately-compound elliptical leaves, which I had not noticed growing out back before. Every ramble through the property reminds me that there is a universe aplenty here, with not a square foot unworthy of respect. Affection returns, and I'm forced to conclude that the whole community's got to be held sacred, even the nattering of my thoughts, or nothing on Earth can be.

On a wet day this past August I bestirred myself to follow a suggestion from Malcolm Margolin's wonderful *Earth Manual*, which is a guide to restoration projects for individuals. He urges the would-be erosion battler "to try to forget everything your mother ever taught you about 'catching your death of cold'" and go out in the rain and see what the water is doing on the land. Despite a downpour it was a mild day, so I figured I could risk a soaking.

I started threading my way back through the thicket and soon was sopping wet, squelching around dreamily sans spectacles, which had been rendered useless by fog and droplets. I looked up and watched the rain falling at me out of a blotter-flat sky; I saw it course in rivulets down the trunks of the chokecherry trees. And by and by I got my new name: Woman Who Rips Up Knapweed In The Rain. Knapweed is one of those nefarious Eurasian invaders. It is a paradigmatic weed, and it has come to pervade the disturbed ground where I live. Periodically I engage in hysterical fits of knapweed pulling, but it's like trying to get rid of oxygen molecules. Every day I pray for a mutant virus to annihilate knapweed.

Emerging from the pines onto the south slope of a little knoll whose north face I regard from my writing studio, I see the makings of gullies. These present a more urgent and slightly more manageable problem than biological pollution. White pine seedlings planted there a couple of years ago weren't quite able to hold the ground.

This summer's periods of drought alternating with slashing rains ensured that until I figured out something to do about that raw hillside, the sound of rain on the roof would just be tributary to a Mississippi of ecoguilt.

A few mornings after the rain-soaked reconnaissance some more of Margolin's advice registered, and I concocted a way to build some brush dams with available materials. On either side of the incipient gullies I pounded in some sharp sticks. Then I lopped branches from the omnipresent pines, thanking the trees for their sacrifice to the cause. I wove the branches crossways against the stakes so as to impede the water's flow and catch the soil on its way down. For now my hope is simply to prevent a blowout, which is how great ugly gashes in the land are referred to in my part of the country. Ecological restoration is yet a ways away.

Henry David Thoreau, who in the mid-nineteenth century hadn't seen nothin' yet, wrote this about ecological loss and change: "I take infinite pains to know the phenomena of the spring, for instance, thinking that I have here the entire poem, and then, to my chagrin, I hear that it is but an imperfect copy that I possess and have read, that my ancestors have torn out many of the first leaves and grandest passages and mutilated it in many places. I should not like to think that some demigod had come before me and picked out some of the best of the stars. I wish to know an entire heaven and an entire earth."

I am not unmindful of my privilege in having an extra-large and rangy backyard peopled with deer, skunk, grouse, jay, knapweed, and revelations un-expected—a fox's skull, an indigo bunting. It is more land than I ever would have dreamed of knowing, but I will never not be cognizant of its woundedness or wondering whether my modest, homespun efforts at restoration won't be obvi-ated by the geophysical and climatic change we humans are producing on the planet.

In an interview with the ecologist Raymond Dasmann, Lewis Mumford al-lowed that he was optimistic about possibilities, if pessimistic about probabilities. Resistance to despair is proper, period. However, I think it's insanity to deny that many of the first leaves, grand passages, and best stars have been picked out and torn away.

I came to ecological concern at a time when the valuing of wild over tame was the rock upon which conservationists built their church. Thoreau's saying, "In wildness is the preservation of the world," was a basic article of faith. I became, in my armchair, a devotee. One of the most startling realizations of the present moment is that the living tissue of wildness—biodiversity—may no longer persist except by human sufferance.

The psychological and moral shock of simultaneously losing that myth, or holy ghost, of autonomous wildness and assuming custodial responsibility for the preservation of species is terrific. To me it seems as though the fundamental ground of mind, god, and being is now shifting.

I subscribe to Gregory Bateson's formulation: "The individual mind is imma-nent, but not only in the body. It is immanent also in pathways and messages outside the body: and there is a larger mind of which the individual mind is only a sub-system. This larger mind is comparable to god and is perhaps what some people mean by god but it is still immanent in the total interconnected social system and planetary ecology."

The places and persons in which that something comparable to god is immanent have never before changed so fast. The relation of humans to the natural world is now the hinge of life's history on Earth.

During our tour of the Somme Woods Preserve, Steve Packard quoted a sage of his discipline as saying, "Ecology is more complex than we think and more complex than we can think." It is both a humbling and challenging observation. Yet I suspect that in the attempts of restorationists to discern what their lands would have of them, there is something of a wish to fathom that larger mind, a joyous wish to become a good part of its imaginings, to make an alliance with wildness. And therein may be the preservation of the world.

—1991

David Brower
It's Healing Time on Earth

There has been, over the past fifty years, no more effective and valiant warrior on behalf of the planet than the articulate, embattled David Brower. He served as the first executive director of the Sierra Club and founded Friends of the Earth, the League of Conservation Voters, Earth Island Institute, and the Brower Fund. The initiator of the biennial Hope of the Earth conferences on peace and environmental and social justice, he is also active in the Restoring the Earth movement and the Global CPR Corps, which promotes conservation, preservation, and restoration. He has written a two-part autobiography entitled *For Earth's Sake* and *Work in Progress* and another book *Let the Mountains Talk, Let the Rivers Run*. He is also the subject of the book *Encounters with the Archdruid* by John McPhee.

Archdruid David Brower begins his lecture mourning the ebbing of democracy in a nation—and a world—run by vast transnational corporations that make us victims, in E. F. Schumacher's phrase, of "an almost universal idolatry of giantism." Brower defines GATT (General Agreement on Tariffs and Trade) as the culmination of that idolatry, as "the end run around the environmental gains of the last century . . . pure gravy for the transnationals." To emphasize the maladaptive folly of corporate domination of global ecosystems, like Thomas Berry he portrays such economic determinism in relation to the cosmic and evolutionary events that have shaped the formation of the Earth. To adapt ourselves to this magnificent living world demands not an ethic of growth but one of restraint in the form of limiting consumption and population growth.

Like Schumacher, Brower castigates nuclear and massive hydro-electrical installations, advocating the development of appropriate-scale, renewable-energy technologies. A life-long activist, he exhorts his listeners and readers to choose a direction, organize, and address an untiring effort toward solving the myriad problems confronting us. Brower himself has been and remains a model of what one human being can do when he dedicates himself to "healing the Earth."

We do not have a democracy in the United States. Any country where only half of the eligible voters are registered and where only half of those who are registered vote and where only half of those who vote like their choice is not a democracy. Any country that isn't ruled by its government, that is ruled instead by the Fortune 500, isn't a democracy. And any world government that is ruled by trans-national corporations isn't a democracy. Yet such is the state of our national and global governments. According to my definition, a corporation is, right now, by law, a lawyer's attempt to create something that can act like a person without a conscience. If you are a CEO or a member of the Board of Directors of a corporation that bypasses an opportunity for profit, you can be sued by the stockholders! There should at least be something written into law that says you can bypass it for sound social or ecological reasons. If you're asked to invest, there should be an Environmental Impact Statement on what your money is going to do to the Earth. If you're going to take over a company because it is trying to operate with a conscience and it's making all the money it can and that's why you're trying to take it over, there should certainly be an Environmental Impact Statement on that. All of these conditions should be required. We should bring this about and see if we can instill ecological conscience into corporate behavior. If that happened, I think we'd be much better off.

I thought it would be useful to do an exercise in perspective relating to time. Squeeze the age of the Earth, four and a half billion years, into the Six Days of Creation for an instant replay. Creation begins Sunday midnight. No life until about Tuesday noon. Life comes aboard, with more and more species, more variety, more genetic variability. Millions upon millions of species come aboard, and millions leave. By Saturday morning at seven, there's enough chlorophyll so that fossil fuels begin to form. At four o'clock in the afternoon, the great reptiles are onstage; at nine o'clock that night they're hauled off. But they had a five-hour run.

Nothing like us appears until three or four minutes before midnight, depending on whose facts you like better. No *Homo sapiens* until a half minute before midnight. We got along as hunter-gatherers pretty well, but the population couldn't have been very big; for those of you concerned about how many hunter-gatherers the Earth can sustain, the range I've heard is between five and twenty-five million people. Then we got onto this big kick: we wanted more of us, we wanted to push forests out of the way so we could feed more people. We wanted to shift from hunting and gathering to starch and thereby start the first big energy crisis (because the greatest energy shortage on Earth is of fuel wood). So we got into agriculture one and a half seconds before midnight. That recently. By the next half-second, we had been so successful that the forests ringing the Mediterranean Sea, for example, were reduced to the pitiful fragments that are the Cedars of Lebanon. That was in one half-second. At about the end of that half-second—we're now one *second* before midnight—after all this time of life being on Earth we began to invent religions.

If I could go back to a point in history to try to get things to come out differently, I would go back and tell Moses to go up the mountain again and get the other tablet. Because the Ten Commandments just tell us what we're supposed to do with one another, not a word about our relationship with the Earth (at least not according to any of the translations I've seen so far). Genesis starts with these commands: multiply, replenish the Earth, and subdue it. We have multiplied very well, we have replenished our population very well, we have subdued all too well, and we don't have any other instruction! The Catholic church just put "steward-ship" in its vocabulary within the past seven or eight years!

So here we are now, a third of a second before midnight: Buddha. A quarter of a second: Christ. A fortieth of a second: the industrial revolution. We began to change ecosystems a great deal with agriculture, but now we can do it with spades—coal-powered, fossil-fuel-powered spades. We begin taking the Earth apart, getting ideas about what we can do, on and on, faster and faster. At one-eightieth of a second before midnight we discover oil, and we build a civilization that depends on it. Then, at two-hundredths of a second, we discover how to split the atom, and we begin the GNP race. (I've been told it was the Soviets who started it, and the United States didn't want anybody to have a grosser national product than ours.) But that's not the race we need; we must change how we think about GNP.

That reminds me of a paradigm shift I've had in mind recently. Through the years I've been quoting Adlai Stevenson in the last speech he gave as our ambassador to the United Nations. It was July 1965 when he said: "We travel together, passengers on a little space ship, dependent upon its vulnerable reserves of air and soil, all committed for our safety to its security and peace, preserved from annihilation only by the care, the work, and I will say the love we give our fragile craft. We cannot maintain it half fortunate and half miserable, half confident, half

despairing, half slave to the ancient enemies of mankind and half free in a liberation of resources undreamed-of until this day. No craft, no crew can travel safely
with such vast contradictions. On their resolution depends the survival of us all."

I wish every person who ever occupied the Oval Office had heard this passage
and committed it to memory and done something about it; it would be a totally
different world right now if that had happened. But I think Adlai Stevenson, if he
were here now, might accept an editorial suggestion or two. One, we have not
liberated resources; we are extirpating resources. Two, let's rethink: our global
conditions are not so clearly defined as to be half one way and half another way;
it's more like 5 percent and 95 percent in the inequity quotient of this Earth.

What is happening? I think we're getting better and better at having despair
and needing to have it. But I'll tell you about the people of Ladakh, the place in
India where Helena Norberg-Hodge has been working for half of every year for
the past seventeen years. She is trying to bring information from Ladakh to us
while also trying to prevent too much of our information from getting to them
(unfortunately, she's losing that struggle a little bit). The Sierra Club has just
come out with a book entitled *Ancient Futures*, with pictures of the people in
Ladakh. When you look at these pictures, there's no great evidence of wealth
there, but there is evidence of something else. You see some of the most beautiful
faces; you see some of the nicest smiles; you see some inner happiness that you
don't see in our supermarkets or on Park Avenue. Where is the despair?

Another thing that is happening is that we are not getting the truth. This is
"the Era of Disinformation." Let me tell you a true story about a Cree Indian who
came down to a city for the first time, to a courtroom, and sat in the witness chair
and was asked, "Do you promise to tell the truth, the whole truth, and nothing
but the truth?" And his answer, according to the interpreter, was: "I can't tell the
truth; I can only say what I know." That is so beautiful. It has something in it that
we, including environmentalists, lack an awful lot of: it has humility. It's true for
all of us, because we don't know the truth.

To take that further, it seems to me that the more we know, the less we know. I
have a beautiful example of that, relating to genetic diversity. When Bernie Frank
was the head of the Division of Forest Influences, United States Forest Service—
and this must have been during the late 1950s—he said, "We know next to
nothing about forest soils." If there was anything known about forest soils, that's
where it would have been known. Four years ago, at a conference in Berkeley on
restoring the Earth, we had some experts on mycorhizal fungi, fungi that live in
symbiotic relationships with the roots of most tree and plant species. These
experts have been studying them for quite awhile, they're learning more and
more about how many different species of mycorhizal fungi there are and how
complicated the relationships are between the fungi and the roots, but they still

can't define these relationships exactly. As for the general knowledge of forest soils in the forest industry, forestry schools, and Forest Service today, it is even less than it was thirty-five years ago!

E. O. Wilson tells us in *The Diversity of Life* that there are something like four thousand different species of bacteria per pinch of soil. We are familiar with the concept of genetic diversity—according to Wilson we have identified 1.4 million species of plants and animals—but we have no idea how many more exist. The estimates I've heard range from five to eighty million. So, as we discover more, we discover that we know less and less about more and more. This is something that should instill some humility into us. It should give us the idea that our agricultural binge—our whole Industrial Age binge—cannot go on. We need to rethink, and our institutions are not ready for it.

To return to the "instant replay": let's back up to a hundredth of a second before midnight. That's when I was born; I mention this for one reason: a huge amount of environmental destruction has taken place since the early 1900s. The population of the Earth has tripled. The population of California has gone up by a factor of twelve. The Earth as a whole has used four times as many resources in those brief eighty years as in all previous history. In our great state of California we had, when I was born, six thousand miles of salmon streams; now we're down to two hundred. We had roughly 80 percent of the original stand of Redwoods, which grow nowhere else but California (except for a few migrants that slipped into Oregon, not knowing what they were doing); that 80 percent is down to 4½ percent. I go into these numbers because all of this has happened in eighty years.

In the past twenty years we have created enough new, man-made deserts—I say "man-made" because women had very little to do with it—to equal the area of cropland in China. We've lost soil through erosion, paving, development, condominiums, suburbia, and inundation by such things as the nonrenewable hydro-electric development in Quebec (the first stage of which, in the James Bay project, inundated four thousand square miles of forest). Now, if you've just learned that there may be four thousand species of bacteria per pinch of soil and you think of the things we're throwing at that soil to get more and faster productivity, you realize that we're on the wrong track. We're wiping out species before we have the foggiest idea that they're there. As Noel Brown of the United Nations Environment Program put it, we may already have destroyed the cure for AIDS. Jay Hair of the National Wildlife Federation tells that when his daughter was three, her doctor said, "She has four days to live," and when he told that story a few years ago, she was then in college, doing all right. The medicine that cured her disease was derived from the rosy periwinkle, which grew only in Madagascar and is now extinct. We're wiping out species everywhere we can possibly reach.

Now it's midnight, and there's a new day coming. What are you going to do with it? You're going to have an important role in what happens in this new day or the next six days or whatever it may be.

But I don't think we're quite ready for it. To begin with, we feel that we have to blame somebody. It's none of us—none of us is guilty for all this, of course—so let's blame the economists. I quote Hazel Henderson: "Economism is a form of brain damage." I heard Fritz Schumacher, when he was lecturing out in Marin County, California, tell this story: There were three people arguing about whose profession was the oldest. The doctor said, "Mine is the oldest, because it took a procedure to get Eve out of Adam." The architect said, "But it took an architect to build a universe out of Chaos." And the economist said, "And who do you think created Chaos?" That is a beautiful story, and it should be carved in stone where the E. F. Schumacher Society has its headquarters. Hazel is an economist, and Fritz was an economist, and even they blame the economists!

Economists are in trouble because they leave out of their calculations two terribly important factors, which they name and do nothing about: the cost to the Earth and the cost to the future. In fact, as David Orr pointed out in his lecture, they discount these factors. That implies they're essentially of no value. Leave out the cost to the Earth, leave out the cost to the future, and whatever your final number is, it's worthless. We're getting worthless advice from those economists who are giving most of the advice about how to run our government, including advice about the General Agreement on Tariffs and Trade. One of my definitions of GATT is that it's the end run around the environmental gains of the past century. It is just pure gravy for the transnationals.

We've got to do something about one of our worst addictions: the addiction to growth. All of the candidates running for office are saying, "We must have a growing economy." If they want to keep it growing the way it has been growing, we absolutely must not have a growing economy. We must have a *sound* economy, a *sustainable* economy. They haven't come up with one single notion of how to move it in that direction. What are we going to do besides grow, grow, grow? In your own body, where the wildness within you puts in a control factor, you have a thymus. Civilization needs a thymus. It needs the word for "enough." But "enough" doesn't sound strong enough—Italian has the right word: *basta*. We must say basta to the kind of growth we've been practicing.

Another thing we need, as Adlai Stevenson pointed out, is love for the fragile craft Earth and all its inhabitants. We haven't been good about that. One small way we could show love would be not just to criticize somebody who's done something we don't like but to thank somebody who's done something we do like. We don't thank the people who deserve it. Think back to Richard Nixon when he first came into office: he made the best speech on population control any president has ever made, before or since. He hedged it a little bit, but nobody has touched what he did. Certainly not Ronald Reagan or George Bush. John Ehrlichman told me in 1969, "That speech was a dud. It bombed at the box office. No support." So I have asked many of my audiences, "How many people think we have a population problem?" Hands up all over the place. "How many people thanked Richard Nixon for what he did?" On the average, only one hand in a thousand.

Jimmy Carter got into the same kind of box on the subject of nuclear power and whether or not to build breeder reactors. Legislation favoring these reactors had passed Congress. He wanted to veto it but felt that he had no alternative except to sign it. I was one of thirteen people who met with Carter to discuss the bind he was in, and eventually it was a letter I signed, which was written by Jeff Knight, Friends of the Earth's energy specialist, that convinced Carter he could veto it. He did! I have also asked audiences, "How many thought that legislation needed to be vetoed?" Almost everybody. "How many thanked Jimmy Carter?" One in a thousand. Yet this is one way we can show a bit of love: by thanking somebody for doing something right. We don't all have to do everything right, but if a person does one thing right, then maybe, with thanks, that person will do something else right. Just think for a moment what might have happened to Richard Nixon if that speech had had the support it deserved—he might have been a completely different Richard Nixon.

I have my own axiom: not to love thy neighbor as much as thyself but to love thy neighbor more than thyself. This might be useful to practice, because out of that behavior something else happens. For example, it could help us be more aware of "the Law of the Minimum": that it doesn't matter how many plants there are if you don't have land; it doesn't matter how much land you have if you don't have soil on it; it doesn't matter how much soil you have if you don't have water for it; it doesn't matter how much water you have if you don't have air; it doesn't matter how much air if you don't have oxygen; or how much oxygen if you don't have judgment; or how much judgment if you don't have love. It would certainly help our transnationals and the "Misfortune 500" if they considered the Law of the Minimum.

This interrelation is terribly important, and there are parts of it that we're not thinking about. I'll just touch for a moment on oxygen. The amount of oxygen on Earth is decreasing because we're getting rid of the world's forests as fast as we can. While a tree is alive, chlorophyll locks up carbon and frees oxygen. But when a great tree falls, it may take two thousand years—or, depending on its chemistry and climate, maybe only two hundred or five hundred or nine hundred years— for it to turn into soil again. During that time it's going to require back all the oxygen it freed so that it can feed Robert Frost's "slow, smokeless burning of decay." This decay is essential to complete the continuing revolution of the cycles of life, particularly the carbon cycle, but it does require oxygen. Simultaneously, we are releasing the carbon that was buried and became fossil fuels over the course of five hundred million years. We have quite a bit of locked-up carbon that could stay locked up, and what do we do with it? We dig it up as fast as we can and put it out as many tailpipes as we can, and we say, "This is jobs." If you're worried about the ozone barrier, then you've got to realize that the damage is going to continue for a long time. CFCs are migrating up; they are destroying ozone now and will continue to destroy it for a hundred years even if their use is stopped today. (I got this number out of the special Fall 1992 issue of *Time*, "Beyond the Year 2000: Preparing for the Next Millennium.")

So for you twenty-year-olds I've got a lot of sympathy. People my age can "check out" reasonably soon, but what is going to happen as this atmospheric imbalance continues to worsen? What is going to heal it? I don't know enough high-school chemistry to know anything but this: if you want O_3 back, you've got to have some O_2 to play with. But we're getting rid of it. So what do we do? People are talking about a carbon tax and other measures, but what we need to do is pay the people who have forests and pay the countries that are storing fossil fuels to keep them where they are. We need to slow down their use as fast as we possibly can, to use every bit of science and technology and humanity we can to slow it down. To say basta to what we've been up to. It's terribly important if you like to breathe. And what are we going to do about the soil if they keep doing to the soil what they've done? There's a big constituency out there of people who like to eat, who like to breathe, and we've got to organize this group.

Where do we start? One opportunity for action is the James Bay situation. I'm on this trip East to try to do something about James Bay, the "thumb" that hangs down from the Hudson Bay. This is my third time here for this purpose. James Bay has a lot of rivers flowing into it, and HydroQuebec—the HydroMafia of Quebec—has been working hard to see if it can get rid of that free-flowing water and turn it into kilowatt-hours for New England. It's going to cost New England fifty billion dollars to finance that operation and receive hydroelectricity. Fortunately, Governor Cuomo—for economic reasons, not for ecological reasons—pulled out of the contract (and I thanked him for it). That puts a big bite of vulnerability into it. Now we've got to get the New England states to pull out. We need to have the people who hold HydroQuebec bonds get rid of them in order to send a signal. We're going to go after universities and other holders of major funds and pressure them to divest themselves of HydroQuebec bonds.

What HydroQuebec wants to do to the Cree Indians is essentially to wipe out their habitat and wipe them out—the same general attitude Henry Kissinger showed toward Micronesia when he said, "There are only ninety thousand people down there; who gives a damn?" HydroQuebec says, "There are seventeen thousand Cree up there, and we do indeed give a dam; we want to build all the dams we possibly can!" Here's a culture that knows its terrain better than we know ours, that has not just a hands-on approach to the Earth but a feet-on approach, and we're trying to destroy it. We're going on the idea—the myth—that hydroelectric energy is renewable, but it's not renewable, because it depends on reservoirs. It's a one-shot thing. It's mining the dam site: you use the dam site up, and that's it. And it's messing around with rivers, taking the meanders out. Rivers know what they're doing—meanders slow the river down, rechanneling it and recharging aquifers. When aquifers can't recharge, what happens? The Kissimmee River in Florida. The Corps of Engineers straightened it out. Now that they've realized their mistake, they're spending fifty million dollars to put the bends back in. Well, that's jobs.

We've got to find alternative forms of energy, certainly alternatives to wasting energy. We've got to cut off these hydroprojects right now. Only God can make a dam site, and we've occupied a lot of them already. We don't need to go on in China, in India, in Japan. I want Hetch Hetchy dam in Yosemite down so that we get another Yosemite Valley in our park. That can be done. Because of the numbers I was throwing at you just now—the resources used up, the population increased—we need a completely new look, a new insight, a new vision of what we're going to do. We've got to worry about numbers. We've got to worry about our appetite, and the best place to start is right at home with our overconsumption. We can stop overconsuming immediately. Just keep your wallet in your pocket, and we'll cut consumption down fast. Yet, as E. O. Wilson says about these numbers, we aren't willing to do anything really drastic. In the case of population control, an acceptable limit is no more than two children per family. I would prefer just one per family, but that means in a short time there will be no cousins anymore. So leave it at two per family, in the families that want them and can take care of them. I firmly believe in life after birth; population control enhances life.

The big pressure is our pressure, our overconsumption. And we have our own Third World, as you know; the homeless aren't using much, people in the ghettoes aren't using much. But those of us who aren't poor are the problem. Buy, buy, buy; consume, consume, consume; toss it away, toss it away. There are people in Massachusetts spending seven million dollars to fight Measure 3, which calls for recycling in an imaginative way; it could be an example for the rest of the United States. We got the governor to say that their arguments were erroneous, but that was boiled down to a tiny piece on page twenty-eight of The Boston Globe.

This brings me to a key area for action: we have to free the media, break the sound barrier. We've got to get the word out, and the media can't get the word out because most of them are indentured. The alternative press, of course, is not indentured, and there are two specific, contrasting examples of fairly "free" magazines. *Ms.* magazine carries no ads. Its editors made that decision because they wanted to be free to speak. They didn't want advertisers looking over their shoulders, and they had firm ideas about what they wanted to tell their audience. They needed to have a circulation of 150,000 to make it work without ads, and as I understand it, they have 250,000, and it's working. That's one way to freedom. But the other way, strangely enough, is in a magazine that is absolutely loaded with ads, and I don't approve of all of them, by any means: *The New Yorker.* Remember what *The New Yorker* did under William Shawn? It gave an entire issue to John Hersey's *Hiroshima.* One issue, maybe it was two, to Rachel Carson's *Silent Spring.* Three issues to *Encounters with the Archdruid,* John McPhee's interview with me. Again and again *The New Yorker* carries pieces that are hard-hitting. I haven't seen it for the past three weeks—I travel too much—but I understand that even with the new editor it's still hard-hitting. In "The Talk of the Town" this week there's apparently an article providing grounds for the impeachment of George Bush. They are bold. They don't give a damn what their advertisers want, but they get them anyway. Their boldness makes them a required medium for advertisers to advertise in. I wish the rest of the media would try that out.

Incidentally, taking a stand on environmental issues is one thing they don't try out. I have been interviewed seven times by *Time* magazine. They have not put in a word of what I said. The first time was when the Alaska pipeline was the cover story. I had a long interview on that. Phil Herrera (he was then the environment editor) put a lot of that material in and submitted it. But *Time* took out everything I'd said (although they left in a picture of me!). Phil said it was taken out by the advertising department. So we *do* have to free the media. They should all be willing and able to say what they think without having to look over their shoulder and wonder, "What will the advertiser think if I do this?" That goes for PBS and NPR as well as anybody else. We *must* free the media, and that will happen only when corporations learn how to operate with conscience. When the corporations do that and the media are free, we'll get our democracy back because the people running for office won't have to go to the Fortune 500 or the transnationals to fund their campaigns. We'll do more of what Jerry Brown was doing with his 800 number: no more than a hundred dollars from anybody. With that 800 number, by that process, he raised twelve million dollars.

What else can we do? Let me tell you about Sam LaBudde. Although you may not know the name, I think you know what he did: he took his courage and his camcorder, which had been given to him by Earth Island Institute and Earth Trust, and he got a job aboard a fishing ship, working as a cook. He said the camcorder was a toy given to him by his father and he wanted to see how it worked. In the July 1989 issue of *The Atlantic* there was a good cover story on him, written by the

writing member of our family, Kenneth Brower. Sam did something that anybody
could do—at least anybody who is thirty-two and a biologist—he got aboard that
ship and took camcorder footage of what was going on in the industry. One of the
best shots shows a set of nets surrounding a school of tuna located under dolphins, and the nets bring in three tuna and kill more than a hundred dolphins.
Hundreds of thousands of dolphins have been killed by the tuna fishing industry.
Sam's tape has been made into specials and news broadcasts around the world.
That tape took a Fortune 500 company—H. J. Heinz—and turned it around. After
a little bit of struggle and some full-page ads Earth Island finally got Bumble Bee
to admit they weren't telling the truth about their fishing methods. So the American tuna industry is now giving you "dolphin-safe" tuna. Mexico's tuna industry
continues to kill dolphins, and they accuse us of eco-imperialism for saying we
will not accept their tuna because it kills dolphins. They're being supported by
the GATT (General Agreement on Tariffs and Trade) philosophy! GATT says
Mexico is right and we are wrong to protect dolphins and to be proud of it. What
we can do now is to help fund—I think Heinz could help fund—research and
development by Mexico that will enable them to catch tuna, as Heinz is doing,
without killing dolphins. They might just as well learn how to do that, because if
they kill all the dolphins, where will they look for their tuna? It's like the old-
growth forests: if we kill all the old-growth forests for the sake of jobs, what will
the lumber industry do when the forests are gone?

Next, Sam LaBudde went aboard a driftnet ship. These ships set out thirty-five
thousand miles of driftnet every night. Then they haul it in. It catches fish that
shouldn't be caught, that need to grow some more and go back to the streams
where they came from. Whales, seals, dolphins, marine birds, and turtles are
killed in the driftnets. Their lives are wasted. Sam sums it up as strip-mining the
high seas. His footage on that has brought changes at the United Nations level. Japan has recently said it will stop using driftnets. Thirty-two years old, bold, with
camcorder. Then Sam went up to Alaska, where young Alaskans were machine-
gunning walruses to trade their ivory tusks for drugs.

By this time Sam was getting a little depressed, so he came up with the idea of
an Earth Corps. I've been working on that: an Earth Corps to take up where the
Peace Corps leaves off. The Earth Corps would be global, whereas the Peace Corps
is just a national thing. It hasn't been interested in the environment. It's more
interested now, but it still isn't willing to displease the transnational corporations.
We want to be able to displease them if necessary. What we want was described
reasonably well, though just briefly, by Mikhail Gorbachev two years ago January
at the Global Forum in Moscow. In that speech Gorbachev called for a "Green
Cross." That's a better name than Earth Corps. The Red Cross takes care of damage
the Earth does to people; the Green Cross will take care of damage people do to

the Earth, to balance things out. It sounded like a great idea, so we started an Earth Island call for the International Green Cross. Then we ran into static from people who were offended by the symbol of the cross. They liked the crescent, they liked the Star of David, but they were offended by the cross. Carl Anthony, the president of Earth Island Institute, suggested that we call it the "International Green Circle." That sounded okay. We thought, "Anyone who lives on a spherical planet and is offended by a circle is in trouble anyway; we'll ignore them." So we called it the International Green Circle, which is a nice, innocuous name, but nobody knows what it is about. Now we're calling it the "Global CPR Corps": Conservation, Protection, and Restoration. A corps like this will help to start the paradigm shift that must come about. We cannot afford to continue feeding our economy, our greed, and what my wife calls our "greedlock." We're running out of some of the things in the Law of the Minimum; we're going over the edge in a so-called Giant Step for Mankind that nobody needs. We've got to avoid it, to do a one-eighty, to make a tire-screeching U-turn and not go over that edge.

I can see no better way to do this than by making a major effort to go back to where we've been, leaving wild the wildness that remains, fulfilling the maxim of Henry David Thoreau, "In wildness is the preservation of the world." We must honor wildness, for as Nancy Newhall writes in *This Is the American Earth*, "The wilderness holds answers to questions we have not yet learned how to ask." It's exciting to discover—it's fun to discover—how nature works, to find out, for example, that *we* have to make cement at 1800° Fahrenheit, while a *hen* can make better cement per unit at 103°, and a *clam* can do it at seawater temperature. What's the trick? We don't know, but I wouldn't mind finding out. Other examples: the bombardier beetle makes actual steam in an internal chamber and fires it at its enemies; another beetle does something to the surface tension of water that makes water skaters sink and become its prey. The giant water bug injects, say, a frog with a chemical that dissolves everything inside the frog's skin—turns it into liquid—and then the bug sucks it out.

So there's all this exciting stuff to discover! We haven't spoiled it all, and we can save all that's left of it. We can go back to where we've been and do better. To science and technology we can add humanity and compassion and go back. Who will do this? Well, we want some teams. We want to build restoration teams on which all the creeds are represented, all the colors, all the ages (I still want something to do), all the classes, and all the sexes. We want to build teams that are willing to put aside their favorite prejudices and get into a symbiotic, instead of an aggressive, relationship with others. We don't agree on a great many things, but we *can* agree that we've got to restore the damage we've done to Earth.

We have examples of restoration work that has been done. Jerry Brown did some when he was the governor of California by supporting the "Investing for

Prosperity" program: when the legislature was cutting every other program, Investing for Prosperity got a hundred and twenty-five million dollars a year to restore forests, wetlands, streams, soil fertility, and other things, and many of these investments have paid off already. Then there's Dan Janzen of the University of Pennsylvania, restoring the dry tropical forest of Costa Rica, and Earth Island Institute, helping to protect Siberia's Lake Baikal. We're trying to get a restoration movement going at Earth Island. We want *every* institution to get involved in it. It's the alternative to war. One of the problems with peace is that it's been rather dull—it's not much fun; if you put restoration into it, it can be great fun, and it can be profit-making. If you don't think so, try taking your car to the shop or your body to your doctor, and find out who's making money. You're glad to pay it: the car works better or your body works better (you hope), so it's a good investment.

There is no better investment, whatever it costs, than getting Earth's life-support systems back in life-supporting, working order. People are worried about the taxing and spending that might be required to pay these costs, but we've been borrowing and spending as well as deferring maintenance and replacement for the past twelve years. We need to pay for restoration because we need to save the wild. If we were to ask the twenty-year-olds and under in the audience, "Would you be willing to pay the bill for restoring the Earth so we can live on it?" I think I know what the answer would be. We have the opportunity now to invest in prosperity, to invest in ecological sanity, and to invest in an understanding of how the Earth works and what we have to do to help it work. We can help nature heal. But we can't be so arrogant as to think we've got all the answers, because we haven't; if we're not careful, we'll make mistakes like bringing more rabbits to Australia or something worse.

We're getting rid of wildness before we have the faintest idea of what we're eliminating. We have got to stop. We can stop by going back to where we've been and doing better there, not by going on further with the idea that we need more and more and more. I think we're getting tired of trashing wildness. It's not making us happy and it's not making us healthy; it's making us miserable and despairing.

So here's a task; it's a challenging one. I've now talked to more than two hundred seventy thousand people. At the end of my pitch for restoration I have asked each audience—and now I'm asking you, "How many people in this audience would be willing to commit at least a year of their lives, out of the next ten, to volunteering for this restoration effort, either getting paid or not?" . . . That's pretty good.

The point is that this is the public wish, as I have seen it represented in these audiences. They haven't just been members and friends of the Schumacher Society: they've been media people; they've been the Physicians for Social Responsibility;

they've been directors, writers, and producers in Hollywood; they've been the audiences I talked to in Japan and in the ex-U.S.S.R. Wherever I have gone, at least two-thirds of the people put their hands up. So the wish is there. The ability to lead needs to be worked on—we need leaders, we need organizers, and of course it would help to have a little money. If you follow through and help organize this and enlist others to help organize it, then it will happen. If it doesn't, we've had it! Civilization as we know it will have had it. We can't continue going that way; we've got to turn around.

We can do it; the talent is here. My old mountaineer friend, William H. Murray, in his book *The Scottish Himalayan Expedition*, expresses his deep admiration for a couplet from Goethe: "Whatever you can do, or dream you can, begin it; boldness has genius, power and magic in it." Do you have magic in you? You bet. Because the minimum of genetic material—the amount necessary to give us all the messages about where our hundred million rods and cones go and about the whole works, conscious mind and unconscious—would fit in a sphere a sixth of an inch in diameter. That sums up the minimum genetic material needed to produce the hundred billion people who have ever lived. That magic, that miracle of life, has been passed on for three and a half billion years. In that time millions of species went by the wayside, but we didn't. From when it began three and a half billion years ago to everyone here: no mistakes, no failure. So a tiny part of each of us is three and a half billion years old, and everything that's alive is related. How did this miracle happen? What shaped it? What informed it? It wasn't civilization, because there wasn't any. It was something else. It was *wilderness*. That's all there was. Trial and error, success and failure, symbiosis; wilderness made it work. Wilderness is the ultimate encyclopedia, holding, just as Nancy Newhall put it, answers to more questions than we have yet learned how to ask.

That's the magic in you. You've got it; let it out.

—1992

Benjamin Strauss
Afterword

Benjamin Strauss began his environmental career in 1989 when he
started a recycling program at his New York City high school. Four
years later, in college, he chaired the Yale Student Environmental Coalition committee that hosted the first Annual E. F. Schumacher Lectures
to be held at Yale. The next winter, in 1994, he was program director
of the Campus Earth Summit, an international conference to advance
environmental education, stewardship, and activism on campuses. In
1995 he researched and wrote for the Nathan Cummings Foundation a
paper on these same topics called "The Class of 2000 Report."

I was born the year before *Small Is Beautiful* came out; however, it
was not until I was twenty years old that I first learned about E. F.
Schumacher. I didn't really have a choice, because I was attend-
ing the Twelfth Annual E. F. Schumacher Lectures. The date was
October 1992. I had driven to Great Barrington with some
college friends to hear David Brower speak, a figure who was
already known to me and who was a hero of mine.

Brower did not disappoint me. What he said that day is now a
part of this book. But I also had the good fortune to make my
first conscious acquaintance with the ideas of Schumacher,

whose philosophy has been an important reference point in my own thinking ever since.

Few of my peers—including those who identify themselves as environmentalists—are familiar with Schumacher or the other thinkers in his tradition who are represented here. Nonetheless, I believe that these lectures speak meaningfully to my generation: they give critiques of the modern era that resonate with discontents we share and yet at the same time offer alternative visions compatible with the various ways many of us choose to serve society.

A particularly important discontent of young people today is embedded in their strong cynicism concerning our government, social institutions, and even, perhaps, possibilities for collective action in general. The Higher Education Research Institute (HERI) at the University of California, Los Angeles, has conducted extensive national surveys of incoming first year college students every year since 1966. For the class entering in 1994 the percentage of students who believed that "keeping up with political affairs" is important in life reached an all-time low, as did the proportion who said they discussed politics regularly.[1] A 1993 survey of undergraduates at all levels showed that most thought "traditional American politics" cannot cause "meaningful social change," that the nation's problems are not being well addressed by the political system, and that Congress does not care about people's best interests.[2]

These attitudes are hardly surprising. They have an all too ample basis in reality, and they are shared by other age groups. They have also been expressed by students before. Americans remember the late 1960s in particular, when students joined in protests, rallies, and sit-ins directed against campus administrations and the national administration. My generation has often been asked, and we have asked ourselves, why do we seem to lack the spirit and energy of the 1960s?

Perhaps we are simply less hopeful of being able to bring about change together—or more aware of the great difficulty of doing so. Efforts to protect the environment make an excellent case study. The first Earth Day, in 1970, was driven largely by students. The principal organizing group, the student-led Environmental Action, went on to become a key player in the passage of the Clean Air Act and in creating the political atmosphere in which the Clean Water Act, Endangered Species Act, and many other major environmental laws were passed.

These laws have resulted in some concrete improvements. Less industrial waste pollutes the nation's rivers now than in 1970, and the bald eagle has been removed from the endangered species list. At the very least most land, resources, and people protected by recent laws are probably better off than they would have been without the laws. But the evidence is compelling that overall our problems are becoming worse and worse. For every endangered species we aid toward recovery, new strip-malls crowd many more toward extinction. Rivers are freer of industrial effluents but suffer increasing damage from more pervasive sources, such as neighboring land development and agricultural run-off. Improved car-engine efficiency has been overwhelmed by the growing number of vehicles on the road and miles driven; consequently, smog is still uncontrolled.

It is plain to see that these problems are tied to deep and powerful forces, not just bad habits correctable by tinkering. Schumacher and the speakers presented here provide insightful analyses that are hard to dispute. On the whole, they agree that the clearest causes of the West's environmental transgressions are economic, especially its high levels of production and consumption and the unrelenting push for growth. Underlying these economic problems are cultural ones: the materialism of Western civilization, its quest for control of nature, and its pursuit of knowledge with little concern for the values it serves.

Related to both the economic and cultural situations is the question of scale, the question that gave its title to Small Is Beautiful. As trade, manufacture, and finance become more interconnected across regions and nations, it becomes more difficult for people to learn how what they purchase or borrow came into being. They cannot easily ascertain the labor conditions or environmental impacts of manufacture, for instance, or the source of funds for a loan. The only value attached to goods or money becomes its material value, which in turn reinforces the materialist values of the culture. Based on the same basic principle of disconnect-edness, large scale encompasses many more effects. As David Ehrenfeld argues, managers of bloated organizations make poor decisions because of their distance from the people or things they are trying to manage. And as Schumacher writes, the integrity of individuals is threatened when they feel as though they have been reduced to small cogs in a vast social machine.

I believe that one of the effects of modern scale, at least in part, is the prevalent cynicism of my generation. Most college students say environmental protection is very important to them—along with a number of other major issues impinging on the public welfare such as crime and racism. But the problems are depressingly large. Government is also large and seems too distant for ordinary citizens to affect in a meaningful way.

The key to change, according to Schumacher, is to decrease the scale of most institutions and the range of most transactions. Local production should provide for local needs as much as possible. A whole section of this book is devoted to filling in the details on how to reconstruct local and regional econo-mies and communities. The speakers share a faith that when we tangibly know the land that gives us food, the people who make our goods, and the bankers who lend us money, the values of responsibility, moral judgment, and environmental stewardship will enjoy a renaissance. They hope that we can return to traditional wisdoms, judged far more essential to happiness and well-being than knowing how to build spacecraft or semiconductors.

Students on the whole are not very aware or supportive of these ideas, although many of the more committed environmentalists among them are. However, students' *actions* do speak to the psychological appeal of the local and the tangible. We have been labeled "Generation X" for our supposed lack of voice, action, or character, but the HERI surveys say we volunteer our time more than any previous cohort in the past thirty years or more. The booming service movement on campus corroborates this finding. When someone volunteers at a homeless shelter or helps with a campus recycling program, the results are visible, gratifying, and encouraging. This contrasts with more ambitious and abstract efforts to reshape politics, laws, or institutions in the hope of improving the social welfare. Performing tangible service actions helps fight the despair over problems that seem insoluble and helps give meaning to individual lives by defining a relationship of responsibility toward society.

We need to do more. A patchwork of individual good works will be overwhelmed by the historical forces now degrading the Earth and the quality of human life upon it. Proposals in this book—to decentralize power, build regional economies, and return to the guidance of traditional wisdoms—are radical and powerful because they address fundamental systemic problems and offer a cohesive vision. They also have a strong appeal because of their emphasis on human scale and community values.

Of course, these proposals cannot fully address the problems we face. They draw upon the better qualities of the remembered past, but the steps of time cannot simply be retraced. As Stephanie Mills points out in her lecture, human restoration of the land cannot return it to an Edenic wilderness, if it ever even existed in such an idealized state. Instead, a new and difficult relationship, unique to this time, has been formed between people and the land, and we must struggle through it.

In a similar vein, pure decentralism or strictly traditional wisdom cannot cope with many of our contemporary problems. The globalization of the economy, for instance, is not likely to reverse itself any time soon. It is still creating new structures, relationships, and dependencies that will expand and endure. This is a practical context which cannot be ignored. Furthermore, many of our problems transcend single regions and call for national, international, or global solutions. If two distant cities provide for their needs locally, they each have structural incentives to treat their own lands and resources well—perhaps as well as any indigenous peoples before them. Both cities, however, still affect and are affected by the same atmosphere and may share a watershed or fishery. Common situations like these require concerted coordination to sort out, especially when the number of cities or towns or nations is large.

Nonetheless, these essays leave little doubt of the many virtues of limiting human organizations to a comprehensible and friendly scale and of bringing a sense of community to all of the interactions that we can. One of our greatest challenges today is to incorporate the beneficial elements of smallness wherever possible into the large-scale coordinating efforts we need. As Schumacher himself wrote, "We always need both freedom and order. We need the freedom of lots and lots of small, autonomous units, and, at the same time, the orderliness of large-scale, possibly global, unity and coordination. When it comes . . . to the indivisibility of peace and also of ecology, we need to recognize the unity of mankind and base our actions upon this recognition."[3]

Notes

1 "The American Freshman National Norms for Fall 1994," Higher Education Research Institute, Graduate School of Education, University of California, Los Angeles, 1994.

2 Alice Dembner, "Poll finds college students ready to make a difference." *The Boston Globe*, September 4, 1994, National/Foreign, p. 1.

3 E. F. Schumacher, *Small Is Beautiful: Economics as if People Mattered* (Harper & Row Publishers, 1989), p. 69.

Chronological List of
E. F. Schumacher Society Lectures

1981	Mt. Holyoke College, South Hadley, Mass.
Wendell Berry	"People, Land, and Community"
Wes Jackson	"Call for a Revolution in Agriculture"*
1982	Cathedral of St. John the Divine, New York City
Elise Boulding	"The Family as a Small Society"*
George McRobie	"The Role of Community in Appropriate Technology"*
1983	Mt. Holyoke College, South Hadley, Mass.
Jane Jacobs	"The Economy of Regions"*
Kirkpatrick Sale	"Mother of All: An Introduction to Bioregionalism"*
1984	Foote School, New Haven
John L. McKnight	"John Deere and the Bereavement Counselor"*
Charlene Spretnak	"Green Politics: The Spiritual Dimension"*
1985	Harvard University, Cambridge, Mass.
Frances Moore Lappé	"Toward a Politics of Hope: Lessons from a Hungry World"*
John Todd	"An Ecological Economic Order"*
1986	Friends' Meeting House, Philadelphia
Wendell Berry	"The City and the Farm Crisis"
Kathryn Waller	"The Farm Crisis"

1987	Monument Mountain High School, Great Barrington, Mass.
Richard Grossman	"Alternative Medicines"
Jeremy Rifkin	"Technology and Its Consequences"
Nancy Jack Todd	"Technology as if the Earth Mattered"
1988	Monument Mountain High School, Great Barrington, Mass.
Alana Probst	"Import Replacement"
August Schumacher	"Linking America's Farmers to Cities: The Farmers-Market Coupon Program"
Robert Swann	"The Need for Local Currencies"*
1989	John Dewey Academy, Great Barrington, Mass.
Hazel Henderson	"Development Beyond Economism"*
Leopold Kohr	"Why Small Is Beautiful: The Size Interpretation of History"*
John McClaughry	"Bringing Power Back Home: Recreating Democracy on a Human Scale"*
1990	John Dewey Academy, Great Barrington, Mass.
David Ehrenfeld	"The Management Explosion and the Next Environmental Crisis"*
Dana Jackson	"Women and the Challenge of the Ecological Era"*
Kirkpatrick Sale	"The Columbian Legacy and the Ecosterian Response"*
1991	John Dewey Academy, Great Barrington, Mass.
Thomas Berry	"The Ecozoic Era"*
Stephanie Mills	"Making Amends to the Myriad Creatures"*
1992	First Congregational Church, Stockbridge, Mass.
David Brower	"It's Healing Time on Earth"*
David W. Orr	"Environmental Literacy: Education as if the Earth Mattered"*
Jakob von Uexkull	"The Right Livelihood Award"*
1993	Yale University, New Haven
George Davis	"Ecologically Sustainable Economic Development"*
Wes Jackson	"Becoming Native to this Place"*
Winona LaDuke	"Voices from White Earth: Gaa-waabaabiganikaag"*
1994	Yale University, New Haven
Gar Alperovitz	"Beyond Capitalism and Socialism: Trajectories to the New Society"
Ivan Illich	"The Wisdom of Leopold Kohr"*

1995	Yale University, New Haven
Paul Hawken	"The Ecology of Commerce: Changing the Nature of Business"
Cathrine Sneed	"The Garden Project: Creating Urban Communities"*
Kent Whealy	"The Seed Savers Exchange"
1996	First Congregational Church, Stockbridge, Mass.
Don Anderson	"The Idea of the Assembly"*
David Morris	"Reclaiming Community"*
Helena Norbert-Hodge	"Moving toward Community: From Global Dependence to Local Interdependence"*

*Available in pamphlet form from the E. F. Schumacher Society, 140 Jug End Road, Great Barrington, MA 01230 (413) 528-1737.

Further Reading

The following list consists of titles by the lecturers as well as titles they have recommended for inclusion here.

Agar, Herbert. *Land of the Free*. Boston: Houghton-Mifflin, 1935.

Annals of Earth. A literary journal of ecological design and philosophy published from One Locust Street, Falmouth, Mass., 02540 by Ocean Arks International and the Lindesfarne Association .

Anderson, Victor. *Alternative Economic Indicators*. London and New York: Routledge, 1991.

Andruss, Van, Christopher Plant, Judith Plant, and Eleanor Wright, eds. *Home! A Bioregional Reader*. Philadelphia, Pa.; Gabriola Island, B.C.; and Santa Cruz, Calif.: New Society Publishers, 1990.

Berry, Thomas. *Befriending the Earth: A Theology of Reconciliation between Humans and the Earth*. Mystic, Conn.: Twenty-Third Publications, 1991.

——. *The Dream of the Earth*. San Francisco: Sierra Club Books, 1988.

Berry, Wendell. *The Agricultural Crisis: A Crisis of Culture*. New York: Myrin Institute, 1977.

——. *Another Turn of the Crank: Essays*. Washington, D. C.: Counterpoint, 1995.

——. *A Continuous Harmony: Essays Cultural and Agricultural*. Magnolia, Mass.: Peter Smith, 1993. First published 1972.

——. *The Gift of Good Land: Further Essays, Cultural and Agricultural*. New York: Farrar, Straus and Giroux, 1981.

——. *Home Economics*. San Francisco: North Point Press, 1987.

——. *Sex, Economy, Freedom and Community*. New York: Pantheon, 1993.

——. *Standing By Words*. San Francisco: North Point Press, 1983.

——. *The Unsettling of America: Culture and Agriculture*. San Francisco: Sierra Club Books, 1977.

———. *What Are People For?* San Francisco: North Point Press, 1990.

Bhave, Vinoba. *Moved by Love: The Memoirs of Vinoba Bhave.* Ed. Kalindi. Trans. Marjorie Sykes. Totnes, Devon, England: Green Books, 1994.

Borsodi, Ralph. *Flight from the City: An Experiment in Creative Living on the Land.* Suffern, N.Y.: School of Living, 1947. First published 1933.

———. *Seventeen Problems of Man and Society.* Anand, India: Charotar Book Stall, 1968.

———. *This Ugly Civilization.* Philadelphia: Porcupine Press, 1975. First published 1929.

Brower, David R. *For Earth's Sake: The Life and Times of David Brower.* Salt Lake City: Peregrine Smith Book, 1990.

———. *Work in Progress.* Salt Lake City: Peregrine Smith Book, 1991.

———. *Let the Mountains Talk, Let the Rivers Run.* New York: Harper Collins, 1995.

Bryan, Frank, and John McClaughry. *The Vermont Papers: Recreating Democracy on a Human Scale.* Post Mills, Vt.: Chelsea Green Publishing Company, 1989.

Buber, Martin. *Paths in Utopia.* Trans. R. F. C. Hull. New York: Collier Books, 1988. First published 1949.

Burns, Scott. *The Household Economy.* New York: Doubleday, 1977.

Capra, Fritjof, and Gunter Pauli. *Steering Business Toward Sustainability.* Tokyo: United Nations University Press, 1995.

Carson, Rachel. *Silent Spring.* Boston: Houghton Mifflin, 1994. First published 1962.

Churchill, Ward. *Struggle for the Land.* Preface by Winona LaDuke. Monroe, Maine: Common Courage Press, 1993.

Cornuelle, Richard C. *Reclaiming the American Dream.* New Brunswick, N.J.: Transaction Pubs., 1993. First published 1965.

Daly, Herman E., and John B. Cobb, Jr. *For the Common Good: Redirecting the Economy toward Community, the Environment, and a Sustainable Future.* Boston: Beacon Press, 1994. First published 1989.

Deloria, Vine, Jr., ed. *American Indian Policy in the Twentieth Century.* Norman: University of Oklahoma Press, 1985.

De Soto, Hernando. *The Other Path.* New York: Harper & Row, 1989.

Douglas, J. Sholto, and Robert A. de J. Hart. *Forest Farming.* With a Foreword by E. F. Schumacher. Emmaus, Pa.: Rodale Press, 1978.

Douthwaite, Richard. *Short Circuit: Building Sustainable Communities as Security in an Unstable World.* Totnes, Devon, England: Green Books, 1996.

Duhl, Leonard, M.D., and Trevor Hancock, M.D. "A Guide to Assessing Healthy Cities." World Health Organization, Feb. 1988.

Eagan, David J., and David W. Orr, eds. *The Campus and Environmental Responsibility.* San Francisco: Jossey-Bass, 1992.

Ehrenfeld, David. *The Arrogance of Humanism.* New York: Oxford University Press, 1978/ 1981.

———. *Beginning Again: People and Nature in the New Millennium.* New York: Oxford University Press, 1993/1995.

Eichstaedt, Peter H. *If You Poison Us: Uranium and Native Americans.* Santa Fe: Red Crane Books, 1994.

Eisler, Riane. *The Chalice and the Blade: Our History, Our Future*. Magnolia, Mass., Peter Smith, 1994. First published 1987.

Fukuoka, Masanobu. *The One-Straw Revolution: An Introduction to Natural Farming*. Ed. Larry Korn. Trans. Chris Pearce, Tsune Kurosawa, and Larry Korn. Preface by Wendell Berry. Emmaus, Pa., Rodale Press, 1978.

Galbraith, John Kenneth. *Money: Whence It Came, Where It Went*. Rev. ed. Boston: Houghton Mifflin, 1995. First published 1975.

Gandhi, Mohandas K. *My Theory of Trusteeship*. Allahabad: Indian Press, 1970.

George, Henry. *Progress and Poverty*. New York: Robert Schalkenbach Foundation, 1992. First published 1879.

Georgescu-Roegen, Nicholas. *The Entropy Law and the Economic Process*. Cambridge, Mass.: Harvard University Press, 1971.

Gesell, Silvio. *The Natural Economic Order*. Trans. Philip Pye. London: Peter Owen, 1958. First published in German 1906 (part one); 1911 (part two); 1916 (parts one and two).

Giarini, Orio. *Dialogue on Wealth and Welfare: An Alternative View of World Capital Formation: A Report to the Club of Rome*. Elkins Park. Pa.: Franklin Book Co., 1980.

Goodman, Percival, and Paul Goodman. *Communitas: Means of Livelihood and Ways of Life*. 2nd ed., rev. New York: Vintage Books, 1960. Reprint of original 1947 edition, New York: Columbia University Press, 1990.

Gregg, Richard. *The Big Idol*. Ahmedabad, India: Navajivan Trust, 1963. First published 1935.

——. *The Power of Nonviolence*. 3rd ed. Canton, Maine: Greenleaf Books, 1984. First published 1944.

Hawken, Paul. *The Ecology of Commerce: A Declaration of Sustainability*. New York: HarperCollins, 1993.

——. *The Next Economy*. New York: Holt, Rinehart and Winston, 1983.

Hayek, Friedrich A. *The Road to Serfdom*. Chicago: University of Chicago Press, 1944.

Henderson, Hazel. *Building a Win-Win World: Life Beyond Global Economic Warfare*. San Francisco: Berrett-Koehler, 1996.

——. *Creating Alternative Futures*. West Hartford, Conn.: Kumarian Press, 1996. First published 1978.

——. *Paradigms in Progress: Life Beyond Economics*. Rev. ed. San Francisco: Berrett-Koehler, 1995. First published 1991.

——. *Politics of the Solar Age*. New York: Doubleday, 1981.

Howard, Albert, Sir. *The Soil and Health: A Study of Organic Agriculture*. New York: Schocken Books, 1972. First published 1947.

Jackson, Wes. *Alters of Unhewn Stone: Science and the Earth*. New York: North Point Press, 1987.

——. *Becoming Native to This Place*. Lexington: University Press of Kentucky, 1994.

——. *New Roots for Agriculture*. New ed. Foreword by Wendell Berry. Lincoln, Neb. and London: University of Nebraska Press, 1981.

Jacobs, Jane. *Cities and the Wealth of Nations: Principles of Economic Life*. New York: Vintage Books, 1985.

——. *The Death and Life of Great American Cities*. New York: Modern Library Edition, 1993. First published 1961.

———. *The Economy of Cities*. New York: Vintage Books, 1970.

———. *The Question of Separatism: Quebec and the Struggle over Sovereignty*. New York: Vintage Books, 1981.

———. *Systems of Survival: A Dialogue on the Moral Foundations of Commerce and Politics*. New York: Vintage Books, 1994.

Kennedy, Margrit. *Interest and Inflation Free Money: Creating an Exchange Medium That Works for Everybody*. Philadelphia: New Society Publishers, 1995.

King, Ynestra. "A Body Politic for a Planet in Crisis: An Ecofeminist Perspective." A paper delivered at the Conference on a Post-Modern Presidency. Santa Barbara: Center for a Post-Modern World, 1989.

Kohr, Leopold. *The Breakdown of Nations*. London and New York: Routledge, 1984. First published 1957.

———. *The City of Man: The Duke of Buen Consejo*. Rio Piedras: University of Puerto Rico Press, 1976.

———. *The Inner City: From Mud to Marble*. Dyfed, Wales: Y Lolfa, 1989.

Kretzmann, John P., and John L. McKnight. *Building Communities from the Inside Out: A Path Toward Finding and Mobilizing a Community's Assets*. Evanston, Ill.: Center for Urban Affairs, 1994.

Kropotkin, Peter. *Fields, Factories, and Workshops*. New Brunswick, N.J.: Transaction Pubs., 1992. First published 1901.

———. *Mutual Aid: A Factor of Evolution*. London: Freedom Press, 1987. First published 1902.

Kumar, Satish. *No Destination: An Autobiography*. Rev. ed. Hartland, Devon, England: Green Books, 1992. First published 1978.

Lappé, Frances Moore. *Diet for a Small Planet*. New York: Ballantine, 1991. First published 1971.

———. *Rediscovering America's Values*. New York: Ballantine, 1989.

Lappé, Frances Moore, and Paul M. Du Bois. *The Quickening of America: Rebuilding Our Nation, Remaking Our Lives*. San Francisco: Jossey-Bass, 1994.

Leopold, Aldo. *A Sand County Almanac*. New York: Oxford University Press, 1989. First published 1949.

Lopez, Barry. *Arctic Dreams: Imagination and Desire in a Northern Landscape*. New York: Charles Scribner's Sons, 1986.

Lovelock, James. *The Gaia Hypothesis*. Oxford: Oxford University Press, 1979.

Mander, Jerry. *In The Absense of the Sacred: The Failure of Technology and the Survival of the Indian Nations*. San Francisco: Sierra Club Books, 1991.

Mander, Jerry, and Edward Goldsmith, eds. *The Case against the Global Economy—and for a Turn to the Local*. San Francisco: Sierra Club Books, 1996.

Margolin, Malcolm. *The Earth Manual: How to Work on Wild Land without Taming It*. San Bernardino, Calif.: Borgo Press, 1991.

Marsh, George Perkins. *Man and Nature*. David Lowenthal, ed. Cambridge: Belknap Press of Harvard University Press, 1965. First published 1864.

McKibben, Bill. *The End of Nature*. New York: Random House, 1989.

———. *Hope, Human and Wild*. New York: Little, Brown & Co., 1995.

McKnight, John. *The Careless Society*. New York: Basic Books, 1995.

Meadows, Donella H., Dennis L. Meadows, Jørgen Randers, and William W. Behrens III. *The Limits to Growth: A Report for the Club of Rome's Project on the Predicament of Mankind.* New York: Universe Books, 1972.

Meadows, Donella H., Dennis L. Meadows, and Jørgen Randers. *Beyond the Limits.* Post Mills, Vt.: Chelsea Green, 1992.

Milbrath, Lester W. *Envisioning a Sustainable Society: Learning Our Way Out.* Albany: State University of New York Press, 1989.

Mills, Stephanie, ed. *In Praise of Nature.* Covelo, Calif.: Island Press, 1990.

——. *In Service of the Wild: Restoring and Reinhabiting Damaged Land.* Boston: Beacon Press, 1995.

——. *Whatever Happened to Ecology?* San Francisco: Sierra Club Books, 1989.

Mischan, E. J. *Technology and Growth: The Price We Pay.* Westport, Conn.: Praeger Publishers, 1970.

Morgan, Arthur E., and Donald Harrington Szantho. *The Small Community: Foundation of Democratic Life.* Yellow Springs, Ohio: Community Service, 1984. First published 1942.

Mumford, Lewis. *The Culture of Cities.* San Diego: Harcourt, Brace and Co., 1996. First published 1938.

——. *The Condition of Man.* New York: Harcourt Brace Jovanovich, 1973. First published 1944.

——. *The Pentagon of Power.* New York: Harcourt, Brace and Co., 1970.

——. *Technics and Civilization.* New York: Harcourt, Brace and World, 1963. First published 1934.

——. *Technics and Human Development.* New York: Harcourt, Brace and Co., 1967.

Nearing, Helen, and Scott Nearing. *Living the Good Life: How to Live Sanely and Simply in a Troubled World.* New York: Schocken Books, 1970. First published 1954.

Nicholls, W. M., and William A. Dyson. *The Informal Economy.* Ottawa: Vanier Institute for the Family, 1983.

Nisbet, Robert A. *Twilight of Authority.* New York: Oxford University Press, 1975.

Nock, Albert Jay. *Our Enemy, the State.* San Francisco: Fox & Wilkes, 1994. First published 1935.

Norberg-Hodge, Helena. *Ancient Futures: Learning from Ladakh.* San Francisco: Sierra Club Books, 1991.

Norberg-Hodge, Helena, Peter Goering, and Steven Gorelick, eds. *The Future of Progress: Reflections on Environment and Development.* Rev. ed. Totnes, Devon, England: Green Books, 1995.

Odum, Howard W., and Harry Estill Moore. *American Regionalism: A Cultural-Historical Approach to National Integration.* Gloucester, Mass.: Peter Smith, 1966. First published 1938.

Orr, David W. "The Campus, the Liberal Arts, and the Biosphere." *Harvard Educational Review,* 1990.

——. *Earth in Mind.* Covelo, Calif.: Island Press, 1994.

——. *Ecological Literacy and the Transition to a Postmodern World.* Albany: State University of New York, 1991.

Orr, David W., and Marvin Soroos. *The Global Predicament: Ecological Perspectives on World Order.* Chapel Hill: University of North Carolina Press, 1979.

Plant, Christopher, and Judith Plant, eds. *Green Business: Hope or Hoax?* Philadelphia; Santa Cruz, Calif.; Gabriola Island, B.C.: New Society Publishers, 1991.

Roberts, Wayne, and Susan Brandum. *Get a Life! How to Make a Good Buck, Dance Around the Dinosaurs and Save the World While You're at It.* Toronto: Get a Life Publishing House, 1995.

Robertson, James. *Future Wealth.* London: Cassell, 1989.

Roche, George C. *The Bewildered Society.* Fairfax, Va., Arlington Book Co., 1972.

Röpke, Wilhelm. *A Humane Economy: The Social Framework of the Free Market.* Trans. Elizabeth Henderson. Lanham, Md: University Press of America, 1986. First published 1958.

Roszak, Theodore. *The Voice of the Earth.* New York: Simon & Schuster, 1992.

——. *Where the Wasteland Ends: Politics and Transcendence in Postindustrial Society.* Garden City, N.Y.: Doubleday & Co., 1972.

Rowe, Stan. *Home Place: Essays on Ecology.* Edmonton, Alberta: NeWest, 1990.

Sachs, Wolfgang, ed. *The Development Dictionary: A Guide to Knowledge as Power.* London and Atlantic Highlands, N.J.: Zed Books, 1992.

Sale, Kirkpatrick. *The Conquest of Paradise: Christopher Columbus and the Columbian Legacy.* New York: Knopf, 1990.

——. *Dwellers in the Land: The Bioregional Vision.* 2nd. ed. Philadelphia: New Society Publishers, 1991. First published 1985.

——. *The Green Revolution: The American Environmental Movement 1962–1992.* New York: Hill and Wang, 1993.

——. *Human Scale.* New York: G. P. Putnam's Sons, 1980.

——. *Rebels Against the Future: The Luddites and Their War on the Industrial Revolution: Lessons for the Computer Age.* Reading, Mass.: Addison-Wesley, 1995.

Schumacher, E. F. *Good Work.* New York: Harper & Row, 1979.

——. *A Guide for the Perplexed.* New York: Harper & Row, 1977.

——. *Small Is Beautiful: Economics as if People Mattered.* New York: Harper & Row, 1989. First published 1973.

Shankland, Graeme. *Wonted Work: A Guide to the Informal Economy.* New York: Bootstrap Press, 1988.

Shkilnyk, Anastasia M. *A Poison Stronger Than Love: The Destruction of an Ojibwa Community.* New Haven: Yale University Press, 1985.

Smith, J. Russell. *Tree Crops: A Permanent Agriculture.* Rev. ed. Greenwich, Conn.: Devin-Adair, 1987. First published 1929.

Snipp, C. Matthew. *American Indians: The First of This Land.* New York: Russell Sage Foundation, 1989.

Snyder, Gary. *The Practice of the Wild: Essays.* San Francisco: North Point Press, 1990.

Solomon, Lewis D. *Rethinking Our Centralized Monetary System: The Case for a System of Local Currencies.* Foreword by Bob Swann. Westport, Conn., and London: Praeger Publishers, 1996.

Spretnak, Charlene. *The Resurgence of the Real: Body, Nature, and Place in a Hypermodern World.* Redding, Mass.: Addison-Wesley, forthcoming.

Spretnak, Charlene, and Fritjof Capra. *Green Politics: The Global Promise.* Sante Fe, N.M.: Bear & Co., 1986.

Steiner, Rudolf. *World Economy: The Formation of a Science of World Economics*. Trans. A. O. Barfield and T. Gordon-Jones. London: Rudolf Steiner Press,1977. First English edition, 1936.

Suzuki, David, and Peter Knudtson. *Wisdom of the Elders*. New York: Bantam Books, 1992.

Swann, Robert S., Shimon Gottschalk, Erick S. Hansch, and Edward Webster. *The Community Land Trust: A Guide to a New Model for Land Tenure in America*. Cambridge, Mass: Center for Community Economic Development, 1972.

Swimme, Brian, and Thomas Berry. *The Universe Story: From the Primordial Flaring Forth to the Ecozoic Era—A Celebration of the Unfolding of the Cosmos*. San Francisco: Harper, 1992.

Thomas, Lewis. *The Lives of a Cell: Notes of a Biology Watcher*. New York: Viking, 1974.

Thoreau, Henry David. *Civil Disobedience*. New York: Dover, 1993. First published 1849.

——. *Walden: Or Life in the Woods*. New York: Knopf, 1992. First published 1854.

Todd, Nancy Jack, and John Todd. *Bioshelters, Ocean Arks, City Farming: Ecology as the Basis of Design*. San Francisco: Sierra Club, 1984.

——. *From Eco-Cities to Living Machines: Principles of Ecological Design*. Berkeley: North Atlantic Books, 1994.

Twelve Southerners. *I'll Take My Stand: The South and the Agrarian Tradition*. New York: Harper & Row, 1962. First published 1930.

Van der Ryn, Sim, and Stuart Cowan. *Ecological Design*. Covelo, Calif.: Island Press, 1995.

Vitek, William, and Wes Jackson, eds. *Rooted in the Land: Essays on Community and Place*. New Haven: Yale University Press, 1996.

Whitmyer, Claude, ed. *In the Company of Others: Making Community in the Modern World*. Los Angeles: Jeremy P. Tarcher / Perigee Books, 1993.

Wilson, E[dward] O. *Biophilia*. Cambridge: Belknap Press of Harvard University Press, 1984.

——. *The Diversity of Life*. New York: W. W. Norton, 1993.

Winner, Langdon. *The Whale and the Reactor: A Search for Limits in an Age of High Technology*. Chicago and London: University of Chicago Press, 1986.

Wood, Barbara. *E. F. Schumacher: His Life and Thought*. New York: Harper & Row, 1984.

Woodhouse, Tom. *People and Planet: Alternative Nobel Prize Speeches*. Hartland, Devon, England: Green Books, 1987.

——. *Replenishing the Earth: Right Livelihood Awards 1986–89*. Hartland, Devon, England: Green Books, 1990.

Worster, Donald. *Nature's Economy: A History of Ecological Ideas*. 2nd ed. New York: Cambridge University Press, 1994. First published 1977.

Wright, Frank Lloyd. *When Democracy Builds*. Chicago: University of Chicago Press, 1945.

Contributors

Thomas Berry	Environmentalist
	Greensboro, North Carolina
Wendell Berry	Farmer, philosopher, poet, author
	Port Royal, Kentucky
David Brower	Chairman of the board, Earth Island Institute
	San Francisco
David Ehrenfeld	Professor of biology, Rutgers University
	New Brunswick, New Jersey
Hunter G. Hannum	German scholar, translator
	Old Lyme, Connecticut
Hazel Henderson	Economist, futurist, author
	Anastasia Island, Florida
Dana Lee Jackson	Associate director, The Land Stewardship Project
	White Bear Lake, Minnesota
Wes Jackson	President, The Land Institute
	Salina, Kansas
Jane Jacobs	Author, regional planner
	Toronto, Canada
Winona LaDuke	Campaign director, White Earth Land Recovery Project
	Ponsford, Minnesota

Frances Moore Lappé	Co-director, Center for Living Democracy
	Brattleboro, Vermont
John McClaughry	President, Institute for Liberty and Community; president, Ethan Allen Institute
	Concord, Vermont
John L. McKnight	Professor of communication studies and urban affairs; co-director, Asset Based Community Development Institute, Northwestern University
	Evanston, Illinois
Stephanie Mills	Author, environmentalist
	Maple City, Michigan
David W. Orr	Chair, Environmental Studies Program, Oberlin College
	Oberlin, Ohio
Kirkpatrick Sale	Author, bioregionalist
	New York City
Cathrine Sneed	Director, The Garden Project
	San Francisco
Benjamin Strauss	Environmental policy researcher, Abt Associates
	Bethesda, Maryland
Robert Swann	President, E. F. Schumacher Society
	Great Barrington, Massachusetts
John Todd	President, Ocean Arks International
	Falmouth, Massachusetts
Nancy Jack Todd	Vice-president, Ocean Arks International; editor, *Annals of Earth*
	Falmouth, Massachusetts
Susan Witt	Executive director, E. F. Schumacher Society
	Great Barrington, Massachusetts

Index

with ecology, 106, 118–19; passive,
108; supply regions, 111–12, 116;
colonial, 112; of rural poverty, 114–
16; urban, 119; industrial, 149; bioregional, 223–24. *See also* Banking;
Barter; Currency; Ecology, use of economic terminology in; Regional economics
Ecosteries, 14, 19, 20, 21
Education, 1, 97, 102, 138, 175, 184,
187, 203, 219, 234, 303; reform of,
236; purpose of, 237; problems of
traditional, 238–44; ecological, 245–
49, 285
E. F. Schumacher Society, 6, 51, 121,
126, 129, 130, 131, 132, 204, 220,
294, 301; Annual Lecture Series, 1, 7,
136, 205, 303; in England, 6; Center,
7; Library, 20. *See also* Schumacher College
Ehrenfeld, David, 7, 49, 236, 305
Eisler, Riane, 46
Ethan Allen Institute, 133
Evans, Terry, 51

Farming. *See* Agriculture
Feedback loops: positive, 65, 66, 67, 93;
negative, 93, 119; currencies provide,
124
Feminine, the: introduced into culture,
40, 46–49; in *The Ring*, 204, 209–13
Feminism, 38–45
Fowles, John, 63
Frank, Bernie, 292
Franklin, Benjamin, 79
Friedman, Milton, 75

Gaia: theory, 3; mentioned, 20, 209,
210, 274; described, 217, 221, 228,
234, 235; worldview, 267
Galbraith, John Kenneth, 124
Galeano, Eduardo, 31
Gandhi, Mahatma, 4, 225

Giarini, Orio, 100–101
Gilligan, Carol, 47
Globalization, 1, 90, 92, 95, 96, 306
Glover, Paul, 131
Goethe, Johann Wolfgang von, 207, 302
Golding, William, 235
Goldsmith, Edward, 223
Goldsmith, Oliver, 65
Gorbachev, Mikhail, 18, 299
Gould, Stephen Jay, 242
Gramdan movement, 33
Grameen Bank, 98, 125
Gray, J. Glenn, 239
Griffin, Donald, 243
Gruchow, Paul, 165

Hardin, Garrett, 36
Havel, Vaclav, 243
Hayek, Friederich A., 125
Henderson, Hazel, 7, 294
Hennessey, Michael, 180, 181, 183
Hill, Stuart, 270
Hobsbawm, Eric, 232
Hooke, Robert, 218
House, Freeman, 282, 284
Howard, Sir Albert, 255

Illich, Ivan, 173
Import replacement, 110, 122, 124, 131
Indigenous peoples, 8, 14, 22, 158,
167, 306; values of, 22–28, 30, 142,
214, 292; land issues, 23, 24, 29–33,
170, 258; definition of, 25; unemployment, 34; elders, 199–200. See
also *Anishinabeg akiing*; Community, indigenous peoples; *Minobimaatisiiwin*
Industrialization, 6, 90–104, 212, 206,
238, 263; worldview of, 20, 25–31
passim, 45, 89, 92, 146, 195; effects on
regions, 109–20; economy of, 123,
143, 165
Institute for Liberty and Community, 133
Intermediate technology, 5, 38, 116